MODULATED-TEMPERATURE DIFFERENTIAL SCANNING CALORIMETRY

Hot Topics in Thermal Analysis and Calorimetry

Volume 6

Series Editor:
Judit Simon, *Budapest University of Technology and Economics, Hungary*

The titles published in this series are listed at the end of this volume.

Modulated Temperature Differential Scanning Calorimetry

Theoretical and Practical Applications in Polymer Characterisation

Edited by

MIKE READING

University of East Anglia, Norwich, U.K.

and

DOUGLAS J. HOURSTON

Loughborough University, Loughborough, U.K.

Springer

A C.I.P. Catalogue record for this book is available from the Library of Congress.

ISBN-10 1-4020-3749-X (HB)
ISBN-13 978-1-4020-3749-X (HB)
ISBN-10 1-4020-3750-3 (e-book)
ISBN-13 978-1-4020-3750-3 (e-book)

Published by Springer,
P.O. Box 17, 3300 AA Dordrecht, The Netherlands.

www.springer.com

Printed on acid-free paper

Printed in the Netherlands.

Contents

Foreword xi

1. Theory and Practice of Modulated Temperature Differential Scanning Calorimetry 1
Andrew A. Lacey, Duncan M. Price, and Mike Reading
1 Introduction 1
2 The Basics of Modulated Temperature Differential Scanning Calorimetry 2
2.1 Some Preliminary Observations on Heat Capacity 2
2.2 The MTDSC Experiment and Deconvolution Procedure 4
2.2.1 The Simple Deconvolution Procedure 6
2.2.2 The Complete Deconvolution Procedure 9
2.2.3 Comments on the Different Deconvolution Procedures 10
2.2.4 Comments on Nomenclature 12
3 Practical Modulated Temperature DSC 13
3.1 The Importance of Linearity 13
3.2 Selection of Experimental Parameters 13
3.3 Common Transformations Studied by MTDSC 16
3.4 Chemical Reactions and Related Processes 16
3.4.1 Characteristics of MTDSC Results for Chemical Reactions and Related Processes 16
3.4.2 Summary 22
3.5 The Glass Transition 23
3.5.1 Characteristics of MTDSC Results for Glass Transitions 23

3.5.2 The Fictive Temperature and Enthalpy Loss on Annealing 34
3.5.3 Summary 37
3.6 Melting 38
3.6.1 Characteristics of MTDSC Results for Polymer Melting 38
3.6.2 The Measurement of Polymer Crystallinity 43
3.6.3 Summary 49
3.7 Calibration 50
3.7.1 Calibration of the Total and Reversing Signals 50
3.7.2 Comments on Methods of Phase Lag Correction 52
3.8 Overview 54
4 Detailed Discussion of the Theory of MTDSC 55
4.1 Introduction 55
4.2 Modulation and Deconvolution 55
4.3 Chemical Reactions and Related Processes 56
4.4 Frequency Dependent Heat Capacity and the Glass Transition 61
4.5 Melting 66
4.6 Calibration 69
References 80

2. The Application of Modulated Temperature Differential Scanning Calorimetry for the Characterisation of Curing Systems 83
Bruno Van Mele, Hubert Rahier, Guy Van Assche, and Steven Swier
1 Introduction 83
2 Polymer Networks and General Nature of Curing 85
2.1 General Aspects of Polymer Network Formation 85
2.2 Cure Reaction Mechanism 87
2.2.1 Step-Growth Polymerisation 88
2.2.2 Chain-Growth Polymerisation without Termination 89
2.2.3 Chain-Growth Polymerisation with Termination 90
2.2.4 Heterogeneous Polymerisation 91
2.3 Mechanistic Versus Empirical Cure Rate Law 91
2.4 Specific Versus Overall Diffusion Control 93
2.5 Glass Transition–Conversion Relationship 94
3 Experimental Procedures to Monitor the Cure Process 94
3.1 Gelation 94
3.2 Vitrification and the Glass Transition 95
3.3 Conversion and Conversion Rate 96
3.4 Diffusion Effects During Cure 97

4 Procedures for MTDSC Cure Measurements 98
4.1 General Considerations for Accurate Kinetic Analysis 98
4.1.1 Sample Preparation, Sample Size and Storage 98
4.1.2 Case of Volatile Reaction Products 99
4.1.3 (Quasi-)isothermal MTDSC Cure Experiments 99
4.1.4 Isothermal or Non-Isothermal Data for Kinetic Analysis? 99
4.1.5 Baseline for Isothermal and Non-Isothermal MTDSC Cure 100
4.1.6 Total Reaction Enthalpy 100
4.2 MTDSC Parameters 101
4.2.1 Modulation Amplitude 101
4.2.2 Modulation Period 101
4.2.3 Temperature-Dependent Heat Capacity Calibration 102
5 MTDSC Characterisation of Cure: Experimental Observations 102
5.1 Experimental Systems 102
5.1.1 Epoxy Systems 102
5.1.2 Polyester–Styrene Systems 103
5.1.3 Melamine–Formaldehyde Resins 103
5.1.4 Inorganic Polymer Glasses 103
5.2 Remarks Concerning MTDSC Signals 103
5.2.1 Non-Reversing MTDSC Heat Flow Versus Conventional DSC Heat Flow 103
5.2.2 Heat Flow Phase 105
5.3 Isothermal Cure with Vitrification 105
5.3.1 Epoxy Systems 105
5.3.2 Unsaturated Polyesters 106
5.3.3 Melamine–Formaldehyde Resins 109
Experimental Requirements and Reproducibility of Condensation Reactions 109
Vitrification During MF Cure 110
5.3.4 Inorganic Polymer Glasses 111
5.4 Non-Isothermal Cure with Vitrification and Devitrification 112
5.4.1 Epoxy–Anhydride 112
5.4.2 Epoxy–Amine 114
5.4.3 Effect of Heating Rate 115
5.5 Combined Cure Paths 116
5.6 Slow Isothermal Cure 118
5.6.1 Unsaturated Polyesters 118
5.6.2 Inorganic Polymer Glasses: Influence of Particle Size 118

5.7 Partial Vitrification 121
5.8 Mobility Factor to Quantify Degree of Vitrification 123
5.9 Heat Capacity Change as a Result of Reaction Before Vitrification 124
5.9.1 Mechanistic Information 124
5.9.2 Step-Growth Epoxy–Amine Polymerisation: Primary and Secondary Reactions 125
5.10 Reaction-Induced Phase Separation 127
6 Modelling the Diffusion-Controlled Overall Kinetics and Cure Rate Law of Epoxy Systems 129
6.1 Proposed Model 129
6.2 Optimised Cure Rate Law 133
6.2.1 Epoxy–Anhydride System 134
6.2.2 Epoxy–Amine System 137
6.3 Remarks Concerning the Proposed Model and Literature Models 139
6.4 Remarks Concerning the Frequency Dependence of Heat Capacity During Cure 140
7 Glass Transition–Conversion Relationship 142
7.1 Validation of the $T_g - x$ Model 142
7.2 ΔC_p and ΔT_g at T_g as a Function of Conversion 144
8 TTT and CHT Cure Diagrams 145
8.1 MTDSC Calculation Procedure 145
8.2 Influence of Chemical Structure on the (De)vitrification Behaviour of the Thermosetting Systems 153
9 Conclusions and Future Developments 154
References 156

3. Applications of Modulated Temperature Differential Scanning Calorimetry to Polymer Blends and Related Systems 161
Douglas J. Hourston and Mo Song
1 Introduction 161
2 Heat Capacity and its Differential with Temperature Signal Over the Glass Transition Region 165
3 Measurements of the Glass Transition Temperature and Increment of Heat Capacity 166
4 Multi-Component Polymer Materials 173
4.1 Improvement in the Measurement of Polymer–Polymer Miscibility 173
4.2 Interface Development Between Compatible Polymer Films 183

4.2.1 Asymmetrical Interdiffusion: Polyepichlorohydrin/Poly(vinyl acetate) 186
4.2.2 Symmetrical Interdiffusion: Poly(methyl acrylate)/Poly(vinyl acetate) 190
4.3 Structured Latex Films 195
4.4 Morphology Analysis of Interpenetrating Polymer Networks 203
4.4.1 Characterisation of Glass Transition Behaviour in Interpenetrating Polymer Networks 204
4.4.2 Model Experiment 207
4.4.3 Analysis of Phase Structure of IPNs 209
5 Conclusions 211
References 212

4. The Application of MTDSC to Polymer Melting 217
Bernhard Wunderlich
1 Introduction 217
2 The Melting and Crystallisation Behaviour of Polymers 219
2.1 Equilibrium Melting 219
2.2 Nucleation of Crystals and Molecules 224
2.3 Irreversible Melting 227
2.4 Local Equilibria 233
2.5 Summary of Thermal Effects 235
3 Instrument and Deconvolution Problems 237
3.1 Evaluation of Heat Capacity by MTDSC as a Baseline for the Study of Melting 237
3.2 Melting and Crystallisation by Standard DSC 245
3.3 Types of Modulation of Temperature 250
3.4 Deconvolution of the Reversing Heat Capacity 252
4 Applications of MTDSC to Polymer Melting 258
4.1 Qualitative and Semi-quantitative Analysis of Polymer Melting by MTDSC 258
4.1.1 Summary 268
4.2 Determination of Heat Capacity of Solids and Melts 269
4.2.1 Summary 274
4.3 Determination of Heat of Fusion, Crystallinity and Kinetics 275
4.3.1 Heat of Fusion and Crystallinity 275
4.3.2 Baseline Fits 276
4.3.3 Quasi-Isothermal Kinetics of the Glass Transition 276
4.3.4 Model Calculation for the Glass Transition with an Underlying Heating Rate 281

4.3.5 Kinetics of Transitions with a Latent Heat 285
4.3.6 Mesophase Transitions 289
4.3.7 Analysis by MTDSC Kinetics of Transitions with a Latent Heat 290
4.3.8 Summary 293
4.4 Determination of Annealing and Reorganisation 294
4.4.1 Annealing of PET 295
4.4.2 Annealing of PTT 297
4.4.3 Annealing and Reversible Recrystallisation in Low Molar Mass PEO 299
4.4.4 Annealing in PEcoO 304
4.4.5 Summary 308
4.5 Reversible Melting 309
4.5.1 Summary 313
5 Recommendations 314
Acknowledgements 316
References 316

Index 321

Foreword

It is always possible to dispute exactly when a new technique was introduced because inventions that bear no resemblance at all to what has gone before are very rare. However, there is a reasonable argument that AC calorimetry (a technique where a measurement is made of the amplitude and phase of the response of a sample to temperature modulation) was first used in the 1960s [1], as was the modern form of the more commonly used technique, differential scanning calorimetry (DSC) [2]. In 1972, a brief paper was published demonstrating the use of sinusoidal temperature modulation with a DSC coupled to an electromechanical lock-in amplifier to obtain the amplitude and phase of the sample's response [3]. Amplitude and phase data (expressed as a complex quantity) were given for four frequencies at one temperature. These authors neither demonstrated nor proposed that anything more than the AC signals could be obtained, and it was not until 1992 that Reading and co-workers introduced a method that provided both the AC signals (amplitude and phase) and the total heat flow signal equivalent to that given by DSC simultaneously in a single experiment [4–8]. The method used a conventional DSC, and the signals were produced by a deconvolution procedure carried out by computer. The name modulated temperature DSC was proposed, but this became copyrighted by TA Instruments, and so the term modulated temperature DSC (MTDSC) was coined. In addition to the technique, these workers introduced a simple theory and method of interpretation that focused on the differences between the AC and the DC signals (often expressed by the 'non-reversing' signal). It was shown how 'hidden' glass transitions could be seen and why this was important for understanding cure behaviour [5] (independently Cassettari

M. Reading and D. Hourston (eds.), Theory and Practice of Modulated Temperature Differential Scanning Calorimetry, xi–xiii.

et al. [9] also developed a technique for doing this on the basis of a modified isothermal calorimeter), how glass transitions could be analysed in far more detail for polymer blends [10,11] and how useful additional information about melting could be gleaned [6]. All of these themes have been much further developed by a variety of workers and the leaders in each of these fields; curing, blends and semi-crystalline polymer systems have contributed to this book.

At the heart of the deconvolution procedure proposed by Reading and co-workers was the idea that it is important to disentangle the sample response that depends on temperature from the response that depends on rate of change of temperature. Others had been exploring a related theme in dielectric response [12]. Since the introduction of MTDSC, exactly the same principles have been applied to thermomechanical analysis by Price [13].

As soon as it was introduced, MTDSC became a controversial technique. This was is large part because, unusually for a new method of characterisation, it was launched as a commercial product with no 'gestation' period in an academic laboratory. Commercial rivalries inevitably led to conflict and a certain amount of misrepresentation. There was also confusion because the first commercial version did not include the ability to use the phase angle to separate the response to the modulation into in- and out-of-phase components. Despite this omission, the fact that this was an option was demonstrated and discussed at the time MTDSC was first described [1–3]. Some workers leapt to the conclusion that this had not been considered and criticised the technique for this reason proposing the 'alternative' method of deriving a complex heat capacity [14] (like Goldbrecht *et al.* [3]). The debate became polarised into advocacy of one or the other approach when, in reality, this conflict was entirely artificial and the use of the phase angle is completely compatible with the practise and theory of MTDSC [15]. Unfortunately, even today this fallacy persists.

Another reason for controversy was the unexpected nature of some of the first results, especially in the melt region. Without going into all the details, the disinclination to accept these data as valid has largely been overcome and important new insights have been gained in the process.

Introducing MTDSC to an often sceptical world was not easy. The authors of these chapters have all played their part. Another notable contributor in this respect was Len Thomas whose tireless efforts made an enormous contribution to wider acceptance and understanding in the early years. Now, the 'dust has settled' and MTDSC has almost passed into routine use, this book is intended to be the collection of all the information an experimenter might need to make the best use of this technique when applying it to polymers. By concentrating it all into one place, we hope to make matters simpler for the user and to promote a better understanding of the benefits and dangers of modulation in DSC.

M Reading and D J Hourston 17/10/2005

References

[1] Y.A. Kraftmakher, Zh. Prikl. Mekh. Tekh. Fiz., 5 (1962) 176.
[2] E.S. Watson, M.J. O'Neil, J. Justin and N. Brenner, Anal. Chem., 36 (1964) 1233.
[3] H. Goldbrecht, K. Hamann and G. Willers, J. Phys. E Sci. Instrum., 4 (1971) 21.
[4] M. Reading, D. Elliott and V.L. Hill, Proc. NATAS (1992) 145.
[5] M. Reading, Trends Polym. Sci., 11(993) 8.
[6] M. Reading, A. Luget and R. Wilson, 238 (1994) 295.
[7] A.A. Lacey, C. Nikolopoulos, M. Reading, 50 (1997) 279.
[8] P.J. Haines, M. Reading and F.W. Wilburn, In Differential Thermal Analysis and Differential Scanning Calorimetry (Chapter 5), M.E. Brown, Ed., Elsevier (1998) p. 321.
[9] M. Cassettari, F. Papucci, G. Salvetti, E. Tombari and S. Verones, Rev. Sci. Instrum., 64 (1993) 1076.
[10] M. Song, A. Hammiche, H.M. Pollock, D.J. Hourston and M. Reading, Polymer, 36, (1995) 3313–3316.
[11] D.J. Hourston, M. Song, A. Hammiche and M. Reading, Polymer, 37 (1996), 243–247.
[12] L.E. Garn and E.J. Sharp, J. Appl. Phys., 53 (1982) 8974.
[13] D.M. Price, Thermochim. Acta, 315 (1998) 11.
[14] J.E.K. Schawe, Thermochim. Acta, 261 (1995) 183.
[15] M. Reading, Thermochim. Acta, 292 (1997) 179.

Chapter 1

THEORY AND PRACTICE OF MODULATED TEMPERATURE DIFFERENTIAL SCANNING CALORIMETRY

Andrew A. Lacey*, Duncan M. Price¶, Mike Reading**
**Department of Mathematics, Heriot-Watt University, Riccarton, Edinburgh EH14 4AS, UK*

¶Institute of Polymer Technology & Materials Engineering, Loughborough University, Loughborough LE11 3TU, UK

***School of Chemical Sciences and Pharmacy, University of East Anglia, Norwich, NR4 7TJ, UK*

1 Introduction

There are two different literatures on the subject of DSC and calorimetry in general. The first deals mainly with its applications, the second primarily with the technique itself. The latter includes, amongst other things, commentary on instrument calibration, the limits of sensitivity and resolution, the details of modelling the response of the calorimeter and separating the effects of the measuring system from those due to the phenomenon being studied. Certainly there is overlap between these two bodies of work. However, it is also true that it is not necessary to understand fully the details of the equations that can be used to model heat flow in a DSC cell in order to measure and interpret a glass transition successfully. In this book, we attempt to strike a balance between satisfying both audiences. In this chapter, in particular, we attempt, in the first part, to provide sufficient information to enable the polymer scientist to interpret correctly his or her results while not burdening the reader with details that might ultimately obscure the central meaning. This is intended for those more interested in

M. Reading and D. Hourston (eds.), Theory and Practice of Modulated Temperature Differential Scanning Calorimetry, 1–82.

the results themselves than the process by which they are derived. There is a discussion of theory, but this is confined to the important results rather than the details of their derivation. In the second part, a more extended discussion is offered on the considerable complexities of understanding the details of modulated temperature calorimetry in its modern form (i.e. an experiment where both the response to the modulation and underlying heat flow are obtained simultaneously and compared for a wide range of transitions). The first part is called 'Practical MTDSC', the second 'Detailed Discussion of the Theory of MTDSC'.

It is not the intention of this chapter to be a review of the literature (if the reader is looking for this, Ref. [1] is a recent example). Its purpose is to serve as an introduction to the technique of MTDSC starting with fairly basic and practical matters than progressing onto more advanced levels. It is also intended to serve as a guide to understanding the remaining chapters that deal with three principal classes of polymeric materials, thermosets, thermoplastic polymer blends and semi-crystalline polymers.

The use of a modulated temperature profile with DSC, combined with a deconvolution procedure in order to obtain the same information as conventional DSC plus, at the same time, the response to the modulation, was first proposed by Reading and co-workers [2–17]. In this section, we describe the basic deconvolution procedure i.e. how that data are processed and presented for a typical polymer sample. We then consider how these data are interpreted.

2 The Basics of Modulated Temperature Differential Scanning Calorimetry

2.1 SOME PRELIMINARY OBSERVATIONS ON HEAT CAPACITY

Heat capacity can be defined as the amount of energy required to increase the temperature of a material by 1 degree Kelvin or Celsius. Thus,

$$C_\mathrm{p} = Q/\Delta T \tag{1}$$

where C_p = the heat capacity
ΔT = the change in temperature
Q = amount of heat required to achieve ΔT.

Often, it would be considered that this is the heat stored in the molecular motions available to the material, that is the vibrational, translational motions etc. It is stored reversibly. Thus, the heat given out by the sample when it is cooled by 1°C is exactly the same as that required to heat it by the same amount. This type of heat capacity is often called *vibrational heat capacity*. Where temperature is changing, the rate of heat flow required to achieve this is given by

$$dQ/dt = C_p dT/dt \tag{2}$$

where t = time

This is intuitively obvious. Clearly, if one wishes to increase the temperature of the material twice as fast, twice the amount of energy per unit time must be supplied. If the sample has twice the heat capacity, this also doubles the amount of heat required per unit time for a given rate of temperature rise. Considering a linear temperature programme, such as is usually employed in scanning calorimetry

$$T = T_0 + \beta t \tag{3}$$

where T = temperature
T_0 = starting temperature
β = the heating rate, dT/dt.

This leads to

$$dQ/dt = \beta C_p \tag{4}$$

or

$$C_p = (dQ/dt)/\beta \tag{5}$$

This provides one way of measuring heat capacity in a linear rising temperature experiment: one simply divides the heat flow by the heating rate. If the temperature programme is replaced by one comprising a linear temperature ramp modulated by a sine wave, this can be expressed as

$$T = T_0 + \beta t + B \sin \omega t \tag{6}$$

where B = the amplitude of the modulation
ω = the angular frequency of the modulation.

The derivative with respect to time of this is

$$dT/dt = \beta + \omega B \cos \omega t \tag{7}$$

Thus, it follows

$$dQ/dt = C_p(\beta + \omega B \cos \omega t) \tag{8}$$

For the special case where β is zero, this yields

$$\mathrm{d}Q/\mathrm{d}t = C_\mathrm{p}\omega B \cos \omega t \tag{9}$$

For the simplest possible case, from Eq. (2), the resultant heat flow must also be a cosine wave. Thus,

$$A_\mathrm{HF} \cos \omega t = C_\mathrm{p}\omega B \cos \omega t \tag{10}$$

where A_HF = the amplitude of the heat flow modulation

It follows that ωB = the amplitude of the modulation in the *heating rate*. Thus,

$$C_\mathrm{p} = A_\mathrm{HF}/A_\mathrm{HR} \tag{11}$$

where A_HR = amplitude of modulation in heating rate ($= \omega B$).

This provides a second method of measuring heat capacity, by looking at the amplitude of the modulation. The same relationship applies even if there is an underlying heating ramp.

In essence, MTDSC is based on simultaneously measuring the heat capacity of the sample using both methods, the response to the linear ramp and the response to the modulation, and comparing them. When the sample is inert and there are no significant temperature gradients between the sample temperature sensor and the centre of the sample, both methods should give the same value. The interest lies in the fact that during transitions, these two methods give different values.

2.2 THE MTDSC EXPERIMENT AND DECONVOLUTION PROCEDURE

Although many different forms of temperature programme are possible, a sinusoidal temperature modulation is most often used, as illustrated in Figure 1.1. Figure 1.2 shows data for amorphous poly(ethylene terephthalate), PET, from below its glass–rubber transition temperature (T_g) to above its melting temperature. The modulation in heating rate and the resultant heat flow is shown as well as one of the signals derived from the deconvolution procedure, the phase lag between the modulation in the heating rate and that in the heat flow. As the first step in the deconvolution process, the raw data are averaged over the period of one oscillation to remove the modulation. This then gives the total signal, which is equivalent to the signal that would have been obtained had the modulation not been used, i.e. a conventional

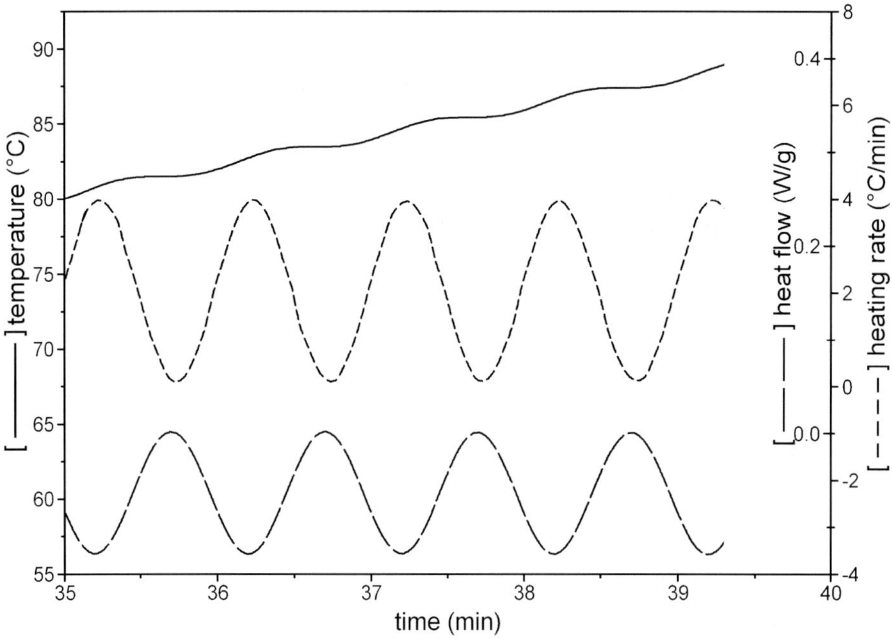

Figure 1.1. Typical temperature–time curve for an MTSC experiment (top) with resultant heating rate modulation and heat flow response (underlying heating rate: 2°C/min, period: 60 s, amplitude: 0.32°C under nitrogen).

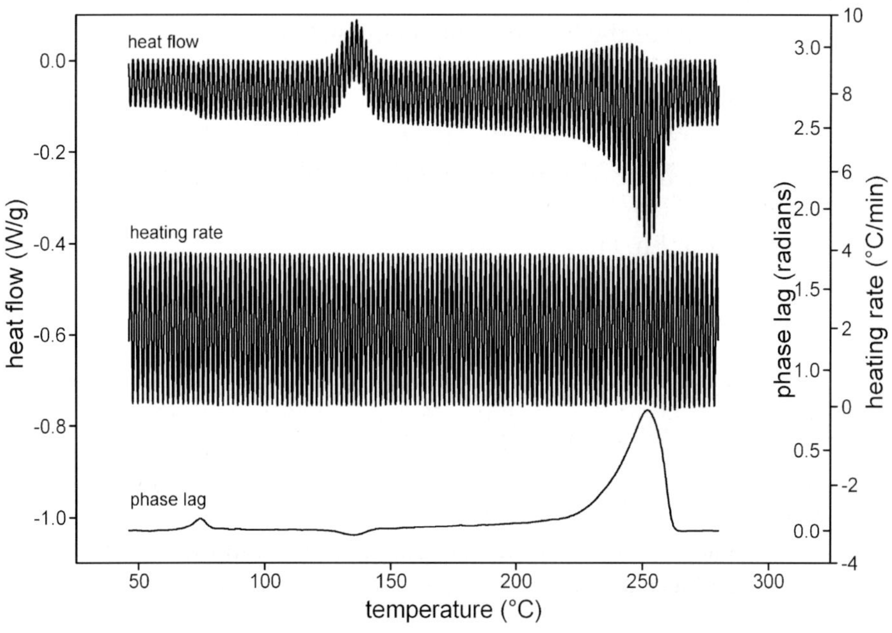

Figure 1.2. Raw data from an MTDSC experiment for quenched PET plus one signal resulting from the Fourier transform, the phase lag (underlying heating rate: 2°C/min, period: 60 s, amplitude: 0.32°C under nitrogen).

DSC experiment (see below). The averaged signal is subtracted from the raw data and the modulation is then analysed using a Fourier transform procedure to obtain the amplitude and phase difference of the heat flow response at the frequency of the imposed modulation.

In contrast to the very simple treatment outlined above in section 2.1, one can allow for the situation that the heat flow modulation might not always follow exactly the cosine modulation in the heating rate (for reasons that will become clear in the discussions on various transitions). Thus, the heat flow may lag behind.

$$\text{The heat flow modulation} = A_{\text{HF}} \cos(\omega t - \phi) \tag{12}$$

where $\phi =$ the phase difference between the modulation in the heat flow and the heating rate, also termed the 'phase lag'.

The basic output from the first stage of the deconvolution procedure is therefore,

$$\langle \mathrm{d}Q/\mathrm{d}t \rangle = \text{the average or total heat flow}$$

where $\langle\rangle$ denotes the average over one or more periods.

$Q =$ heat,
$A_{\text{HF}} =$ amplitude of the heat flow modulation,
$A_{\text{HR}} =$ amplitude of modulation in the heating rate,
$\phi =$ the phase lag.

Having obtained the amplitudes of the modulations in heating rate and heat flow, the next step is to use these quantities to calculate a value for the heat capacity as in Eq. (11).

$$A_{\text{HF}}/A_{\text{HR}} = C^* \tag{13}$$

where $C^* =$ the reversing heat capacity (also called the cyclic heat capacity or modulus of complex heat capacity. See below).

There are then two alternative ways of proceeding with the deconvolution – both of which were originally proposed by Reading and co-workers [2–5].

2.2.1 The Simple Deconvolution Procedure

If the results are to be expressed as heat capacities, then the average total heat flow is divided by the underlying heating rate β. Thus,

$$\langle \mathrm{d}Q/\mathrm{d}t \rangle/\beta = C_{\text{pT}} = \text{the average or total heat capacity} \tag{14}$$

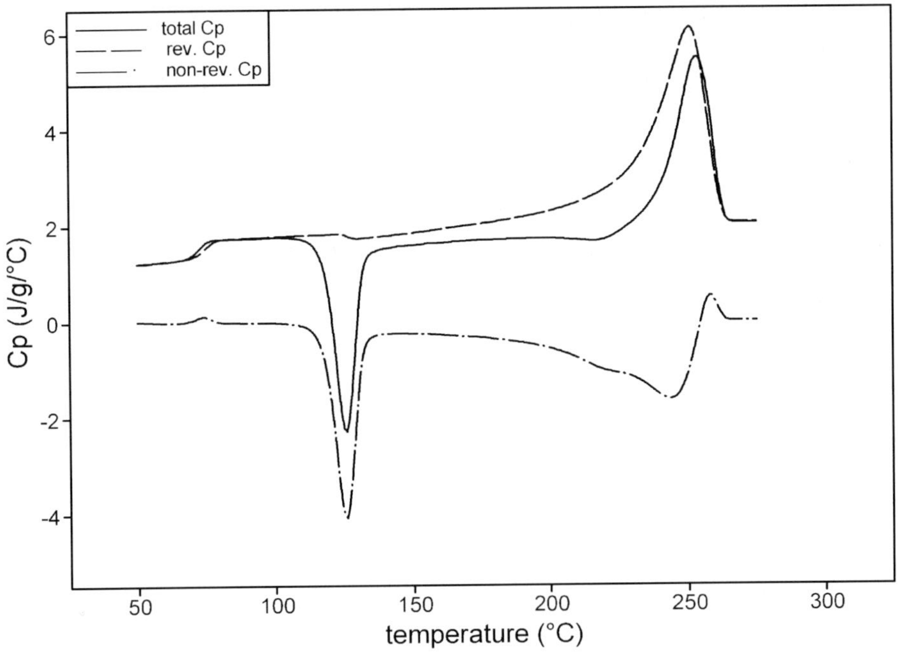

Figure 1.3. Results from simple deconvolution procedure for the data shown in Figure 1.2.

Having obtained the reversing heat capacity, then one can calculate the non-reversing heat capacity. Viz

$$\langle \mathrm{d}Q/\mathrm{d}t \rangle/\beta - C^* = \text{the non-reversing heat capacity} = C_{\mathrm{pNR}} \qquad (15)$$

This is illustrated in Figure 1.3 using the data shown in Figure 1.2. Note that in non-transition regions, for example below the glass transition and in the molten state, the reversing and total heat capacities are the same. As should be clear from the discussion in section 2.1, and the theoretical arguments advanced below, this is exactly what we would expect. If measurements were made on an inert material such as sapphire, then the reversing and total signals should be coincident and the reversing signal would be zero. However, all measurements contain errors and so exact agreement is difficult to achieve. It requires careful calibration (see below) and good experimental practice. Where there are minor discrepancies, it is useful to use non-transition regions as a kind of internal calibration and use a linear baseline correction such as is illustrated in Figure 1.4. The two signals are forced to be the same where it is known that they should be. Whether the total or the reversing heat capacity is taken to give the 'correct' value is a matter of

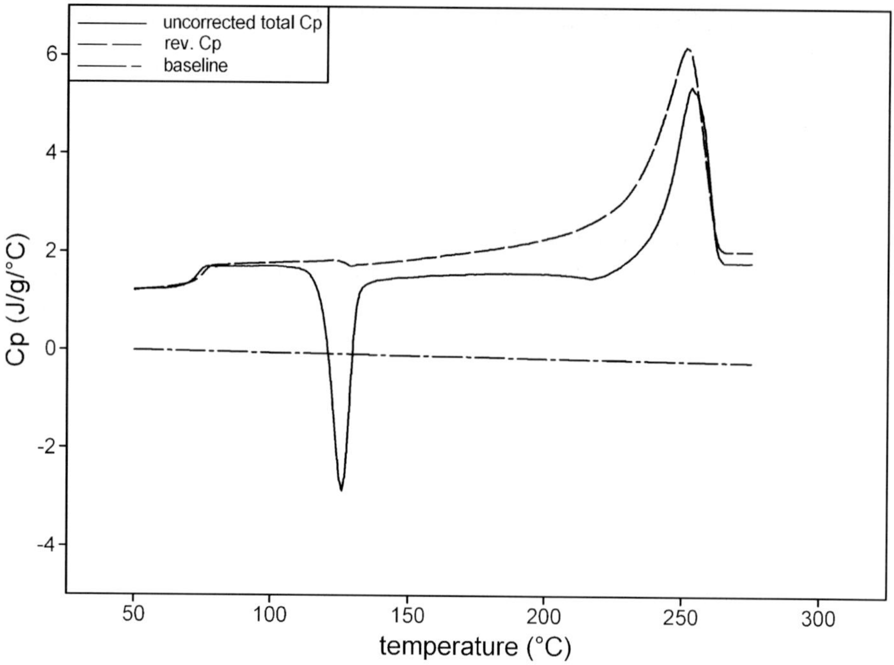

Figure 1.4. Total and reversing C_p before baseline correction applied to data in Fig. 1.3.

judgement given the experimental conditions used (and may well be irrelevant depending on what information is being sought from the experiment). This is discussed below in the sections dealing with selection of experimental conditions and calibration. The non-reversing signal is calculated after any shift to make the non-transition reversing and total signals the same.

It can be argued that enthalpies associated with, for example, crystallisation, should not be expressed as changes in heat capacity in the way shown in Figure 1.3. Perhaps this term is best reserved for the reversible storage of heat in the motions of the molecules such as we see in the non-transition regions. This is a moot point. In practice, results are often expressed in terms of heat capacity, regardless of any transitions that occur, and this convention is followed in this book.

Although all of these signals in Figure 1.3 are expressed as heat capacities, they can equally well be expressed as heat flows.

$$\langle \mathrm{d}Q/\mathrm{d}t \rangle = \text{average or total heat flow}$$

$$C^* \beta = \text{reversing heat flow} \tag{16}$$

This is then subtracted from the total heat flow to obtain the non-reversing heat flow. Viz

$$\langle \mathrm{d}Q/\mathrm{d}t\rangle - C^{*}\beta = \text{non-reversing heat flow} \tag{17}$$

The convention often adopted for heat flux DSCs means that exotherms go up and so, in addition to changing the units on the y-axis, expressing the signals as heat flows also sometimes means inverting the curves compared to Figure 1.3. However, it is not uncommon to express exotherms as going down even when plotting the data as heat flow. The reader simply needs to be careful in regard to what units and conventions are being used.

Note that the simple deconvolution procedure makes no use of the phase lag signal.

2.2.2 The Complete Deconvolution Procedure

In this procedure, the phase lag is used to calculate the in- and out-of-phase components of the cyclic heat capacity. Viz

$$C^{*}\cos\phi = \text{phase-corrected reversing heat capacity} = C_{\mathrm{pPCR}} \tag{18}$$

$$C^{*}\sin\phi = \text{kinetic heat capacity} = C_{\mathrm{pK}} \tag{19}$$

In reality, the phase angle cannot usually be used directly. A baseline correction is required. This is dealt with in the calibration section below.

The complete deconvolution then proceeds in the same way as for the simple deconvolution, except that the phase-corrected reversing heat capacity is used instead of the reversing heat capacity. Thus

$$\langle \mathrm{d}Q/\mathrm{d}t\rangle/\beta - C_{\mathrm{pPCR}} = \text{phase-corrected non-reversing heat flow} = C_{\mathrm{pPCNR}} \tag{20}$$

The results of this deconvolution applied to the data in Figure 1.2 are given in Figure 1.5.

Again, all of the signals can also be expressed as heat flows.

$$\langle \mathrm{d}Q/\mathrm{d}t\rangle = \text{average or total heat flow}$$

$$C_{\mathrm{pPCR}}\beta = \text{phase-corrected reversing heat flow} \tag{21}$$

$$\langle \mathrm{d}Q/\mathrm{d}t\rangle - C_{\mathrm{pPCR}}\beta = \text{phase-corrected non-reversing heat flow} \tag{22}$$

$$C_{\mathrm{pK}}\beta = \text{kinetic heat flow} \tag{23}$$

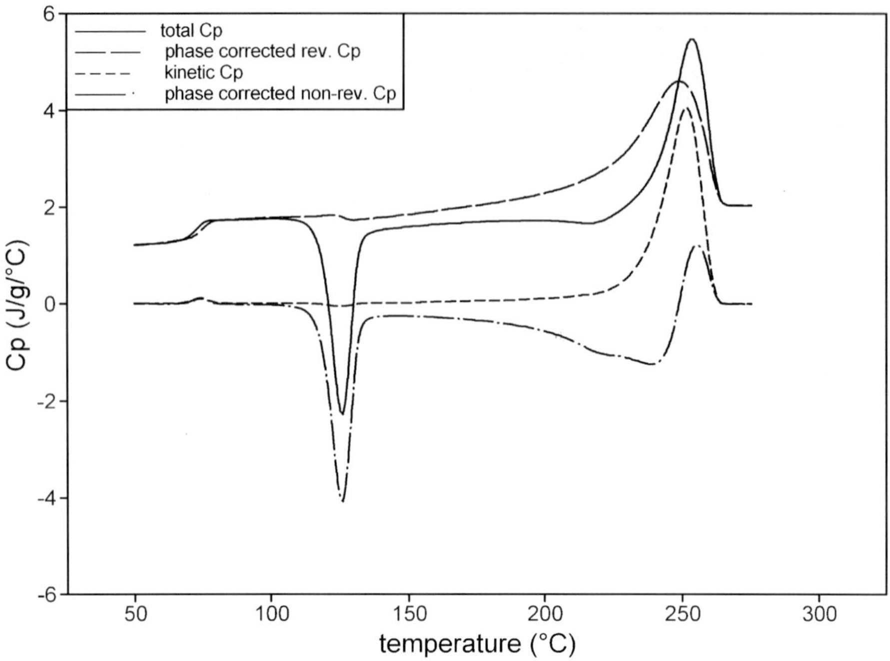

Figure 1.5. Results of complete deconvolution procedure for the data shown in Figure 1.2.

Thus, in general, all signals can equally well be expressed as heat capacities or heat flows simply by multiplying or dividing by the underlying heating rate, β, as appropriate. Often both types of signals are mixed, so reversing heat capacity is co-plotted with non-reversing heat flow.

2.2.3 Comments on the Different Deconvolution Procedures

In Figure 1.6, a comparison is made between the reversing and the phase-corrected reversing and non-reversing signals. It can be seen that there is only a significant difference in the melt region. In reality, the simple deconvolution is an approximate form of the complete deconvolution procedure. The phase correction is, in most polymer transitions, except melting, negligible, as illustrated in Figure 1.6. Thus,

$$C^* \approx C_{\mathrm{pPCR}} \tag{24}$$

A quantitative interpretation of results in the melt region, with or without the use of the phase lag, is often problematic. As a consequence of this, it often does not matter whether the phase correction has been applied or not

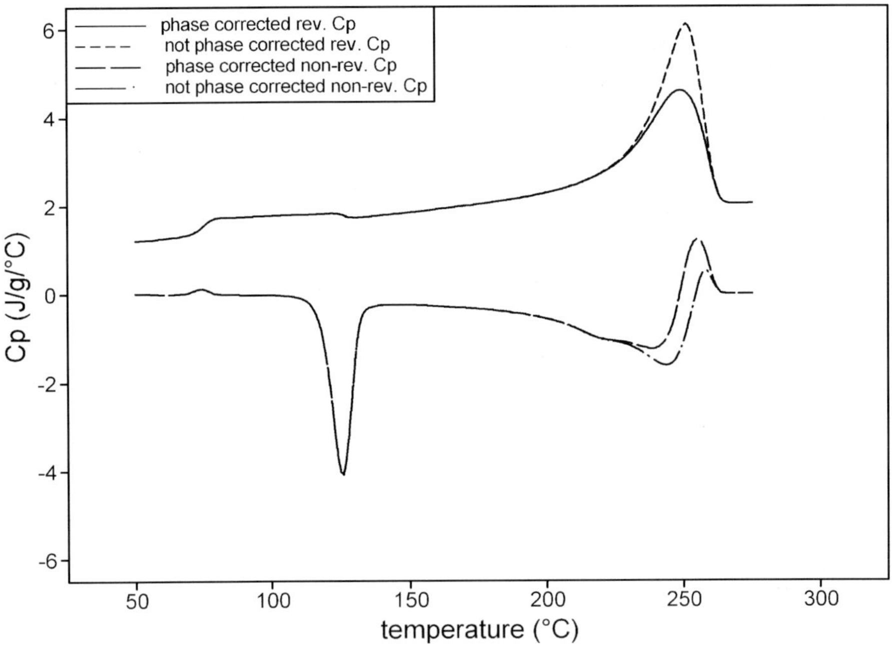

Figure 1.6. Co-plot of reversing and non-reversing heat capacity arising from the simple and full deconvolution procedures applied to the raw data from Figure 1.2.

unless the kinetic heat flow is specifically of interest. In many of the applications described in this book, no phase correction has been used. However, it must be stressed that there is no conflict between the simple and complete deconvolution procedures. Perhaps because of an initial misunderstanding in the literature [2,15,17,18], even today authors present the deconvolution into reversing and non-reversing as an alternative to using the phase correction (to derive the phase-corrected reversing and kinetic heat capacities or complex heat capacity. See below). It is sometimes even presented as a rival method. This confusion in the literature is an obstacle to a proper understanding of the technique and it is to be hoped that in future it will cease. The use of the phase lag is an optional refinement that has always been part of MTDSC from the time it was first introduced [2]. The full deconvolution does provide the maximum information and workers who prefer this are encouraged to pursue it. If it is not used routinely, it is simply because the phase lag is sensitive to non-ideal behaviour of the combination of the sample, pan and measuring system and correcting for this requires additional effort (see the calibration section below) often with little practical benefit. However, improvements in instrumentation and software will probably make the full deconvolution routine in future.

2.2.4 Comments on Nomenclature

The reason for the nomenclature reversing and non-reversing will be given below as part of the discussion on practical MTDSC in the section on chemical reactions and related processes. It was the original intention of Reading and co-workers that the term reversing should mean what is referred to above as the phase-corrected reversing [19], while accepting that in most cases the phase lag correction would not be used because it is very small. However, the de facto current practice is that reversing applies to the non-phase-corrected signal and this is the convention that we use in this book.

It is also possible, and often helpful, to use complex notation. The ratio of the amplitudes of the modulations of the temperature rise and heat flow gives one useful piece of information: $C^* = A_{\mathrm{HF}}/A_{\mathrm{HR}}$. The phase lag gives another. These two bits of information are equivalent to knowing both C_{pPCR} and C_{pK}, or the single complex quality $\hat{C} = C_{\mathrm{pR}} - iC_{\mathrm{pK}}$ where $i =$ the square root of -1. Since the temperature rise and heat flow modulations can be written as $\mathrm{Re}\{\omega B \mathrm{e}^{i\varpi t}\}$ and $\mathrm{Re}\{A_{\mathrm{HF}}\mathrm{e}^{i(\omega t - \phi)}\} = \mathrm{Re}\{(A_{\mathrm{HF}}\cos\phi - iA_{\mathrm{HF}}\sin\phi)\mathrm{e}^{i\omega t}\}$, respectively, $(A_{\mathrm{HR}} = \omega B)$, the complex heat capacity can be defined directly.

$$\hat{C} = \frac{A_{\mathrm{HF}}\mathrm{e}^{-i\phi}}{A_{\mathrm{HR}}} = C' - iC'' \tag{25}$$

where $C_{\mathrm{pPCR}} = C' =$ the real component,
$C_{\mathrm{pK}} = C'' =$ the imaginary component.

Manipulations needed to relate heat flow A_{HR} to temperature changes through theoretical models for transitions, or through properties of calorimeters, are usually more conveniently done via such complex qualities. The value $\hat{C}$ can then lead directly to evaluations of real specific heat and parameters controlling kinetics. However, the use of complex notation does not imply a different theoretical treatment or method. It is simply a more convenient mathematical formalism. The terms 'real' and 'imaginary' heat capacity and 'phase-corrected reversing' and 'kinetic' heat capacities are interchangeable.

It is regrettable that such a proliferation of names is in common use and this must be confusing to many workers. However, by paying close attention to the above text, it should be possible to deduce the correct signal in almost all cases.

3 Practical Modulated Temperature DSC

3.1 THE IMPORTANCE OF LINEARITY

One point that needs to be mentioned is that the analysis described above assumes that the sample's response to the modulation can be approximated as linear. Clearly, the processes such as those that follow Arrhenius kinetics or related kinetics of a glass transition are not linear with temperature. However, over a small temperature interval they can be approximated as linear. Where this cannot be said to be true, the above analysis fails because it assumes a linear response.

Where a multiplexed sine wave or saw-tooth modulation is used the deconvolution procedure can be used to extract the response at a series of frequencies [4,10,19,20]. However, current commercial products restrict themselves to using the first component of the Fourier series, which is then, with the assumption of linearity, equivalent to using a single sinusoidal modulation. It is true that looking at the whole Fourier series, rather than just the first component, offers scope for increasing the amount of information that can be obtained from an MTDSC experiment. This applies even to single sinusoidal modulations (because non-linearities produce harmonics) as well as multiple simultaneous sine waves or saw-tooth modulations. This will be considered in greater detail below in the section on advanced MTDSC.

3.2 SELECTION OF EXPERIMENTAL PARAMETERS

A fundamental consideration that always applies is the requirement that there be many modulations over the course of any transition. Stated simply, the deconvolution procedure described above can only make sense if the underlying heat flow is changing slowly and smoothly under the modulation. If this is true, averaging the modulated signal over the period of the modulation will provide, to a good approximation, the same information as an un-modulated experiment. The averaging will usually mean the modulated experiment looks 'smoothed' to some extent. Thus, the tops of peaks may be a little 'rounder', but the areas under the peaks and all of the essential features will be the same. If a significant part of a transition occurs over the course of a single modulation, this invalidates the assumptions behind the use of the Fourier series. As the reader proceeds through the sections on theory and typical results, it is hoped that these points will become intuitively obvious. As a general rule of thumb for most polymer applications, where the transition is a peak in $\mathrm{d}Q/\mathrm{d}t$, then there should be at least five modulations over the period represented by the width at half height. Where the transition

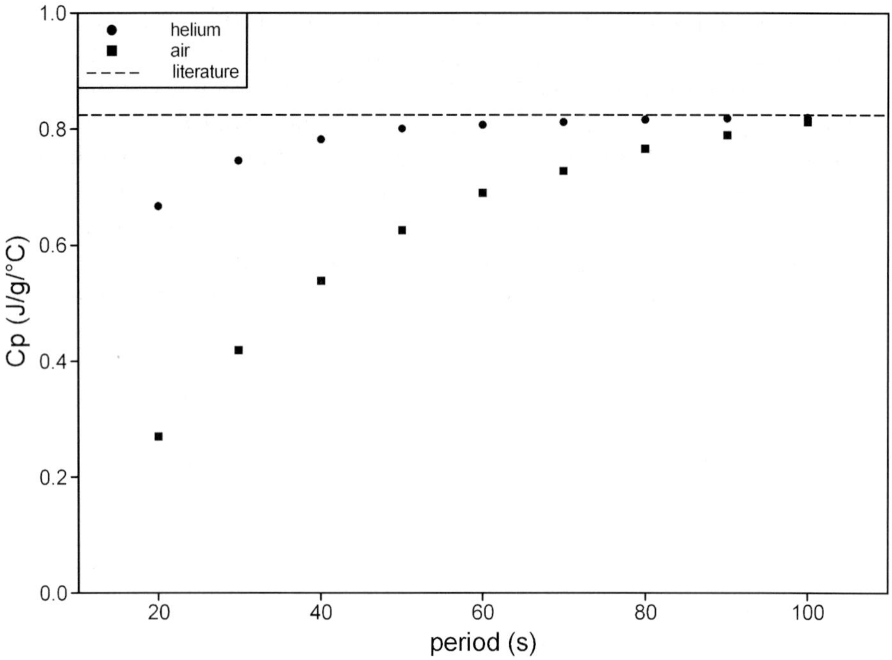

Figure 1.7. Reversing C_p for sapphire as a function of period in helium and air (quasi-isothermal measurement at 50°C, amplitude 1°C). Literature value from ref. 38.

is a step change, there should be at least five modulations over that part of the transition where change is most rapid. Where there is doubt, the number of modulations should be increased by reducing the underlying temperature ramp to check whether this significantly changes the reversing signal.

There is the question of what period should be used. As mentioned above, for an inert material the reversing heat capacity should provide an accurate measure of the specific heat capacity (= heat capacity/mass) of the sample when the calorimeter is calibrated in the conventional way (see below). This is true when the period is long, typically over 100 s or more. As the period becomes shorter, the apparent reversing heat capacity becomes smaller as illustrated in Figure 1.7. This happens because there are thermal resistances between the pan and the temperature sensor, the pan and the sample and within the sample itself. A long period implies a slow underlying heating rate that is undesirable because this means a long time for the experiment and a reduction in the signal-to-noise in the total signal. A typical compromise is 60 s used with a calibration factor determined using a calibrant with an accurately known heat capacity. (This is described in the calibration sections below). In Figure 1.7, it can be seen that the effects of the thermal resistances are smaller when helium is used and a reasonable compromise is 40 s (again with a calibration factor). While it is true that considerable progress has been made by some workers in characterising and compensating for these non-ideal experimental conditions [12,21–27], for most experimentalists, the best approach is to use longer periods that avoid the complications engendered by these thermal resistances. It should be noted that, if helium is used, the concentration of helium in the actual cell will generally not be 100% and will vary with flow rate. This means that the flow rate must be accurately controlled (usually with a mass flow controller).

Once the period is chosen, the requirement that there be many modulations over the course of all transitions then sets limits on the maximum heating rate that can be used. A typical heating rate with a 60-s period would be 2°C/min, or 3°C/min for a 40-s period. A lower rate might be used if a transition is particularly sharp or more resolution is required. Alternatively, there will be circumstances when a faster underlying heating rate might be used. Generally, in current instruments, which usually use a nitrogen purge, a 60-s period with a 2°C/min ramp is a reasonable starting point, but as in conventional DSC the conditions will vary according to the sample and the specific information being sought.

The choice of modulation amplitude is firstly governed by the signal-to-noise ratio. If the amplitude is too small, then it will be difficult to detect and so the signal-to-noise will degrade. A few tenths of a degree should normally be sufficient. If the amplitude is too large then this will 'smear' the transition. Consider a glass transition that is 10°C wide. If the modulation amplitude is also 10°C, then when the average temperature is 5°C below its onset, the modulation will already be significantly influenced by the transition. There is also the problem of linearity. If the amplitude it too large, then the response will be significantly non-linear. A check is to change the amplitude and it should be possible to find a range of values where the result remains invariant. An amplitude of 0.5°C will often give satisfactory results for the kinds of applications considered in this book.

It is possible to select a programme for a rising temperature experiment such that the minimum heating rate is always positive or zero (this is the case in Figure 1.2), or the heating rate is sometimes negative. In the next section, the various different types of transition that can be studied by MTDSC are discussed. In general, any type of heating programme can be used except when it is the melting behaviour that is of interest. In the case of melting, it has been shown that the material that melts while the temperature is increasing will not crystallise when the temperature is decreased [8,10]. This then gives rise to a highly asymmetrical and, therefore, non-linear response to the modulation. Consequently, when melting is being studied, conditions should be chosen so that the heating rate is never negative. (This is sometimes referred to as 'heat-only' conditions).

In reality, it is not possible to recommend experimental conditions that will apply very generally to a wide range of materials and types of study. The above comments are intended as a simple guide for the novice. These guidelines are often contravened in this this book! There is no substitute for gaining a good understanding of the basic theory of MTDSC and then building experience through practical study.

3.3 COMMON TRANSFORMATIONS STUDIED BY MTDSC

In the next part of this chapter, we will consider the most commonly encountered types of processes that are studied by MTDSC in polymeric materials. The types of results they give and the appropriate specific kinetic functions will be discussed. The categories are as follows.
1) Chemical reactions and related processes.
2) Glass transitions.
3) Melting.

3.4 CHEMICAL REACTIONS AND RELATED PROCESSES

3.4.1 Characteristics of MTDSC Results for Chemical Reactions and Related Processes

In this section, the discussion will begin with the simplest case that can realistically be considered—a zero-order irreversible chemical reaction. In this example, the reaction rate is a function only of temperature until all reactant is consumed and the reaction stops. The exact function governing the temperature dependence of the reaction rate is not defined in this initial analysis, but it can be, it is assumed, approximated to be linear over the small temperature interval of the modulation. The more general case where the chemical reaction can be considered to be a function of time (and therefore conversion) and temperature is then treated. Finally, the Arrhenius equation is dealt with, as this is the most relevant case to the subject of this book.

In the case of a zero-order reaction, the rate of the reaction is dependent only on the temperature. Thus, it produces heat at a rate given by some function of temperature. Taking the heating programme given above in equation 6,

$$\mathrm{d}Q/\mathrm{d}t = C_\mathrm{p}(\beta + \omega B\cos\omega t) - h(T_0 + \beta t + b\sin\omega t) \tag{26}$$

where $h(T)$ = some function that determines how the heat output from the reaction changes with temperature.

Note that the contribution to the heat flow from the sample's heat capacity is included. As discussed above, the heat capacity can be considered as the energy contained in the various vibrational, translational etc. modes available to the sample. In this section, these processes are considered to be very fast and can normally be treated as instantaneous when compared to the frequency of the modulation that typically has a period of several tens of seconds. This means that any heat flow deriving from the heat capacity will not depend on the heating rate or frequency of the modulation. The energy contained in these molecular motions is stored reversibly. This can be contrasted with the enthalpy associated with the zero-order chemical reaction being considered in this case, which is irreversible.

It can be shown (see section 4.3 on detailed MTDSC theory) that to a good approximation under realistic conditions

$$\mathrm{d}Q/\mathrm{d}t = C_\mathrm{p}\beta - h(T_0 + \beta t) + \omega B C_\mathrm{p} \cos \omega t + \mathrm{C} \sin \omega t \qquad (27)$$

For clarity, this can be rewritten as:

$$\mathrm{d}Q/\mathrm{d}t = C_\mathrm{p}\beta - h(T_0 + \beta t) \qquad \ldots \text{the underlying signal}$$
$$+\,\omega \mathrm{B} C_\mathrm{p} \cos \omega t + C \sin \omega t \quad \ldots \text{the response to the modulation} \qquad (28)$$

where $C = B\mathrm{d}h(T_0 + \beta\mathrm{t})/\mathrm{d}T$ = the derivative of $h(T_0 + \beta t)$ with respect to temperature.

Note that the underlying signal is the same as would be obtained in a conventional non-modulated experiment. Averaging over the period of a modulation will suppress the modulation. Thus,

$$\langle \mathrm{d}Q/\mathrm{d}t \rangle = C_\mathrm{p}\beta - h(T_0 + \beta t) \qquad (29)$$

Also,

$$C_\mathrm{pPCR} = C_\mathrm{p} \qquad (30)$$

$$C_\mathrm{pK} = (\mathrm{d}h(T_0 + \beta t)/\mathrm{d}T)/\omega \qquad (31)$$

Thus, it follows that

$$\langle \mathrm{d}Q/\mathrm{d}t \rangle - C_\mathrm{pPCR}\beta = h(T_0 + \beta t) = \text{the phase-corrected non-reversing heat flow} \qquad (32)$$

In other words, it is possible to separate the contribution in the total heat flow from the heat capacity and which arises from the zero-order reaction. It is this ability that is one of the main advantages of MTDSC. In most cases, it is not necessary to use the phase lag correction in order to achieve this. So, the simple deconvolution procedure is adequate.

The above is intuitively satisfactory when one considers that, in a zero-order reaction, the reaction rate will change only with temperature and will thus follow the $B \sin \omega t$ of the modulation. The contribution from the heat capacity, on the other hand, follows the derivative of temperature and thus follows $\omega B \cos \omega t$. The in-phase contribution arises from a signal that depends only on the heat capacity. Thus, this provides a means of separating or deconvoluting these two different contributions to the heat flow.

We now consider a more general process that gives rise to a heat flow and is governed by a kinetic function that is dependent on temperature *and* time, $f(t, T)$. The derivation of this result given below is provided in section 4.3 on advanced theory. In effect, we come to essentially the same conclusion as for the zero-order case.

$$\begin{aligned} \mathrm{d}Q/\mathrm{d}t = \beta C_\mathrm{p} + f(t, T_0 + \beta) \quad & \ldots \text{the underlying signal} \\ + \omega B C_\mathrm{p} \cos \omega t + C \sin \omega t \quad & \ldots \text{the response to modulation} \end{aligned} \tag{33}$$

where $C = B(\partial f/\partial T)$ (as defined above) to some approximation, but may be considered to include other terms depending on the experimental conditions and the nature of the $f(t,T)$ term. (See the section on detailed theory).

By analogy with the case considered above,

$$\begin{aligned} \langle \mathrm{d}Q/\mathrm{d}t \rangle - C_{\mathrm{pPCR}}\beta &= f(t, T_0 + \beta t) \\ &= \text{the phase-corrected non-reversing heat flow} \end{aligned} \tag{34}$$

Thus, as also demonstrated in Eq. (32), by carrying out this deconvolution procedure it is possible to separate the two fundamentally different contributions to the total heat flow: the reversible contribution that derives from the heat capacity (the phase-corrected reversing heat flow) and the contribution that derives from $f(t,T)$ which is, on the time-scale of the modulation, irreversible. In most cases, the phase-corrected reversing heat flow will be the same as the reversing heat flow to an accuracy greater than that of the measurement being made.

In the description given above, essentially represented in Eqs. (26)–(34), the 'reversing' signal was truly reversible and the 'non-reversing' signal came from a nominally irreversible process. However, the non-reversing signal can also be the heat from a crystallisation or from the loss of volatile material. Both of these processes are reversible in the sense that, with large-scale temperature changes, crystals can be melted and, on cooling, moisture can be reabsorbed. For this reason, the term non-reversing was coined to denote that at the time and temperature the measurement was made the process was not reversing although it might be reversible.

Most of the transitions being considered in this section will follow, to some approximation, the Arrhenius equation, viz

$$\mathrm{d}x/\mathrm{d}t = f(x)A\mathrm{e}^{-E/RT} \tag{35}$$

where x = the extent of the reaction,
t = time,
$f(x)$ = some function of the extent of reaction,
A = the pre-exponential constant,
E = the activation energy,
R = the gas constant,
T = absolute temperature.

This type of behaviour is associated with the well-known energy barrier model for thermally activated processes. In this model, a material changes from one form to another more thermodynamically stable form, but must first overcome an energy barrier that requires an increase in Gibbs free energy. Only a certain fraction of the population of reactant molecules have sufficient energy to do this and the extent of this fraction and the total number of reactant molecules determine the speed at which the transformation occurs. The fraction of molecules with sufficient energy is dependent upon the temperature in a way given by the form of the Arrhenius equation. Thus, this must also be true for the transformation rate. The types of process that can be modelled using this type of expression include chemical reactions, diffusion controlled processes such as the desorption of a vapour from a solid and some phase changes such as crystallisation. There will be some constant of proportionality, H, such that the rate of heat flow can be directly related to the rate of the process, viz

$$(\mathrm{d}Q/\mathrm{d}t)_\mathrm{r} = H\mathrm{d}x/\mathrm{d}t = Hf(x)A\mathrm{e}^{-E/RT} \tag{36}$$

One can derive the following equation. (See the advanced theory section)

$$\mathrm{d}Q/\mathrm{d}t = \beta C_\mathrm{p} + Hf(\langle x\rangle)A\mathrm{e}^{-E/RT} \ldots \text{the underlying signal}$$
$$+ B\omega C_\mathrm{p}\cos\omega t + C\sin\omega t \quad \ldots \text{the response to modulation} \tag{37}$$

where $C = Bf(\langle x\rangle)\mathrm{d}(HA\mathrm{e}^{-E/R\langle T\rangle})/\mathrm{d}T = Bf(\langle x\rangle)(HAE/R\langle T\rangle^2)\mathrm{e}^{-E/R\langle T\rangle}$.

A typical form of $f(x)$ might be

$$f(x) = (1 - x)^n$$

where $n =$ the reaction order.

However, there are many other possibilities that are already well established in the literature 28. Some of these are considered in detail in Chapter 2. Again, one can say

$$\langle \mathrm{d}Q/\mathrm{d}t\rangle = \beta C_\mathrm{p} + Hf(\langle x\rangle)A\mathrm{e}^{-E/RT} \tag{38}$$

$$\begin{aligned}\langle \mathrm{d}Q/\mathrm{d}t\rangle - C_\mathrm{pPCR}\beta &= Hf(\langle x\rangle)A\mathrm{e}^{-E/RT}\\ &= \text{the non-reversing signal}\end{aligned} \tag{39}$$

Thus, it is possible to conclude that the non-reversing heat flow contains that part of the underlying signal that comes from the chemical reaction. In most cases, it is also true to a very good approximation that $C^* = C_\mathrm{pPCR}$.

Thus, it is not necessary to use the phase correction in order to measure the heat capacity and then calculate the non-reversing signal. So, the simple deconvolution can bc uscd.

Also since $C_\mathrm{pK} = (HAEf(\langle x\rangle)/\omega R\langle T\rangle^2)\mathrm{e}^{-E/R\langle T\rangle}$ and a comparison with the phase-corrected non-reversing signal shows that the activation energy is given by

$$E = (\omega R\langle T\rangle^2 C_\mathrm{pK})/(\langle \mathrm{d}Q/\mathrm{d}t\rangle - C_\mathrm{pPCR}) \tag{40}$$

Toda *et al.* have shown that Eq. (40) can be used to determine E [29].

Above, the simplest possible case (a zero-order reaction) has been considered. Here, the results are intuitively easy to understand. The general case, $f(t, T)$, where the kinetics are a function of both time and temperature is then considered and essentially the same result is achieved. Finally, for completeness, the most commonly encountered case (the Arrhenius equation) is dealt with. In all of these examples, we came to the same conclusions (mathematical details are given in section 4.3 on MTDSC theory).

Figure 1.8 shows results for a curing sample. In the reversing signal, a glass transition is observed during the course of the cure reaction, which provides the enthalpy change that appears in the non-reversing signal. Clearly, it is not possible to obtain the same information from a conventional DSC experiment, which would not be able to separate these two contributions to

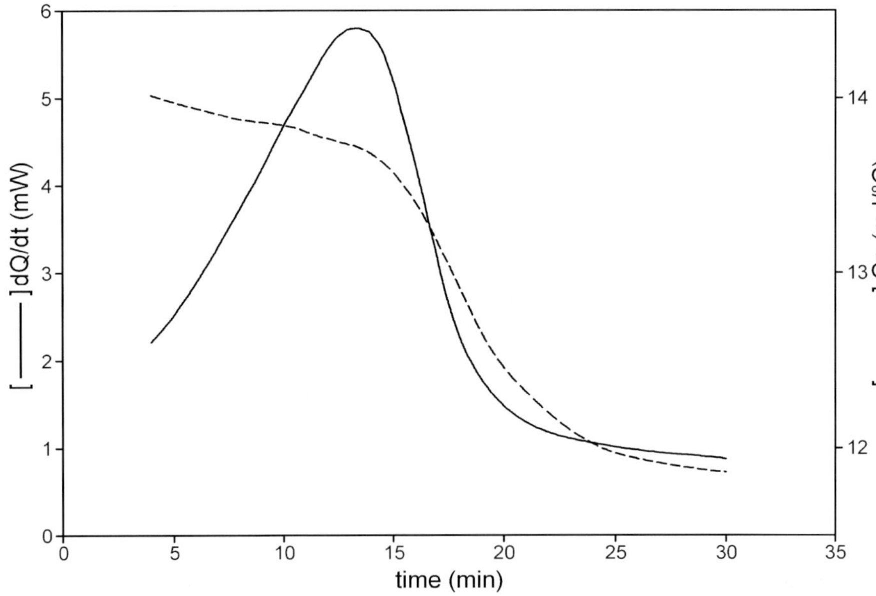

Figure 1.8. Isothermal cure of an epoxy resin showing a glass transition during cure. Data from Ref. [5].

the total heat flow. The advantages that this affords for the study of reacting systems are illustrated extensively in Chapter 2.

Figure 1.9 shows an example of detecting a glass transition beneath a cold crystallisation exotherm. The total heat flow corresponds to a conventional DSC experiment. It is not possible from inspection of the distorted peak in this curve to conclude that it is formed from an exotherm (from the crystallisation of PET) superimposed on a glass–rubber transition (from the polycarbonate). The additional signals of MTDSC make this interpretation clear. In this case, the crystallisation acts like a chemical reaction: once formed the crystals remain as the temperature increases through the peak. Thus, the process is non-reversing.

Inspection of Figure 1.3 shows there is a decrease in reversing heat capacity as initially purely amorphous PET crystallises. This effect is present, but cannot be seen easily in Figure 1.9, in part because the change is correspondingly smaller in this sample as there is a large amount of second amorphous material present and also due to the increase in heat capacity through T_g. The results in Figure 1.3 are an accurate reflection of the fact the crystals have a lower heat capacity than the amorphous material that proceeded them. Note also that, during the cold crystallisation, the peak in the phase lag is negative (and so, therefore, is the kinetic heat capacity). This is exactly what theoretical analysis predicts.

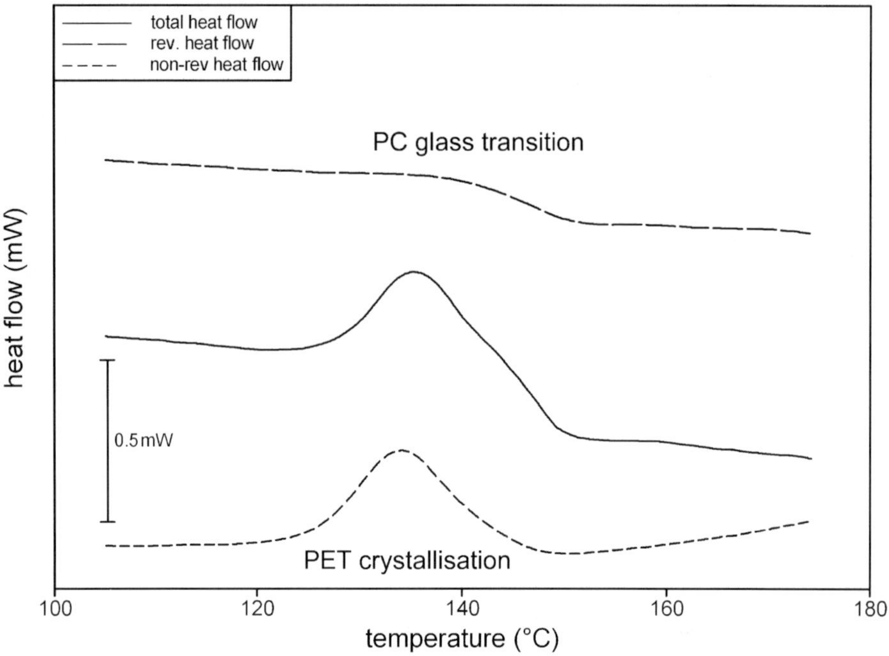

Figure 1.9. Crystallisation of PET:PC bilayer film showing detection of PC glass transition during crystallisation of PET. Data from Ref. [3].

This ability to measure changes in vibrational heat capacity that occur during the course of a process that gives rise to a heat flow such as a chemical reaction or crystallisation is a very useful aspect of MTDSC. It applies equally well to the loss of volatile material, for example, that can mask a glass transition.

Often the deconvolution into reversing and non-reversing is most useful when there is a 'hidden' glass transition such as in Figures 1.8 and 1.9. For reasons that are discussed in the section below on glass transitions, the presence of a glass transition in the reversing signal implies an error in the non-reversing signal. This is because not all of the energy changes associated with a glass transition is to be found in the reversing signal. At T_g, there is always a (usually) small non-reversing contribution. In most cases, this can be neglected. Where it is important to account for this, it can be done by measuring the non-reversing signal of the relevant glass transition when other processes are not present (see Ref. [38]).

3.4.2 Summary

- By averaging the modulated heat flow signal, one can recover results that are equivalent to conventional DSC. This is important because DSC is a highly successful technique for the good reason that the information it provides is very useful.

- One can measure the sample's vibrational heat capacity independently of any other process that is occurring, such as a chemical reaction, by looking at the in-phase response to the modulation. This signal gives C_p directly.
- The out-of-phase response can be expressed as the kinetic or, in complex notation, the imaginary heat capacity or simply as C in many of the above equations. It can take a variety of forms depending on the details of the experiment conditions and the form of $f(t, T)$. However, it is generally approximated by taking the derivative with respect to temperature of the heat flow generated by the reaction or other process. This signal can be used to determine the activation energy for a reaction.
- Very often the out-of-phase component C is small, so the reversing heat capacity (modulus of the complex heat capacity) is the same as the in-phase component (phase-corrected reversing or real heat capacity). So, the phase correction can be neglected. This means that the simple deconvolution defined above can be used.
- The non-reversing signal gives a measure of the energy that arises from the chemical reaction.
- Where a glass transition is present underneath a non-reversing peak due to a cure reaction or a similar transformation, then this does imply an error in the non-reversing signal because there is a non-reversing component arising from devitrification. This can usually be neglected or corrected for.

3.5 THE GLASS TRANSITION

3.5.1 Characteristics of MTDSC Results for Glass Transitions

Figure 1.10 shows typical MTDSC results for a glass transition for a polystyrene sample that has been annealed for different lengths of time. It can be seen that, as expected, the total signal is the same as that observed for a conventional DSC experiment. As annealing increases, the characteristic endothermic peak at the glass transition increases. At low levels of annealing, there are noticeable changes in the total signal as the characteristic relaxation peak is seen to develop. However, the changes in the reversing and kinetic signals are small. It follows that the non-reversing signal shows an increasing peak with annealing time. The use of MTDSC seems to eliminate the influence of annealing and enables the relaxation endotherm to be separated from the glass transition itself. To a first approximation, this is true, but this must be understood within the context of the frequency dependence of the glass transition.

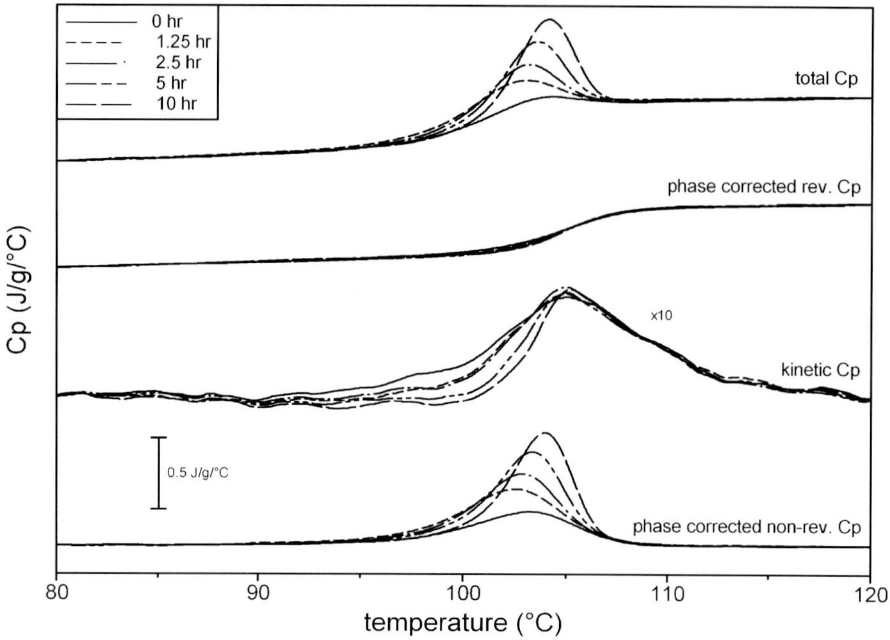

Figure 1.10. Typical results for a glass transition with different degrees of annealing (polystyrene annealed at 90°C and re-heated at 2°C/min, period: 40 s, amplitude 0.21°C under helium).

It is well known that the temperature of the glass transition is frequency dependent from measurements made with dynamic mechanical and dielectric measurements. This same frequency dependence is seen in MTDSC [30]. Figure 1.11 shows the results for polystyrene at a variety of frequencies. For a cooling experiment with MTDSC, there is both a cooling rate, β, and a frequency (the frequency of the modulation, ω). If the cooling rate is kept the same and the frequency is varied, the underlying signal remains constant, while the reversing signal changes. The underlying signal will always give a lower T_g than the reversing signal because the underlying measurement must, in some sense, be slower (i.e. on a longer time-scale) than the reversing measurement. This is because of the requirement that there be many modulations over the course of the transition. As the cooling rates become slower, in other words as the time-scale of the measurement becomes longer, T_g moves to a lower temperature. Similarly, as the frequency decreases, T_g moves to lower temperatures. As a consequence of this, there is a peak in the non-reversing signal as the sample is cooled that is clearly not related to annealing, but is a consequence of the difference in effective frequency between the average measurement and that of the modulation.

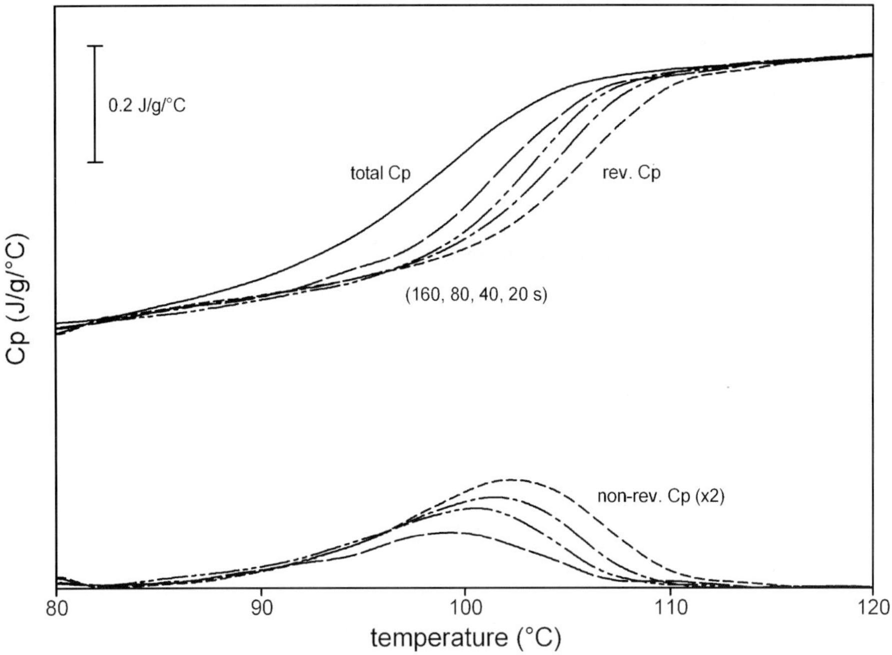

Figure 1.11. Experimental results that illustrate the effect of frequency on the total, reversing and non-reversing C_p for the glass transition of polystyrene in cooling (period: 20, 40, 80 and 160 s, underlying rate: 1°C/min, 'cool-only' under helium).

Thus, the non-reversing signal changes with cooling rate and modulation frequency. This is shown in Figure 1.11.

On heating, the non-reversing signal, as can be seen from Figure 1.10, is related to the amount of annealing and also must contain the effects of the different effective frequencies used in the measurement. These effects can be treated as additive. Thus, the non-reversing signal gives a measure of the enthalpy loss on annealing with an offset due to the frequency difference. This is intuitively satisfactory, as the enthalpy that is regained by the sample on heating after annealing cannot be lost again on a short time-scale at the time and temperature at which the measurement is made. In this sense, it is non-reversing in the same way as a chemical reaction or crystallisation event. This simple picture is only a first approximation, but it will be adequate in many cases. In particular, the non-reversing peak at the glass transition can be used to rank systems in terms of degree of annealing.

In the discussion above it is assumed that, as indicated in Figure 1.10, the reversing signal is not affected by annealing. In reality, this is not correct. At higher degrees of annealing, the reversing signal becomes sharper. Thus, the simple relationships outlined in the previous paragraph break down. This

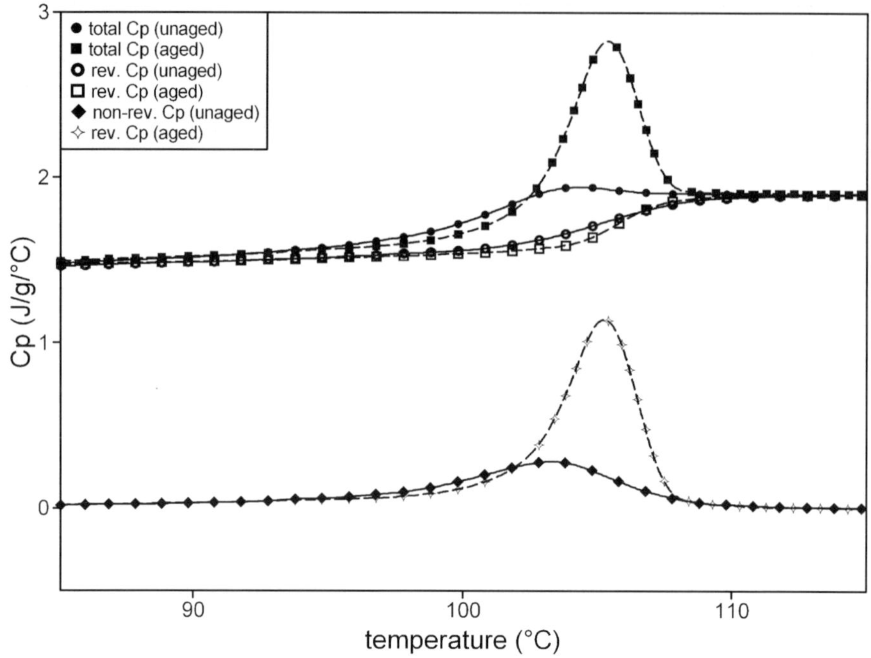

Figure 1.12. Effect of long annealing times on the total and reversing signals (polystyrene unaged and aged at 90°C for 40 h, re-heated at 2°C/min, 40s period, amplitude: 0.21°C under helium).

is illustrated in Figure 1.12, which compares the behaviour of a sample of polystyrene that has been subjected to a low and a high level of annealing. Figure 1.13 shows that slower cooling rates also lead to sharper reversing transitions. In both cases, the sample is closer to equilibrium when it undergoes the transition in the reversing signal and this leads to a narrowing of the temperature range over which it occurs. How this can be allowed for is discussed below.

At first sight, the step change in C_p that occurs at the glass transition might be interpreted as a discontinuity: that would mean that it would be a second-order transition. In fact, the transition is gradual as it occurs over about 10°C or more. Its position also varies with heating rate (and with frequency in MTDSC), which reveals that it is a kinetic phenomenon. The co-operative motions that enable large-scale movement in polymers have activation energies in a way that is similar to (but not the same as) the energy barrier model mentioned above for Arrhenius processes. Thus, as the temperature is decreased, they become slower until they appear frozen. There is a contribution to the heat capacity that is associated with these motions. Therefore, as the temperature is reduced, these large-scale motions are no longer possible and consequently the material appears glassy (rigid)

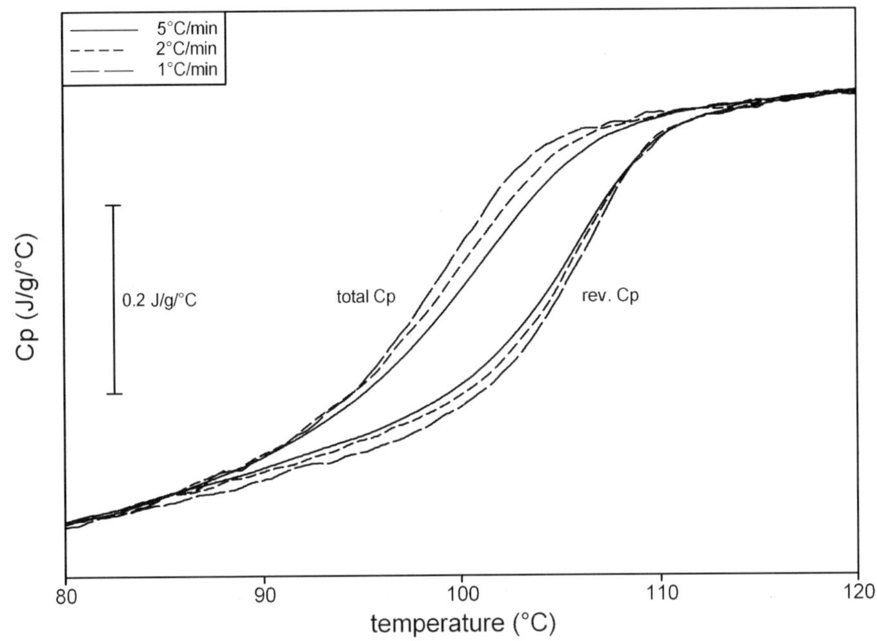

Figure 1.13. Experimental results that illustrate the effect of cooling rate on the total and reversing C_p for the glass transition of polystyrene (period: 20s, underlying rate: 1, 2 and 5°C/min, 'cool-only' under helium).

and the heat capacity decreases. In reality, whether a polymer appears glassy or rubbery depends the time-scale of the observation. Thus, if the polymer is being vibrated at a frequency of several times a minute, it may be springy and return to its original shape when the stress is removed. If it is being deformed and released over a period of a year, it may well behave like a pliable material that creeps under load, thus retaining a permanent distortion in dimensions when unloaded. There is a parallel dependence of the heat capacity on how rapidly one is attempting to put heat into or take it out of the sample. Thus, the position of T_g changes with heating and cooling rates.

Figures 1.14 and 1.15 give the enthalpy and heat capacity diagrams for glass formation. The enthalpy gained, or lost, by a sample is determined by integrating the area under the heat capacity curve. Above T_g, the sample is in equilibrium (provided no other processes such as crystallisation are occurring). Consequently, this line is fixed regardless of the thermal treatment of the sample and a given temperature corresponds to a unique enthalpy stored within the sample. As the sample is cooled, there comes a point at which the C_p changes as it goes through the glass transition. Thus, dQ/dt changes and so does the slope of the enthalpy line. At different cooling rates, the temperature at which this occurs changes. Thus, a different

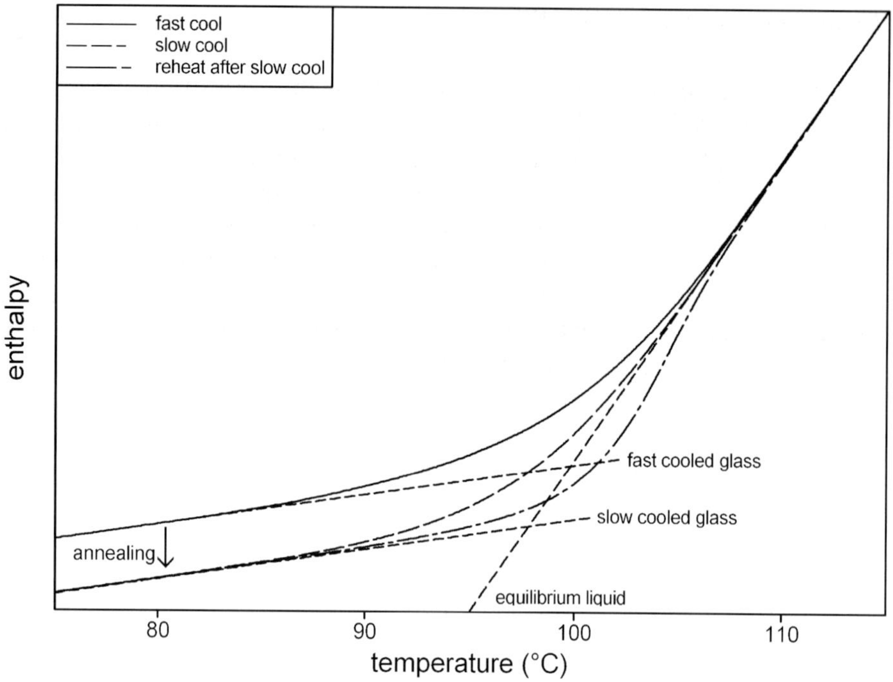

Figure 1.14. Enthalpy diagrams for the glass transition – original data obtained on polystrene. The change in C_p between the glass and rubber has been exaggerated for clarity.

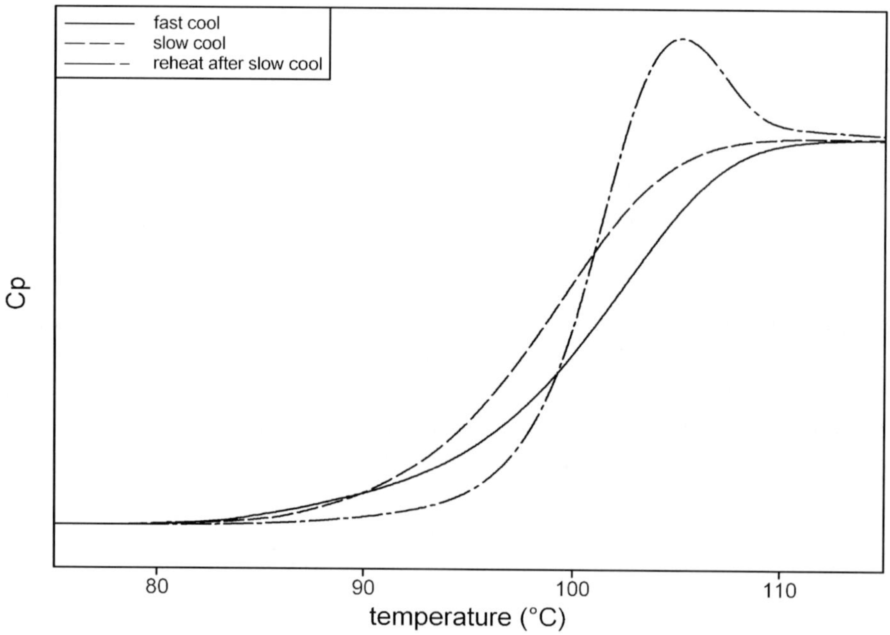

Figure 1.15. Schematic heat capacity plot that corresponds to the enthalpy diagram shown in Figure 1.14 showing the peak in C_p arising from annealing.

glass with a different enthalpy is created. Above the transition, the sample is at equilibrium. Below T_g, it is at some distance from this equilibrium line, but is moving towards it very slowly. Thus, glasses are metastable. If the glass is annealed at temperatures a little below the glass transition, it looses enthalpy relatively rapidly and becomes a different glass as it moves toward the equilibrium line. At temperatures far below T_g, the rate of enthalpy loss becomes very slow and effectively falls to zero. When the sample is heated, the enthalpy lost on annealing must be regained and this gives rise to the characteristic peak at the glass transition as seen in Figure 1.10.

From a simple model [31] of the glass transition, it is possible to derive approximate analytical expressions that model the response to the modulation at the glass transition. (See the discussion in section 4.4). Viz

$$\Delta C_{\mathrm{pPCR}} = \Delta C_{\mathrm{p}}/\left(1 + \exp\left(-2Q_{\omega}\Delta h^{*}(T - T_{\mathrm{g}\omega})/RT_{\mathrm{g}\omega}^{2}\right)\right) \tag{41}$$

$$\begin{aligned} C_{\mathrm{pK}} = q\,\Delta C_{\mathrm{pPCR}} \exp\left(-Q_{\omega}\Delta h^{*}(T - T_{\mathrm{g}\omega})/RT_{\mathrm{g}\omega}^{2}\right)/ \\ \left(1 + \exp\left(-2Q_{\omega}\Delta h^{*}(T - T_{\mathrm{g}\omega})/RT_{\mathrm{g}\omega}^{2}\right)\right) \end{aligned} \tag{42}$$

where ΔC_p = the change in heat capacity at the glass transition,
Q_ω and q = shape factors related to the distribution of relaxation times and mechanism of the relaxation process,
Δh^* = the apparent activation energy,
$T_{g\omega}$ = the glass transition temperature (at half height) at frequency ω.

The $T_{g\omega}$ is given by relating the period of oscillation to the time-scale associated with the Arrhenius relationship, viz

$$\omega = A\mathrm{e}^{-\Delta h^{*}/RT_{\mathrm{g}\omega}} \tag{43}$$

The change in the average or total signal for heating or cooling rate β can be approximately modelled using the following relationship which simply combines a step change (first term on rhs of the equation which is based on the equation for the reversing signal, see equation 41) with a peak (second term on rhs of the equation and which is zero on coling) viz:

$$\begin{aligned} \Delta C_{\mathrm{p}T\beta} = & \frac{\Delta C_{\mathrm{p}}}{\left(1 + \exp\left(-2Q_{\beta}\Delta h^{*}(T - T_{\mathrm{g}\beta})/RT_{\mathrm{g}}^{2}\right)\right)} \\ & + \frac{h\exp(T - T_{\mathrm{gp}})}{(1 + \mathrm{a}\exp(T - T_{\mathrm{gp}}))^{n}} \end{aligned} \tag{44}$$

Here, $T_{g\beta}$ = the apparent glass transition temperature for the total signal,
Q_{β} = shape factor for the underlying measurement at heating/cooling rate, β, where this would generally be different on heating and cooling.
T_{gp}, h and n are all fitted parameters that change with the degree of annealing (see Fig. 17).

Equation (44) is an ad-hoc model that is used here for illustrative purposes because it is often useful to think of the glass transition as a combination of a step change in heat capacity with an additional peak that increases in size with increasing enthalpy loss. This is illustrated in Fig. 17. However, it must be stressed that at higher levels of annealing this model cannot be applied. There is no simple analytical expression that can be used and one is forced to use numerical solutions to models such as that given in equation 94.

$T_{g\beta}$ would normally show an Arrhenius dependence on cooling rate:

$$\beta = zA\mathrm{e}^{-Dh^{*}/RT_{g\beta}} \tag{45}$$

where z is some constant with units of °C^{-1}. In fact, this pre-exponential factor can be considered to be a function of heating rate, but this is beyond the scope of this discussion. For any frequency, there must be a cooling rate that would give the same transition temperature (taken at the half height of the step change) and so there should be a frequency–cooling rate equivalence.

These must of course be obtained from two separate experiments as these signals can never give the same T_g in a single experiment. One way of looking at this is to think in terms of the time taken to traverse the transition as (with suitable weighting) a measure of the time-scale of the linear cooling rate measurement. This then is related to the period that gives a measure of the time-scale of the cyclic measurement. Thus, β and ω can be related by

$$z\omega = \beta \tag{46}$$

The concept of a reversing response can be extended to the total signal by considering a heating and cooling experiment at the same rate. The vibrational heat capacity of a purely inert sample should be exactly the same at any temperature and so completely reversing (and reversible). An experiment where the sample is cooled through a glass transition, and then heated at the same rate, will give a similar, but not identical, result in both directions. Because of this, it is convenient to define a hysteresis factor $h_{\beta}(t, T)$ describing this difference

$$h_{\beta}(t, T) = \Delta C_{\mathrm{p}T\beta\mathrm{H}} - \Delta C_{\mathrm{p}T\beta\mathrm{C}} \tag{47}$$

where $\Delta C_{pT\beta C}$ = the change in heat capacity on cooling

$\Delta C_{pT\beta H}$ = the change in heat capacity on heating without annealing

It is possible to raise objections to this approach on the basis that change below T_g never ceases. Thus, there is no end to the transition region on cooling and so any choice of temperature at which to reverse the cooling programme is arbitrary. This implies that the shape of the heating curve cannot be fixed. However, in reality, the rate at which the sample approaches the equilibrium line decreases very rapidly below the glass transition. Thus, a few tens of degrees below the mid-point of the step change, the transition can be said to have come to an end.

To a reasonable approximation

$$\Delta C_{\mathrm{pT}\beta\mathrm{C}} = \Delta C_{\mathrm{p}}/\left(1 + \exp\left(-2Q_{\mathrm{C}\beta}\Delta h^*(T - T_{\mathrm{gC}\beta})/RT^2_{\mathrm{gC}\beta}\right)\right) \quad (48)$$

$$\Delta C_{\mathrm{pT}\beta\mathrm{H}} = \Delta C_{\mathrm{p}}/\left(1 + \exp\left(-2Q_{\mathrm{H}\beta}\Delta h^*(T - T_{\mathrm{gH}\beta})/RT^2_{\mathrm{gH}\beta}\right)\right) + \frac{\mathrm{h}\exp(T - T_{\mathrm{p}})}{(1 + \mathrm{a}\exp(T - T_{\mathrm{p}}))^{\mathrm{n}}} \quad (49)$$

Where $\Delta C_{\mathrm{pT}\beta\mathrm{H}}$ = the change in heat capacity on heating with or without annealing.

It is also convenient to define a function for the enthalpy recovery at T_g due to any annealing.

$$N\beta(t, T) = \Delta C_{\mathrm{pT}\beta\mathrm{H}} - \Delta C_{\mathrm{pT}\beta\mathrm{H}\omega} \quad (50)$$

so that

$\int N_\beta(t, T)\,\mathrm{d}t$ = enthalpy loss on annealing.

Combining these equations, we obtain

$$\Delta C_{\mathrm{pT}\beta\mathrm{H}} = \Delta C_{\mathrm{pT}\beta\mathrm{C}} + h_\beta(t, T) + N_\beta(t, T) \quad (51)$$

Note that Eqs. (41), (42), (44), (47)–(49) and (51) give the behaviour of the step change at the glass transition. For a more complete model, the change in heat capacity as a function of temperature must be taken into account above and below T_g. To deal with this, all equations that feature ΔC_p can be adapted to follow real world behaviour by assuming a linear function above and below the glass transition. Taking Eq. (41) as an example, yields

$$C_{\mathrm{pPCR}} = ((a_2 - a_1)T + (b_2 - b_1))/\left(1 + \exp\left(-2Q_\omega\Delta h^*(T - T_{\mathrm{g}\omega})/RT^2_{\mathrm{g}\omega}\right)\right) + (a_1T + b_1) \quad (52)$$

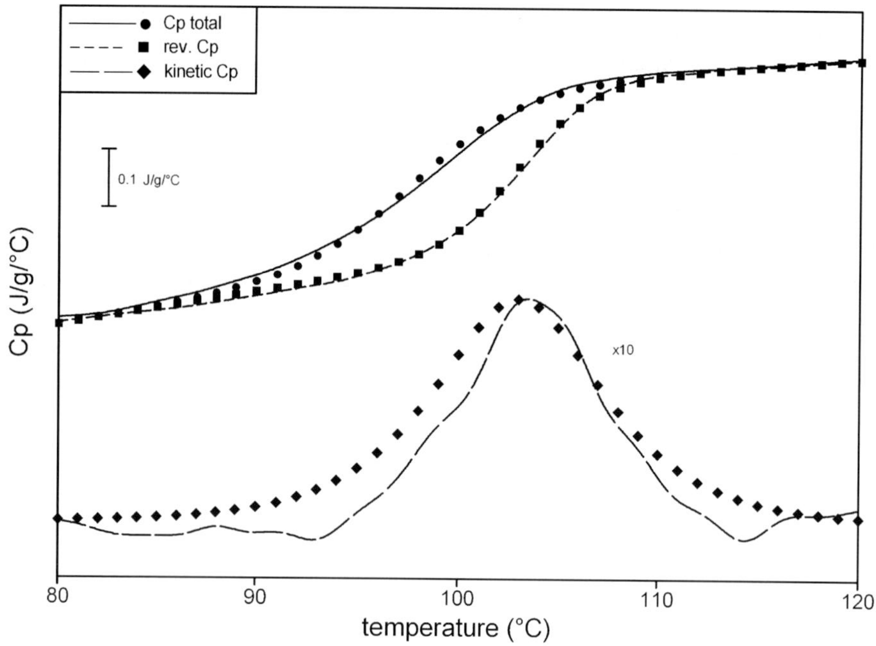

Figure 1.16. Results of modelling the glass transition behaviour of polystyrene in cooling at 80 s period shown in Figure 1.11 by applying Eqs. (41), (42) and (44) ($\Delta h^* = 690$ kJ/mol, $T_{g\omega} = 105.5°$C, $Q_\beta = 0.3$, $T_{g\beta} = 93°$C, $Q_\omega = 0.45$, $q\Delta C_p = 0.0078$ J/°C/mol).

where $C_{pg} = a_1 T + b_1$ below T_g, a_1 and b_1 being constants and
$C_{pl} = a_2 T + b_2$ above T_g, a_2 and b_2 being constants.

This modification can also be applied to Eqs. (42), (44), (47)–(49) and (51). Figure 1.16 provides an example of fitting with this expression for all of the signals on cooling, while Figure 1.17 gives examples of fitting to heating curves with different degrees of annealing.

If we return to the general expression for heat flow for MTDSC, we can express the response at the glass transition as follows.

$$\begin{aligned} dQ/dt = {} & \beta C_{pC\beta} + \langle f(t,T)\rangle && \ldots \text{the underlying signal} \\ & + C_{p\omega} B\omega \cos\omega t + C \sin\omega t && \ldots \text{the response to modulation} \end{aligned} \tag{53}$$

where $C_{p\omega}$ = the heat capacity at the frequency ω, given approximately by Eq. (41). The transition temperature and shape factor change slightly with high levels of annealing or very slow cooling.

$C_{pC\beta}$ = the 'reversing' heat capacity implied by the heating or cooling rate, β, given approximately by Eq. (51).

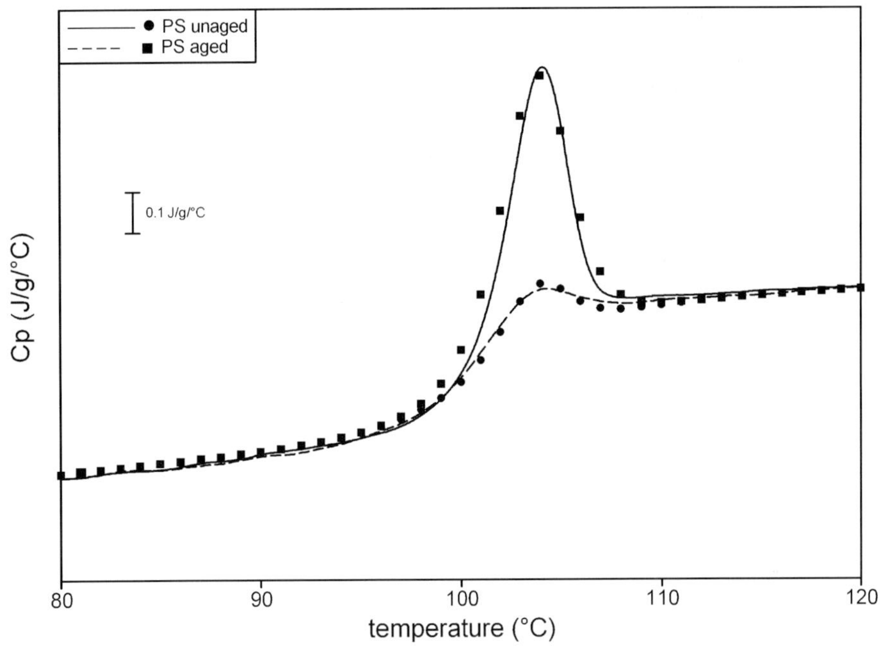

Figure 1.17. Results for modelling the glass transition for low and high levels of annealing by combining a step change in C_p with a peak (first and second terms, respectively, of equation 44). The difference between these peaks arising from the second term of equation 44 is a measure of the enthalpy loss. This approach breaks down at long annealing times.

C = the kinetic response = $B\omega C_{pK}$ where C_{pK} is given by Eq. (42). This signal becomes higher and narrower and occurs at a higher temperature with high levels of annealing.

$\langle f(t, T)\rangle = \beta h_\beta(t, T) + \beta N_\beta(t, T)$ This expresses changes below T_g that give rise to the hysteresis when heating with no annealing plus the enthalpy recovery at T_g caused by any annealing. This can be represented by equations (47) and (50).

The value of the approximate analytical expressions given in Eqs. (41)–(52) is that they enable the experimenter to gain an intuitive understanding of the phenomenology of the glass transition simply by inspection. The form of C_{pK} in Eq. (29) may be very different from the case of a chemical reaction as given in Eqs. (28), (33) and (37). However, it is still basically a manifestation of the kinetics of the glass transition. Thus, the concept that this signal is a measure of the kinetics of the transition, remains valid. An exact description of the non-reversing signal at T_g is complex because of the influence of the time-scale dependence of all measurements at the glass transition. However, for a sample cooled at a certain rate, annealed then heated at the same rate, the non-reversing signal contains the enthalpy recovery necessitated by the annealing. Although reversible on a sufficiently long time-scale, the enthalpy recovery due to annealing is non-reversing under the conditions of the measurement. In this way, it is similar to the non-reversing signal obtained during, for example, a reversible chemical reaction. In the discussion on advanced theory, models are discussed that are also phenomenological, but they have fewer variables and provide for a more fundamental insight into the underlying mechanisms governing the glass transition. However, they have to be solved numerically and thus cannot by simple inspection provide a guide to thought. The model expressed in Eqs. (41)–(52) is in part based on these models, but it is principally designed as an aid to understanding the behaviour (rather than its causes). A detailed discussion of the fundamental nature of the glass transition is beyond the scope of this chapter.

MTDSC has several significant practical advantages for studying glass transitions. The first is that the limit of detection is increased. The effect of using a Fourier analysis to eliminate all responses not at the driving frequency of the modulation reduces unwanted noise. The second is that it increases resolution. A high signal from the heat capacity is assured by a high rate of temperature change over the course of a modulation a high resolution can be assured by using a low underlying rate of temperature increase. The third is that it makes the correct assignment more certain. When a glass transition is weak, and set against a rising baseline due to the gradual increase in heat capacity of other components, the presence of a relaxation endotherm can give the impression of a melt or some other endothermic process rather than a glass transition. A clear step change in the reversing signal makes a correct assignment unequivocal in most cases. A fourth advantage is that quantification of amorphous phases is made more accurate. The increase in signal to noise already discussed above is obviously helpful in this respect. In addition, the suppression of annealing effects makes it easier to quantify the increase in heat capacity correctly. Examining the derivative of the reversing heat capacity with respect to temperature is the best approach to doing this. This approximates very well to a Gaussian distribution and numerical fitting procedures can be used to quantify multiple phases. This is explored in detail in Chapter 3 on polymer blends.

3.5.2 The Fictive Temperature and Enthalpy Loss on Annealing

The fictive temperature (T_{gf}) can be obtained by extrapolation of the linear portions of the enthalpy lines above and below the glass transition as illustrated in Figure 1.18. This can be calculated in the case of MTDSC from the following approximate relationship.

$$T_{gf} = T_{gr} + \Delta H_{NR}/\Delta C_p \tag{54}$$

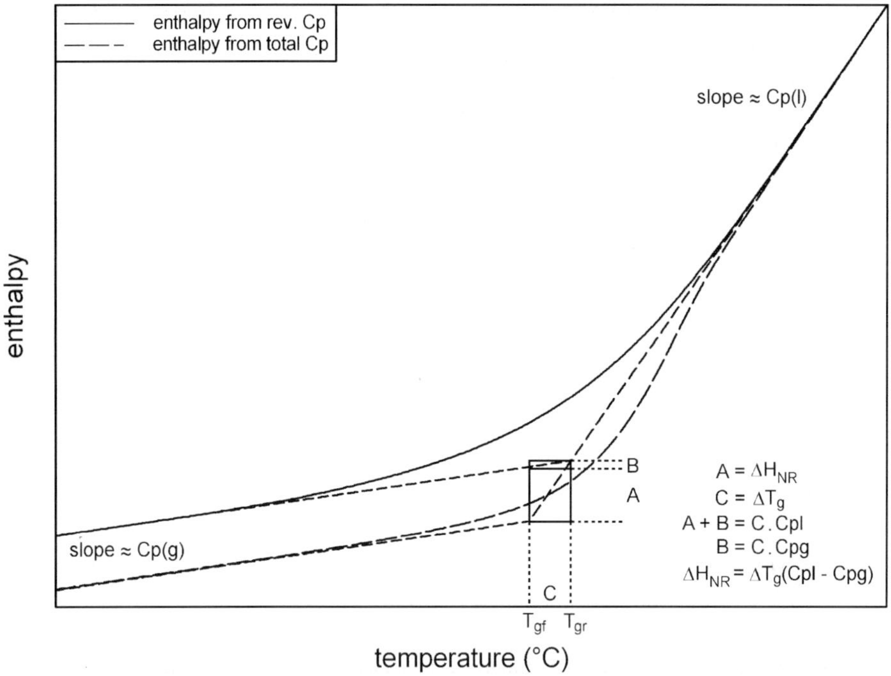

Figure 1.18. Schematic diagram illustrating the relationships between fictive temperatures, enthalpy and heat capacity.

where T_{gf} = the fictive temperature
T_{gr} = the glass transition at the mid-point of the reversing signal
ΔH_{NR} = the area under the non-reversing curve (i.e. the area between the reversing and total curves)
ΔC_p = the heat capacity change at the glass transition.

The geometric relations illustrating this equation are given in Figure 1.18. If enthalpies are required relative to some reference glass, then one approach is to use the following equation.

$$\Delta H = (T_{gfr} - T_{gfm})\Delta C_p \quad (55)$$

where ΔH = difference in enthalpy between reference state and the measured sample
T_{gfr} = fictive temperature of the reference state
T_{gfm} = fictive temperature of the measured sample.

Equations (54) and (55) can be criticised because they assume a unique value for ΔC_p whereas this varies slightly as the liquid and glass heat capacities have different slopes. (For highest accuracy ΔC_p should be determined for the mean of T_{gf} and T_{gr}). Alternatively, enthalpy loss on annealing is often measured by using a result from a sample with low annealing (say

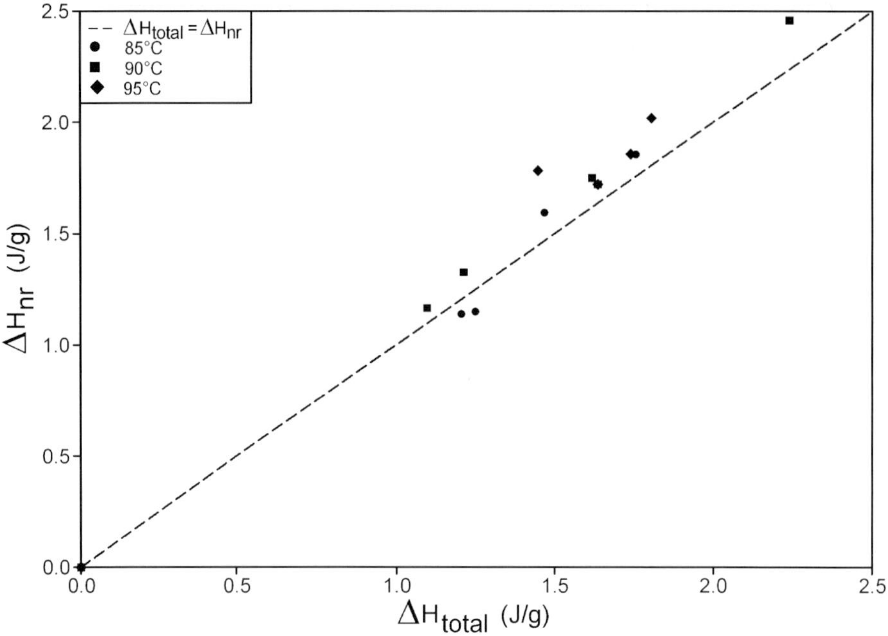

Figure 1.19. Relationship between enthalpy change on annealing obtained from total C_p and non-reversing C_p for polystyrene aged for different times at 85, 90 and 95°C.

cooled at a specified rate then immediately heated again at that rate) as the baseline that is subtracted from an annealed sample. At low degrees of annealing there should be an approximately linear relationship between this measurement and the area under the non-reversing signal because the reversing signal is not greatly affected by low small amounts of annealing. This is illustrated in Figure 1.19.

The early points show that the scatter in the data is greater than the deviation from the linear relationship, then there is a clear positive deviation as annealing increases, which can exceed 20% [16], as we would expect. This observation has also been made by Hutchinson [32,33] and Monserrat [34] who confirmed the earlier work of Reading *et al.* [16], but drew the overly pessimistic conclusion that the non-reversing signal could not be used for measuring enthalpy loss. Figure 1.19 here, Figure 4 in [34] and [35] demonstrate that, while there are problems for highly annealed samples, for low degrees of annealing a linear relationship can be assumed. In reality, a deviation of the order of 5–10%, which is what is found at moderate annealing, is within the scatter that would typically be expected with two different operators making ostensibly the same measurements. Experimenters must judge for themselves whether this is adequate for their needs. Certainly it is good enough to make comparisons between samples.

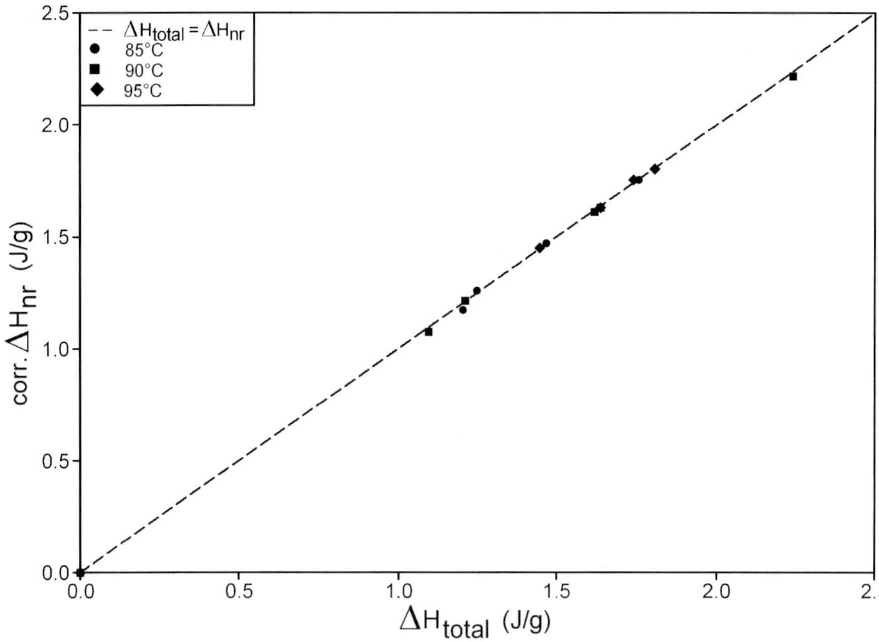

Figure 1.20. As Figure 1.19 following correction due to change in reversing T_g.

However, any debate on this subject is redundant for two reasons. The first is that the changes in the reversing signal can easily be compensated for using the following correction.

$$\Delta H = \Delta H_{\mathrm{NR}} + \Delta T_{\mathrm{gr}} \Delta C_{pa} \tag{56}$$

where ΔT_{gr} = the change in the reversing glass transition temperature.

Figure 1.20 illustrates how applying this correction excellent agreement with the more conventional approach is achieved. The second is that, whilst it is useful to understand the relationships between the results given by MTDSC and the kind of parameters often determined by conventional DSC (such as fictive temperature and enthalpy loss), MTDSC does not afford any advantages over conventional DSC for such studies. Conventional DSC measurements are to be preferred in this case due to shorter measurement times and less data processing [36].

3.5.3 Summary

In summary, the important concepts that should be born in mind when considering glass transitions are as follows.

- The glass transition temperature (T_g), as measured by the reversing heat capacity, is a function of frequency.

- T_g as measured by the total heat capacity on cooling is a function of cooling rate.
- Broadly, there is equivalence between these two observations because both changing the frequency of the modulation and the cooling rate changes the time-scale over which the measurement is made. This means that there is always, in the non-reversing signal, a contribution from $\beta(C_{p\beta} - C_{p\omega})$ which is present regardless of annealing. (For example, it is present when cooling.)
- Ageing below the glass transition produces an enthalpy loss that is recovered as a peak overlaid on the glass transition. However, this ageing does not, at low degrees of annealing, have a great effect on the reversing signal and this is intuitively satisfactory as the ageing effect is not reversible on the time-scale of the modulation. This means that the non-reversing signal includes a contribution from the different time-scales of the cyclic and underlying measurements, plus a contribution from annealing expressed as $N(t, T)$ in Eq. (50). This implies that the relationship between the enthalpy loss on annealing and the area under the non-reversing peak should be linear.
- At high degrees of annealing, the reversing signal is affected and the non-reversing signal no longer increases linearly with enthalpy loss. However, this can be compensated for by use of the fictive temperature and associated equations such as Eq. (56).
- The fact that the reversing signal is largely unaffected by annealing and its derivative provides an approximately Gaussian peak makes it a much better signal for assessing the structure of blends as described in Chapter 3.

3.6 MELTING

3.6.1 Characteristics of MTDSC Results for Polymer Melting

A first-order phase transition is characterised as a change in specific volume accompanied by a latent heat. The most common example studied by DSC is melting. Typically, at the melt temperature, the sample will remain isothermal until the whole sample has melted. The factor that determines the speed of the transition is the rate at which heat can be supplied by the calorimeter. Normally this is fast compared with the overall rate of rise of temperature so the transition is very sharp with a little 'tail', the length of which is determined by the speed with which the calorimeter can re-establish the heating programme within the sample. The area under the peak is a measure of the latent heat of the transition.

Pure, low molecular weight organic materials generally produce very narrow melting peaks. Because this narrow temperature range inevitably lies, either entirely or to a significant extent, within the course of only one modulation, it means that the response to the modulation will not be linear and the deconvolution procedure we have described above cannot be used. It is possible to obtain useful information by looking at the Lissajous figures generated by the modulation. This is dealt with in Chapter 4, which covers melting.

Polymers, in contrast, produce a range of crystallites with different melting temperatures [44]. Typically, semi-crystalline polymers will contain a distribution of crystals with differing degrees of perfection and thus different melting temperatures. The melting transition in these materials is broad, as a succession of crystallite populations melts one after the other, as the sample temperature reaches their melting temperatures. The amount of energy required to melt these crystallites is fixed as is their melting temperature. This means that if one wants to melt them twice as fast (i.e. the heating rate is doubled) the rate of energy input must be twice as fast. It follows from this that the heat flow required to melt the crystallites is a linear function of heating rate. Therefore, the enthalpy of melting will be seen in the reversing signal. In a simple case, this type of melting behaviour closely mimics heat capacity. This is discussed in more detail in [12] and the advanced section.

It should be noted that this simple picture breaks down if cooling occurs during the modulation. As we can see from Figure 1.1, it is not necessary to have a negative heating rate at any point in an MTDSC experiment: there can simply be faster and slower rates of heating. Having cooling at any point is an option. If cooling does occur then, to maintain linearity, the crystallites must crystallise instantly to form the same structure as before, something that is generally unlikely both because super-cooling is common and crystallisation to form exactly what was present before is uncommon.

Figure 1.21 shows some typical results for a semi-crystalline PET. It can be seen that there is a strong frequency dependence of the results in the melt region. The simple model discussed above (from [12]) does not predict that this will occur. Figure 1.22 shows how the peak in the melt region is also significantly affected by the underlying heating rate which is again in contradiction to the simple model. In both cases, the simple model predicts that the reversing signal should be invariant.

In Figure 1.3 it can be seen that, above the cold crystallisation temperature, the reversing signal is greater than the average until very near the end of the melting peak. This means that the non-reversing signal is exothermic over most of the melt region. This can be observed in more detail in Figure 1.23, which is an enlargement of a selected region of the raw data shown in Figure 1.2. Here, at the lowest heating rates (approximately equal to zero), an exotherm is observed within the modulation along with an endotherm at the highest heating rates. At zero heating rate, where the contribution to heat flow from the vibrational heat capacity must be zero, the heat flow is

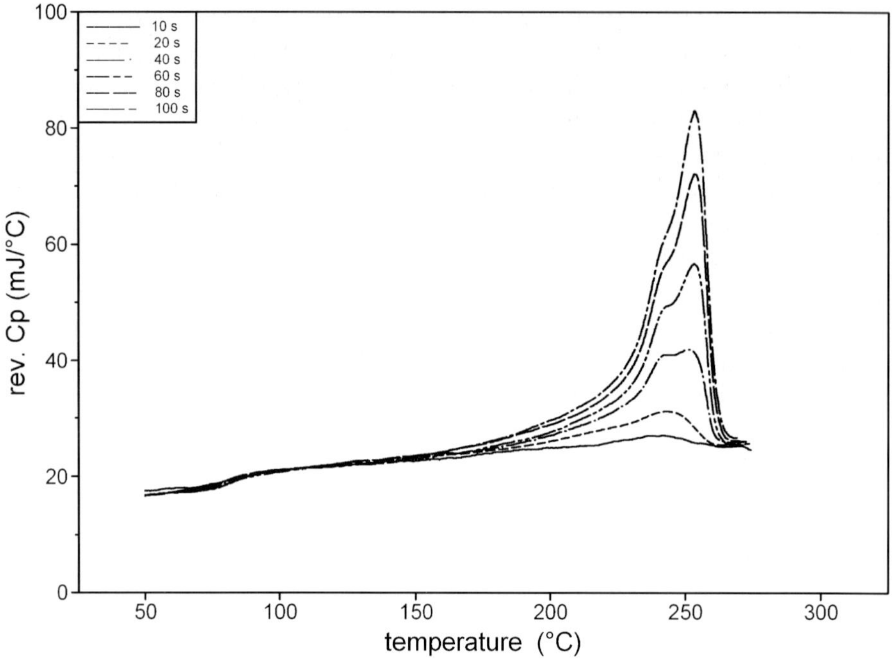

Figure 1.21. Effect of period (heating rate 2°C/min) on reversing heat capacity for semi-crystalline PET ('heat-only' conditions under helium).

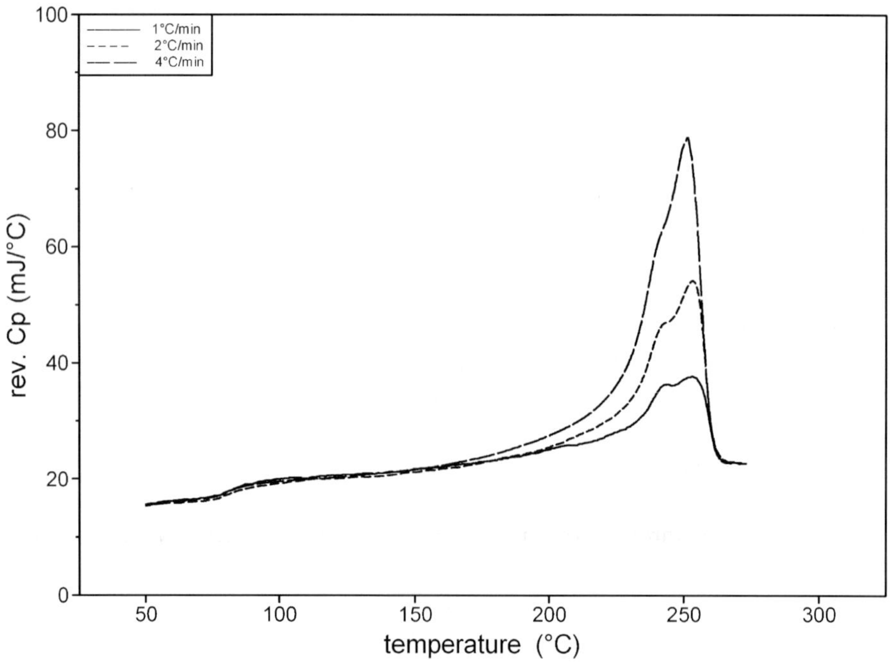

Figure 1.22. Effect of heating rate (period: 40 s, 'heat-only' under helium) on reversing heat capacity for semi-crystalline PET.

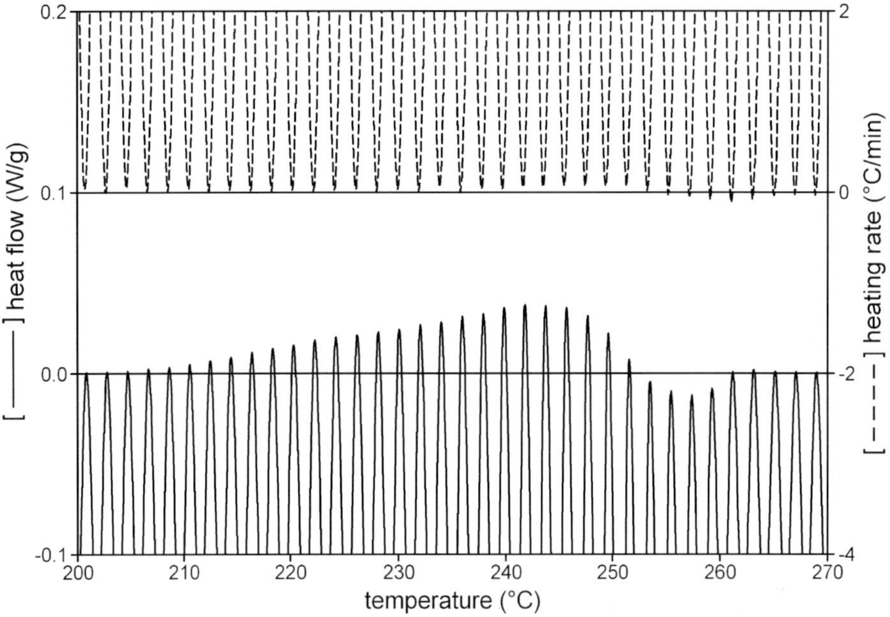

Figure 1.23. Enlargement of raw data from Figure 1.2 illustrating the exothermic crystal perfection during melting when the heating rate is zero.

exothermic. This is symptomatic of a rearrangement process. The molten material produced by melting the crystallites with lower melting temperatures can crystallise to form more perfect crystals with a higher melting temperature. This is seen because, at the lower heating rate, the rate of melting is lowest, thus the exothermic process can predominate. At the higher heating rates the reverse is true. In conventional DSC, which provides the same curve as the average signal, there is little or no indication that this rearrangement process is occurring—as the exothermic and endothermic processes cancel each other out. Thus, one benefit from using MTDSC is simply the qualitative one that it can make the occurrence of this phenomenon far more apparent.

Turning to some simple mathematical representation of melting behaviour, we can express this as follows.

$$dQ/dt = \beta(C_p + g(t, T)) \qquad (57)$$

where $g(t, T)$ = some function that models the contribution to the heat flow from the melting process.

When the melting is rapid with respect to the measurement, $g(t, T)$ will be simply a function of temperature, $g(T)$. This means that, in the case of the distribution of crystallites, the melting contribution to heat flow will scale with heating rate exactly like heat capacity if no other process occurs. Taking this simple model gives

$$\begin{aligned} \mathrm{d}Q/\mathrm{d}t &= \beta[C_\mathrm{p} + \langle g(T)\rangle] \quad \ldots \text{ the underlying signal} \\ &+ (C_\mathrm{p} + E)B\omega\cos\omega t + C\sin\omega t \quad \ldots \text{ the response to modulation} \end{aligned} \tag{58}$$

where, approximately, $E = \langle g(T)\rangle$ and $C = 0$ for this simple case.

Frequently, what is encountered is a complex process that involves melting a population of crystallites with a range of melting temperatures to form molten material which then recrystallises (following some kinetics, thus involving some $f(t, T)$) to form a further population of crystallites which then, in their turn, melt and possibly undergo further rearrangement. The data shown in Figure 1.23 illustrate this process. To complicate matters further, some workers have suggested that melting is often not rapid with respect to the frequency of the modulation. Thus, there is a time dependency in $g(t, T)$ [25] and C is not zero even without taking account of crystallisation (see the advanced theory section). To allow for this complex range of possibilities it is convenient to define a composite kinetic function that includes all terms other than the heat capacity and models both melting and the kinetics of crystallisation, viz

$$f(t, T) = g(t, T)\,\mathrm{d}T/\mathrm{d}t + f_2(t, T) \tag{59}$$

Under modulated conditions *with no cooling*

$$f(t, T) = \langle f(t, T)\rangle + D\sin\omega t + EB\omega\cos\omega \tag{60}$$

Equation (58) now becomes

$$\begin{aligned} \mathrm{d}Q/\mathrm{d}t &= \beta C_\mathrm{p} + \langle f(t, T)\rangle \quad \ldots \text{ the underlying signal} \\ &+ B\omega(C_\mathrm{p} + E)\cos\omega t + D\sin\omega t \quad \ldots \text{ the response to modulation} \end{aligned} \tag{61}$$

Note that the ‘reversing’ signal during the melt no longer has the same meaning as for an Arrhenius process and the glass transition because it contains a contribution, E, from the melting of the crystallites which will typically not be fully reversible due to super-cooling. As noted above, this gives rise to the requirement that there be no cooling at any point during the modulation so that the response does not become asymmetric and thus strongly non-linear.

The question arises as to what might be the form of $f_2(t, T)$ and hence E and D. At the current state of development, there are no well-established candidates although this may well change in the near future. This point is further discussed in the section dealing with details of the theory.

The situation becomes even more complex when we consider that, even for Eq. (61) to be true, there must be no significant temperature gradients in the sample. We can reasonably expect that, in the melt region, this will generally not be true. Taking all of these factors into consideration, the melt region is significantly more complex than the other transitions we have considered and it is generally true that a quantitative interpretation of melting behaviour, particularly during experiments with a non-zero underlying heat rate, is not generally possible at present.

3.6.2 The Measurement of Polymer Crystallinity

A problem encountered frequently in determining the crystallinity of a polymer using DSC is that, as illustrated in Figures 1.3 and 1.22, the sample changes its crystallinity during the experiment. The problem becomes one of establishing the initial crystallinity before the experiment started.

An understanding of the problem is best approached from the perspective of enthalpy diagrams [31]. Figure 1.24 shows an enthalpy-temperature diagram for completely amorphous PET, 100% crystalline PET and a 50% crystalline PET. In the molten state, all of these samples must have the same enthalpy, so the curves obtained for each example are aligned to make this the case. As the diagram indicates, below the melting temperature, the enthalpies are different due to contributions from the latent heat of fusion and the different vibrational heat capacities of the crystal and/or glass compared to that of the liquid.

The distance between the 100% amorphous and 100% crystalline enthalpy line is the enthalpy required to melt a 100% crystalline sample. This changes with temperature. Consequently there is no unique enthalpy of fusion for a given degree of crystallinity. This must be considered to be function of temperature. A 50% crystalline material will follow an enthalpy curve approximately half way between the lines defined by 0 and 100% crystallinity. If one measures the enthalpy change between the equilibrium melting temperature and just above the glass transition temperature, this can be broken down into two contributions: one derived from the latent heat, ΔH_{m}, and a contribution from the integral of the vibrational heat capacity, ΔH_{vib}. The total change in enthalpy can always be measured. If one can estimate the contribution for the vibrational heat capacity of the sample, the difference between this value and the total change in enthalpy will be a measure of the latent heat of melting at the equilibrium melting temperature.

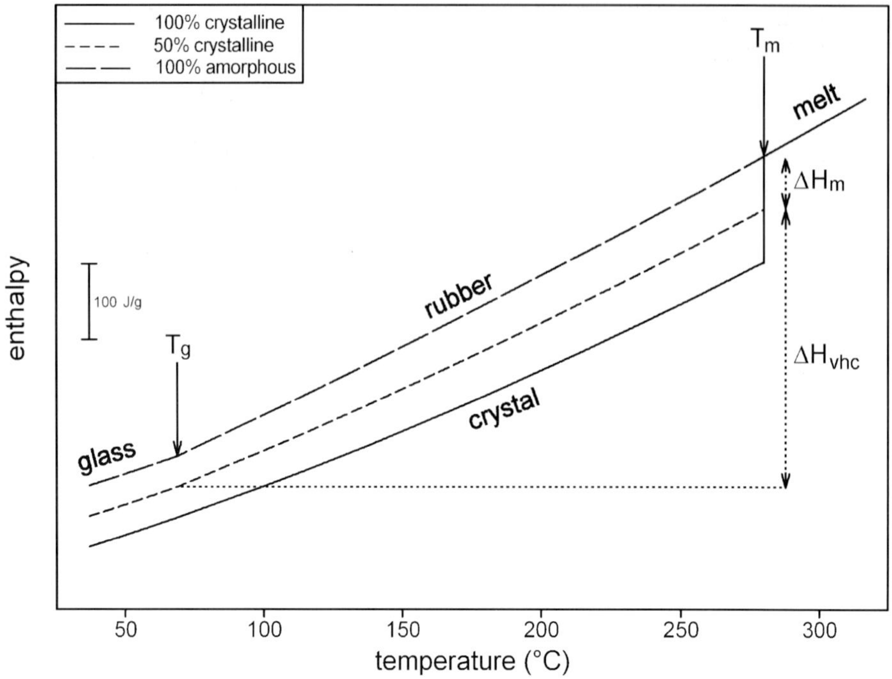

Figure 1.24. Enthalpy diagram for 100% amorphous, 100% crystalline and 50% crystalline PET (data from AThAS databank: http://web.utk.edu/~athas/).

The reversing signal can be used to estimate this quantity. It should be noted that this approach solves the problem of the temperature dependence of enthalpy of melting. This is because the total enthalpy (equals latent heat of melting plus the enthalpy required to account for the vibrational heat capacity) must be the same when integrating over the whole of the relevant temperature interval regardless of at which temperature the melting (or crystallisation on cooling) occurs.

Starting by considering the simplest case of a purely amorphous polymer, Figure 1.25 shows again the results for quenched PET. The simple deconvolution procedure has been used (thus, the phase lag has been neglected) and the non-reversing signal has not been calculated. One can consider the reversing signal in isolation as shown in Figure 1.26. The broad peak that is seen from about 150°C is not due to vibrational heat capacity, but arises from the contributions made by the melting and rearrangement processes that occur as the sample is heated as discussed above. In Figure 1.26, an attempt is made to correct for this by interpolating a baseline to approximate the vibrational heat capacity that would have been measured had crystallisation not occurred. This 'corrected' signal can then be re-plotted with the total signal and difference between them (Figure 1.27). This difference, when

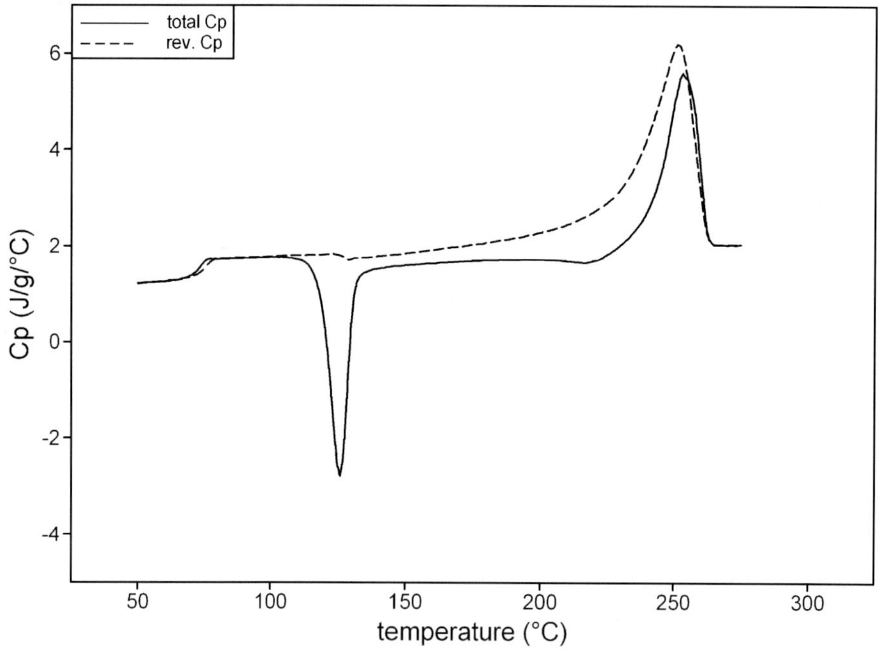

Figure 1.25. MTDSC results for quenched PET showing the peak in the reversing signal that comes from the reorganisation process that occurs after the cold crystallisation (underlying heating rate: 2°C/min, period: 60 s, amplitude: 0.32°C under nitrogen).

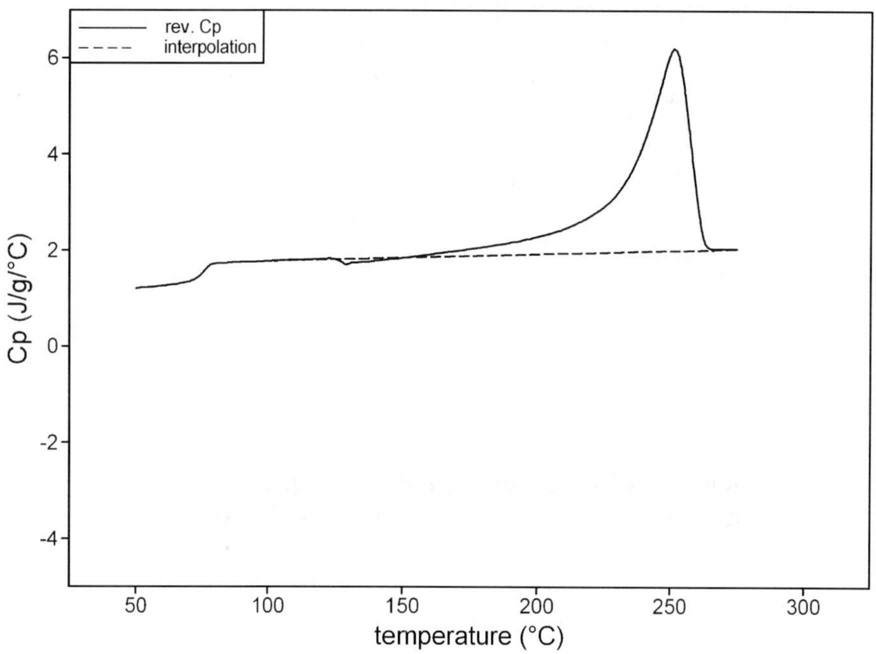

Figure 1.26. The reversing C_p from Figure 1.25 with the interpolation that seeks to approximate the true vibrational heat capacity of the material before crystallisation and rearrangement.

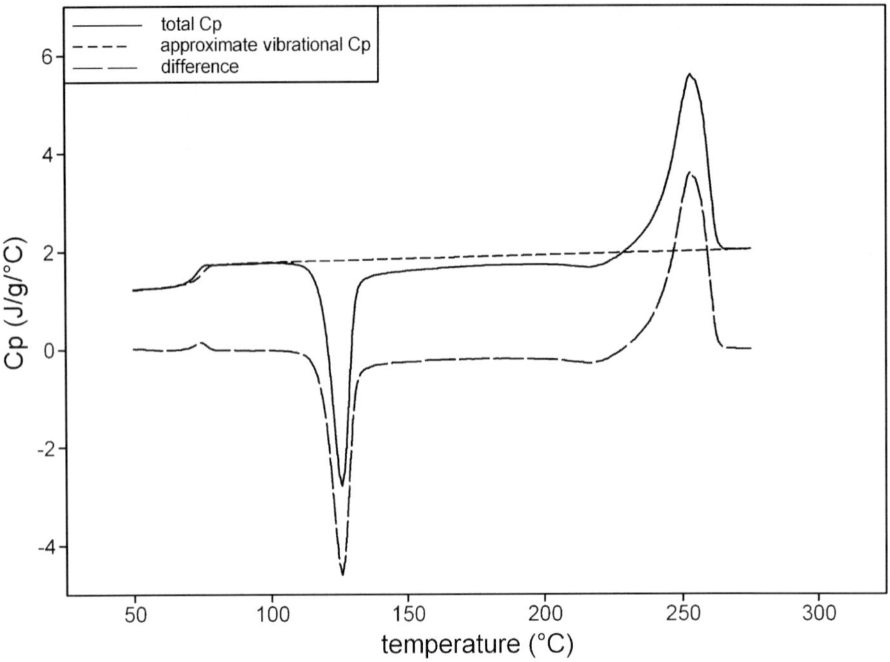

Figure 1.27. Co-plot of total heat capacity with the approximated vibrational heat capacity obtained from the reversing C_p together with the difference between these two signals. The integral of the difference gives the enthalpy associated with the latent heat of melting.

integrated, provides a measure of the enthalpy of melting which is, in this example, zero (to within experimental error). This simple case does not require the use of modulation to estimate the appropriate baseline heat capacity. However, in more complex cases the use of modulation can provide a distinct advantage. Note that, because the peak in the reversing signal is eliminated, the use of the phase lag is irrelevant. Furthermore, had a different frequency or heating rate been used, thus changing the area under the reversing peak during melting, this would also have made no difference to the calculation of crystallinity for the same reason.

Mathematically, we can express these measurements as follows.

$$\Delta H_u = (1 - X_c(T_1)) \int C_{p,a} dT + X_c(T_1) \int C_{p,x}\, dT + X_c(T_1) \Delta H^o(T_2) \tag{62}$$

where T_1 = some temperature before melting begins

T_2 = the equilibrium melting temperature of a 100% crystalline sample

ΔH_u = the enthalpy of melting of the unknown sample

X_c = fraction crystallinity before heating.

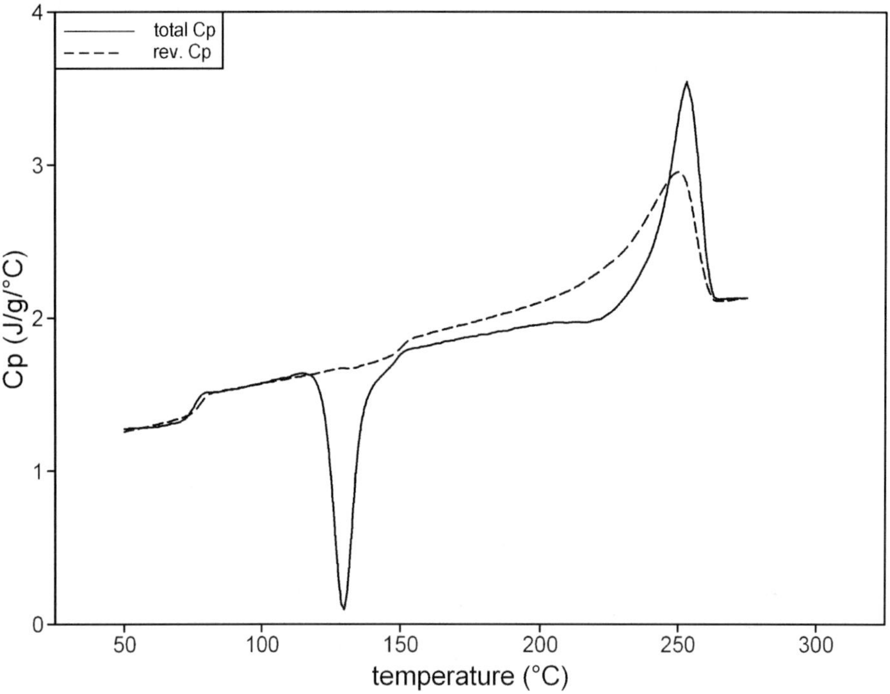

Figure 1.28. MTDSC results for a blend of PET and polycarbonate (underlying heating rate: 2°C/min, period: 60 s, amplitude: 0.32°C under nitrogen).

$C_{\mathrm{p,a}}$ = the heat capacity of the amorphous material
$C_{\mathrm{p,x}}$ = the heat capacity of the crystalline sample
ΔH^{o} = the enthalpy of fusion of a 100% crystalline sample at T_2.

With the interpolation procedure, we are attempting to estimate the quantity $\Delta H_{\mathrm{vib}} = (1 - X_{\mathrm{c}}(T_1)) \int C_{\mathrm{p,a}}\mathrm{d}T + X_{\mathrm{c}}(T_1) \int C_{\mathrm{p,x}}\mathrm{d}T$. When this is subtracted from the total signal (i.e. to calculate the non-reversing signal), this gives $X_{\mathrm{c}}(T_1)\Delta H^{\mathrm{o}}(T_2)_{\mathrm{f}}$ and so, provided $\Delta H(T_2)_{\mathrm{f}}$ is known, the crystallinity at the start of the experiment.

Figure 1.28 shows a blend of polycarbonate (PC) with PET. The glass transition of the amorphous PC occurs beneath the crystallisation peak of the PET and this complicates the interpretation of the data considerably. Again, the contribution from the rearrangement and melting of the PET can be removed by interpolation to produce the approximated vibrational heat capacity baseline shown in Figure 1.29. The difference signal estimates the crystallinity of the PET as 2.5%, close to the correct value of zero. There is an error that arises from the difference between the reversing and total signal at the glass transition (see above). If this is corrected for, then

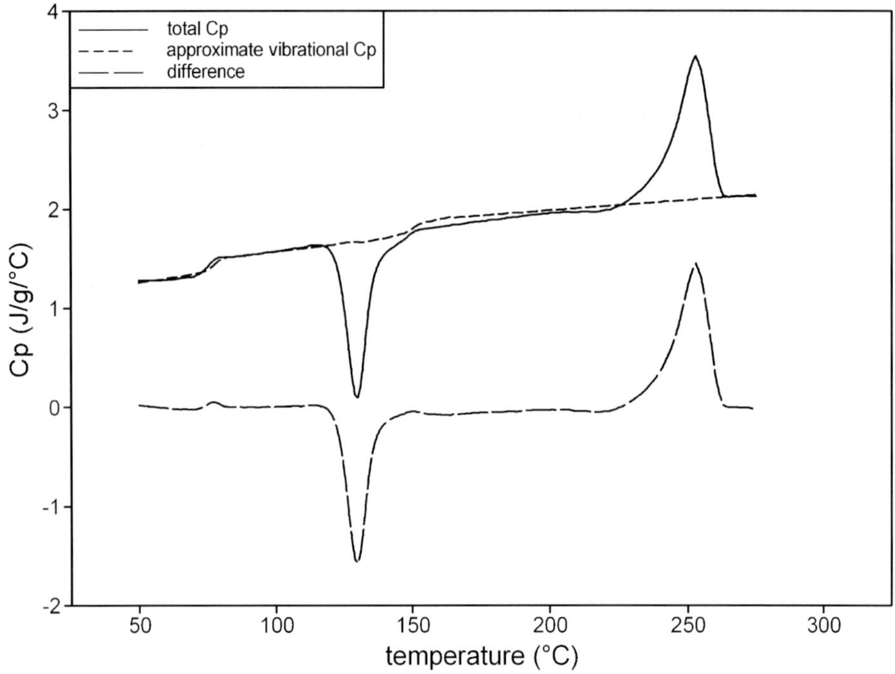

Figure 1.29. Data from Figure 1.28 with interpolated approximation for the vibrational heat capacity and the difference between them.

the apparent crystallinity becomes 1% i.e. the correct answer to within the typical accuracy for this kind of measurement. In this example, ΔH_{vib} is the vibrational heat capacity of everything except the component that can crystallise and/or melt in the temperature range of interest. For an absolute value for crystallinity the mass fraction of this component must be known. Failing this relative crystallinity can be assessed.

There are assumptions in this approach. Probably the most important is that the interpolation assumes the value for the heat capacity of the sample at the upper temperature is the same as the amorphous polymer. For a 50% crystalline sample, for example, the vibrational heat capacity at T_2 would not be the same as a 100% amorphous sample, which is, in effect, the assumption that is made in this method. When the level of crystallinity is low, then the error from this source will be small. As crystallinity increases the potential error increases, but the fact that, for most polymers, the liquid and crystalline vibrational heat capacities converge around the equilibrium melting temperature tends to reduce this problem. For PET at 48% crystallinity the error from this source has been estimated to be 1.5%. A fuller description of this method can be found in [37].

There are other approaches to measuring polymer crystallinity based on conventional DSC [31]. They use database values for the crystalline and amorphous heat capacity sometimes in combination with extrapolation procedures. Where they can be applied, these methods might be preferred as they make fewer assumptions than the MTDSC approach described above. However, industrially useful systems are often blends and/or contain fillers or other additives, so complex samples are frequently encountered. When this is the case, the more conventional approaches will either produce significant errors or will simply be inapplicable. Under these circumstances MTDSC can offer very real advantages.

3.6.3 Summary

- Because polymer melting is the result of a distribution of species all melting at their equilibrium melting temperature, the enthalpy of melting is found in whole or in part in the reversing signal. In principle, where there is no cooling during the modulation cycle, this type of melting behaves in a similar manner to heat capacity. Consequently, the reversing signal no longer has the same meaning as it does when considering chemical reactions and glass transitions because the reversing signal contains a contribution from an essentially non-reversing process. Clearly, it could be argued that this means the nomenclature for the different signals should be changed. However, this nomenclature is now so well established that it seems better to accept it, while bearing in mind that it is somewhat misleading in the case of melting and rearrangement.
- When, during an experiment that has an underlying heating rate, part of the modulation cycle causes cooling to occur, super-cooling will often mean that the response to the modulation is distorted. Consequently, this type of experiment is generally inadvisable. Conditions should be selected that are 'heat-only'.
- Quasi-isothermal experiments in the melt region can often reveal a wealth of information and this is dealt with in detail in Chapter 4.
- When rearrangement is occurring as the sample is heated, the lower heating rates cause an exothermic response balanced by the endothermic response at the higher heating rates. This causes the reversing heat capacity to be higher than the total heat capacity. This behaviour is a good indication that rearrangement is occurring whereas conventional DSC often provides little indication that this is happening. This is because the endothermic and exothermic processes largely cancel each other out.
- The additional information given by MTDSC, compared to conventional DSC, can help in quantifying crystallinity more accurately.

3.7 CALIBRATION

3.7.1 Calibration of the Total and Reversing Signals

The calibration of the average, or total heat flow, or heat capacity is carried out in the same way as conventional calorimetry because this signal is the same as that normally obtained in conventional un-modulated DSC (see [19]). Briefly, an empty-pan baseline run is carried out where the DSC contains empty pans. This is then subtracted from a run in which sapphire is used as the sample. Sapphire is a good calibrant because it undergoes no transitions in the temperature range covered by most DSC's used to study polymers. The apparent heat capacity of the sample can then be calculated by dividing the empty-pan corrected heat flow by the average heating rate. The heat capacity of sapphire as a function of temperature has been established to high accuracy [38]. Thus, these known values can be compared with the apparent values calculated in the foregoing experiment. Correction factors can then be calculated as a function of temperature. In this way, errors that arise from an imperfect baseline and those derived from inaccuracies in the calibration can be corrected for. In subsequent experiments, the sapphire is replaced by a sample in a pan of matched weight. Note that the weight of the pan is matched to that used during the calibration, not necessarily that of the reference pan (see below the discussion on calibrating the reversing signal). The same empty-pan baseline is subtracted and the correction factors are applied to calculate the sample heat capacity as a function of temperature.

The calibration of the reversing heat capacity is also carried out using a heat capacity standard such as sapphire. The procedure is basically the same as that used for calibrating for heat capacity in a conventional DSC experiment described above. A baseline run is carried out with empty pans, then the sapphire calibrant is used. After applying a correction for the empty-pan baseline (see below) to the heat capacity measurements, comparisons are made between the apparent values for the heat capacity calculated from Eq. (13) and the known heat capacity of sapphire. Correction factors can then be derived as a function of temperature. Sometimes a one point (i.e. at one temperature) calibration is offered in commercial software. This then applies the same correction factor over the whole temperature range. This approach should be used with caution as the correction factors can change significantly with temperature. Note, as shown in Figure 1.7, correction factors can be strongly influenced by the period of the modulation. The reasons for this and methods of overcoming them are dealt with in the discussion on advanced theory. Here one is considering the best, simple approach that will work with comparatively long periods. In MTDSC, both of the above calibrations are carried out simultaneously using the same empty-pan baseline and sapphire experiments. Any change to the experimental conditions such as modulation period, heating rate etc. necessitates re-calibration.

The empty-pan baseline correction is somewhat more complex for the reversing measurement than for the conventional measurement. This is because, if the cell-plus-pan asymmetry is such that the reference side has an apparent heat capacity that is greater than the sample side, then this represents a heat capacity deficit that must be added to, not subtracted from, the measured value for the sample heat capacity. Because the cyclic heat capacity is derived from a ratio of amplitudes there can never be a negative value. Thus, the problem is posed, how does one know when to subtract and when to add the empty pan baseline? In principle, the phase lag will be 180° out-of-phase when the reference pan has a higher heat capacity than the sample pan. However, many factors can affect the phase lag and, in general, relying on this signal is inadvisable. Probably the simplest approach is to use a lighter-than-average pan in the reference position and measure the sample pan weight in order to ensure that it is heavier. In this way, the empty-pan baseline can be systematically subtracted from any measurement. Note that the sample pan weight must be matched to that used during the calibration.

Sometimes the empty-pan baseline correction for heat capacity is omitted for the reversing signal because, when closely matched sample and reference pans are used, it is usually small. Whether this is adequate depends on the type of information being sought. For example, if all that is required is the glass transition temperature, then a full heat capacity calibration may be excessive. However, as an absolute minimum, a calibration must be performed to obtain a correction factor for the cyclic heat capacity at one temperature in the range of interest.

It has become common practice to present MTDSC results both as heat flow and heat capacity within the same data set. Typically, the greatest source of error is that from the empty-pan baseline for the average signal. Many workers, in order to save time, choose not to make this correction and then, almost by default, the total signal is presented as heat flow rather than heat capacity, because this correction has not been made. This is not necessarily bad practice. The experimenters may well be able to obtain the information they require without this additional calibration step. Consequently, one must take care when reading the literature to ascertain what types of data reduction and calibration have been carried out. The questions that must be asked are as follows.

- Has an empty-pan baseline correction been carried out on the reversing and/or total measurement?
- Has the heat capacity calibration constant been determined at a single temperature or as a function of temperature?

- Has the phase lag correction been applied when calculating the reversing signal?
- What type of baseline correction has been applied to the phase lag signal?

If some, or all, of the answers to the above questions are absent, this implies that a less than optimal procedure has been adopted. One must form one's own judgement as to whether this lack undermines the conclusions that are drawn for a particular case.

It has been discussed in section 2.1.1 and illustrated in Figure 1.4, how it can be useful to force the reversing signal to be equal to the total signal in non-transition regions. When doing this, one must decide which signal to take as the accurate one. If no empty-pan correction has been made on the total signal, then the reversing signal would usually give the more reliable value. However, it should be noted that an empty-pan correction is generally desirable to avoid the effect of baseline curvature which can significantly confuse interpretation even when forcing agreement, as illustrated in Figure 1.4, is used because this method assumes a linear offset.

Where empty-pan corrections have been made, and long periods (60 s or more) are used, this implies low underlying heating rates – as there must be many modulations over each transition. Generally these conditions would still mean the reversing signal is more accurate, as low heating rates give poorer quantification in the total signal. However, a very thick sample would mean the reversing signal could be significantly in error, because of the effects of sample thermal conductivity (see the advanced theory section), while the accuracy of the total signal is improved because the sample is large. Shorter periods make the reversing signal less quantitative and mean higher heating rates can be used – thus making the total signal more reliable. Where accurate values for the heat capacities are needed (rather than differences between the total and the reversing) there is, in reality, no simple answer to suit all cases. Experimenters are encouraged to gain experience with their particular materials. Making accurate heat capacity measurements in non-transition temperature regions using long periods (typically 100 s) can do this. Although there is evidence that MTDSC can be used to obtain more accurate heat capacity measurements than conventional methods (see Chapter 4), the benefits of MTDSC are mostly achieved through the additional signals it provides through the reversing and non-reversing signals.

3.7.2 Comments on Methods of Phase Lag Correction

In much of the practical section on MTDSC it is assumed that the calorimeter behaves ideally. In other words, the sample and sensor can respond instantaneously and there are no significant thermal lags in any part of the system including the sample. In reality, this is clearly not true and there is a

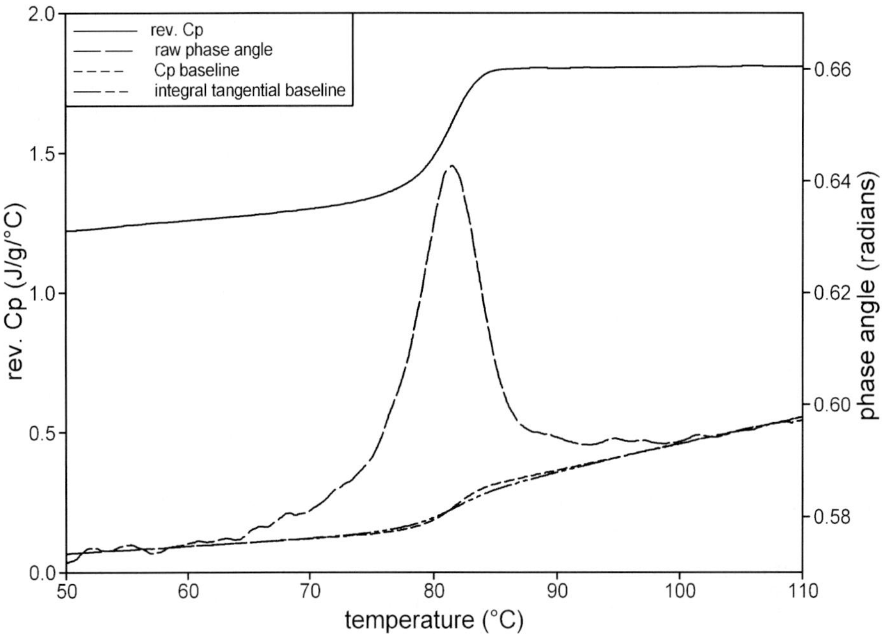

Figure 1.30. Constructions showing baseline correction of phase angle during glass transition (PET).

discussion of this in the advanced section. The signal that is, in some ways, most affected by non-ideality is the phase signal. In an ideal calorimeter, it would be zero except in a transition region. However, this is never the case in practice. The first and simplest solution to this was first proposed by Reading who used a simple interpolation between the start and end of a transition region to provide the baseline that is subtracted from the phase angle to provide the 'true' phase angle [16].

In modelling the non-ideal calorimeter, the simplest non-ideality that we might consider is the thermal resistance that exists between the sensor and the sample pan. It has been shown that when this resistance is significant, the phase angle is affected by heat capacity changes in the sample. Thus, not only is the phase non-zero, it changes with changing sample heat capacity [12]. This poses a problem that can most clearly be seen in the glass-rubber transition. The baseline phase lag before and after the transition are not the same. After the transition, the phase can be greater because the heat capacity is greater. See Figure 1.30. One solution is to construct a baseline that takes its shape from the reversing heat capacity [39] using the idea that it tracks the phase shift due purely to changes in heat capacity rather than the kinetics of the transition. It is these kinetics which, as discussed below, are what dictates the 'true' value of the phase angle during a transition under ideal conditions. This method of baseline construction is well founded except that in 'real' samples other things are often changing during the course of the transitions. For example, changing contact resistance between the sample and the pan. Consequently it is often observed that the change in the phase angle between the start and end of the transition goes the opposite way to that dictated by this simple model [16].

By extension of the original method proposed by Reading, the construction of an interpolated sigmoidal baseline, using the well-established integral tangential method normally adopted for constructing baselines under peaks in DSC experiments [40], can also be applied to the phase angle. This has been found to give reliable results even in cases where the phase shift 'goes the wrong way'. The two different approaches to deriving a baseline are shown in Figure 1.30 for the case of a 'well behaved' glass transition. In this case, the shift in-phase lag baseline follows the expected trend. The two baselines are almost identical. Consequently, either approach can be used. The interpolation method has the advantage that it is more robust to non-ideal behaviour. Both methods are, in fact, equivalent if it is considered that the fractional area under the phase angle peak is a measure of the extent to which the transition is complete. In some cases, the phase angle correction is applied over a large temperature range, as, for example in Figure 1.2. In such circumstances, a simple linear interpolation is advisable.

The simplest approach, and one that in practice is often justified, is to not carry out the phase lag correction and make the approximation as has already been mentioned above, i.e. $C^* = C_{\mathrm{pR}}$. In many practical cases, knowing C_{pK} or C $(= C_{\mathrm{pK}}\beta)$ is not of any value. Thus, it can simply be neglected.

3.8 OVERVIEW

The results of the deconvolution process shown in Figure 1.4 can now be discussed in terms of the simple theory offered above. Outside of transition regions, the total signal and the reversing signal should be the same and the kinetic heat capacity should be zero. At the cold crystallisation, the reversing signal is not greatly affected, thus the non-reversing peak contains the enthalpy of crystallisation and one observes a negative peak in the kinetic heat capacity. This is in accordance with the simple theory outlined above. At the glass transition, the reversing signal shows the step change in heat capacity expected at T_{g}. The non-reversing signal shows a peak. This is, at first glance, in accordance with the simple theory because the enthalpy loss on annealing is recovered during the transition, but not quickly lost again. Thus, this aspect of the glass transition is non-reversing. However, this interpretation must be made within the context of the frequency dependence of the glass transition as would be expected from dynamic mechanical and dielectric measurements. Over the melt region there is a complex behaviour that cannot be accounted for by a simple theory for melting. However, a qualitative description allows one to identify crystalline rearrangement when it occurs in the reversing signal, even when it is not easily discerned in the total signal (equivalent to conventional DSC). One can also exploit the additional information provided by the modulation to enable a more accurate measure of initial crystallinity to be made in complex samples. In all cases, whether studying kinetic processes like cure or crystallisation, investigating blends by looking at their glass transitions or measuring polymer crystallinity, the modulation offers significant advantages. The remaining chapters of this book illustrate this point very well.

4 Detailed Discussion of the Theory of MTDSC

4.1 INTRODUCTION

Under the heading of this more detailed discussion of the theory of MTDSC, we will again consider the following types of transitions and phenomena.

- Chemical reactions and related processes.
- Frequency-dependent heat capacity and glass transitions.
- First-order phase transitions.

In each case, the full derivations of the expressions used in the more practical exposition above will be given. The discussion will also look more closely at some of the simplifying assumptions and the problems that arise when these no longer apply. Before this some comments are made on alternative modulations and deconvolution methods.

4.2 MODULATION AND DECONVOLUTION

It is possible to use multiple sine waves [10] and so extract as a Fourier series (or other deconvolution procedure) the response to several frequencies simultaneously, as illustrated in Chapter 4. An extension of this is the use of saw-tooth temperature modulations [20]. These can be considered to be a combination of an infinite series of sine waves (though only a limited range will be available in practice). A symmetric saw-tooth (same heating and cooling rate) only has odd harmonics, but an asymmetric saw-tooth (different heating and cooling rates) is equivalent to a broad range of frequencies.

The use of averaging combined with a Fourier transform is by no means the only possible deconvolution procedure [17]. Details of a linear fitting approach have been published that could easily be adapted to deal with some forms of non-linear behaviour [10]. Other multi-parameter fitting approaches are possible. However, overwhelmingly, the current practice is to use a single sinusoidal modulation and a Fourier transform so this will be the focus of this Chapter and this book.

4.3 CHEMICAL REACTIONS AND RELATED PROCESSES

For a zero-order chemical reaction (including a term for the contribution to the heat flow from heat capacity), the rate of the reaction is dependent only on temperature. Thus, it produces heat at a rate given by some function of temperature $h(T)$. Taking the heating programme given in Eq. (6)

$$\mathrm{d}Q/\mathrm{d}t = C_{\mathrm{p}}(\beta + \omega B \cos \omega t) - h(T_0 + \beta t + b \sin \omega t) \tag{63}$$

For an exactly linear case, $h(T) = h_1 + h_2 T$

$$\frac{\mathrm{d}Q}{\mathrm{d}t} = (C_p\beta - h_1 - h_2 T_0 - h_2\beta t) + B(C_p\omega \cos \omega t - h_2 \sin \omega t). \tag{64}$$

More generally, $h(T_0 + \beta t + B \sin \omega t)$ can be expanded as a power series

$$h(T_0 + \beta t) + Bh'(T_0 + \beta t) \sin \omega t + \frac{B^2}{2} h''(T_0 + \beta t) \sin^2 \omega t + \cdots$$

So,

$$\frac{\mathrm{d}Q}{\mathrm{d}t} = (C_{\mathrm{p}}\beta - h(T_0 + \beta t)) + B(C_{\mathrm{p}}\omega \cos \omega t - h'(T_0 + \beta t) \sin \omega t) - \frac{1}{2} B^2 h''(T_0 + \beta t) \sin^2 \omega t + \cdots \tag{65}$$

Should β be small enough for terms in B^2, B^3, to be negligible in comparison with B, the response of the heat flow is effectively linear, and $\mathrm{d}Q/\mathrm{d}t$ agrees with that for the exactly linear case. In the event that the amplitude of the temperature modulation B is not so small, the terms in $B^2 \sin^2 \omega t$, $B^3 \sin^3 \omega t$ are significant and the higher harmonics $\cos 2\omega t$, $\sin 3\omega t, \ldots$ appear in the modulation and the heat flow $\mathrm{d}Q/\mathrm{d}t$. These can, in principle, give information about the kinetic law h through its derivatives $h', h'' \ldots$

Returning to the linear case

$$\langle \mathrm{d}Q/\mathrm{d}t\rangle = C_p\beta - h_1 - h_2T_0 - h_2\beta t \tag{66}$$

Modulation of the heat flow $= B(C_p\omega\cos\omega t - h_2\sin\omega t) = A_{\mathrm{HF}}$ $\cos(\omega t - \phi)$

where $A_{\mathrm{HF}} = BC_{\mathrm{p}}\omega\sqrt{1 + h_2^2/C_{\mathrm{p}}^2\omega^2}$, $\quad \phi = -\tan^{-1}(h_2/C_{\mathrm{p}}\omega)$ and $\mathrm{A}_{\mathrm{HF}} = \omega B$

In the absence of the kinetic process ($h = 0$) a measurement of the heat capacity is

$$C_{\mathrm{p}} = A_{\mathrm{HF}}/A_{\mathrm{HR}}. \tag{67}$$

This indicates that

$$A_{\mathrm{HF}}/A_{\mathrm{HR}} = C^*$$
$$C^*\cos\phi = C_{\mathrm{pR}}$$
$$C^*\sin\phi = C_{\mathrm{pK}}$$

could all be useful measured quantities as indicated above. For the general zero-order linear reaction, $C^* = C_{\mathrm{p}}\sqrt{1 + h_2^2/C_{\mathrm{p}}^2\omega^2}$ differs from the true heat capacity, although not significantly if the frequency is high enough. The phase-corrected reversing heat capacity, however, is given by the following relation.

$$C_{\mathrm{pPCR}} = C^*\cos\phi = C_{\mathrm{p}}. \tag{68}$$

At the same time, the non-reversing heat flow is simply

$$\left\langle\frac{\mathrm{d}Q}{\mathrm{d}t}\right\rangle - \beta C^*\cos\phi = \text{Average heat flow} - \beta C_{\mathrm{pPCR}} = h(T) \tag{69}$$

Thus, it is possible to separate the contribution in the total heat flow from the heat capacity and that which arises from the zero order reaction.

Considering a more general process that gives rise to a heat flow, and is governed by a kinetic function that is dependent on temperature *and* time, viz for some process;

$$dQ/dt = f(t, T) \tag{70}$$

By analogy with the simpler case considered above

$$\begin{aligned} dQ/dt &= \beta C_p + BC_p\omega\cos\omega t + f(t, T) \\ &\simeq (C_p\beta - f(t, T_0 + \beta t)) \\ &+ B\left(C_p\omega\cos\omega t - \frac{\partial f}{\partial T}(t, T_0 + \beta t)\sin\omega t\right) \end{aligned} \tag{71}$$

neglecting the non-linear (higher order) terms. Following the above procedure leads to an average heat flow

$$\begin{aligned} \langle dQ/dt \rangle &= C_p\beta + f(t, T_0 + \beta t) + O(B^2) \\ &= C_p\beta + f(t, T) + O(B^2) \end{aligned} \tag{72}$$

an amplitude of the heat flow modulation

$$A_{HF} = BC_p\omega\sqrt{1 + \left(\frac{\partial f}{\partial T}\Big/ C_p\omega\right)^2} + O(B^2) \tag{73}$$

and a phase lag

$$\phi = \tan^{-1}\left(\frac{\partial f}{\partial T}\Big/ C_p\omega\right) + O(B) \tag{74}$$

Neglecting the higher order terms

$$C_{pPCR} = C_p, \quad C_{pK} = \frac{\partial f}{\partial T}\Big/\omega \quad \text{and} \quad C^* = C_p\sqrt{1 + \left(\frac{\partial f}{\partial T}\Big/ C_p\omega\right)^2} \tag{75}$$

where $\partial f/\partial T$ means $(\partial f/\partial T)(t, T_0 + \beta t)$. So, it is the average value.

Also, using the complex notation, the heat flow modulation is just

$$B\omega\mathrm{Re}\left\{\left(C_\mathrm{p}-i\frac{\partial f}{\partial t}\Big/\omega\right)\mathrm{e}^{i\omega t}\right\}=B\omega\mathrm{Re}\{\hat{C}\mathrm{e}^{\mathrm{i}\omega\mathrm{t}}\} \tag{76}$$

where $\hat{C}=C_\mathrm{p}-i\dfrac{\partial f}{\partial T}/\omega$.

Thus,

$$C_\mathrm{pPCR}=C_\mathrm{p}\quad\text{and}\quad C_\mathrm{pK}=(\partial f/\partial T)/\omega. \tag{77}$$

Considering the Arrhenius equation, viz

$$\frac{\mathrm{d}Q}{\mathrm{d}t}=\beta C_\mathrm{p}+Hf(x)A\mathrm{e}^{-E/RT}+C_\mathrm{p}B\omega\cos\omega t, \tag{78}$$

where

$$HA\mathrm{e}^{-E/RT}=HA\mathrm{e}^{-E/(RT_0+\beta t)}+\left(\frac{\mathrm{d}}{\mathrm{d}T}\left(HA\mathrm{e}^{-E/RT}\right)\right)\Big|_{T=T_0+\beta t}B\sin\omega t$$
$$+\text{higher order terms}$$

and the reaction extent satisfies Eq. (79)

$$\frac{\mathrm{d}x}{\mathrm{d}t}\Big/f(x)=A\mathrm{e}^{-E/R(T_0+\beta t+B\sin\omega t)} \tag{79}$$

This last equation, because the temperature variation is taken to be sufficiently small, leads to

$$x=\langle x\rangle+\text{oscillatory term}$$

where $\langle x\rangle$ is the same as the reaction extent during conventional DSC and the oscillatory term takes the form

$$\frac{x_\mathrm{c}(t)B}{\omega}\cos\omega t+\text{higher order terms and terms in }\frac{B}{\omega}$$

the coefficient x_c being independent of the modulation.

The combination of all of these eventually results in a total heat flow signal

$$\frac{dQ}{dt} = C_p\beta + \left\langle Hf(x)Ae^{-E/RT}\right\rangle \quad \text{... the underlying signal}$$

$$+ B\left((\omega C_p \cos\omega t + f(\langle x\rangle)\left(\frac{d}{dT}\left(HAe^{-E/RT}\right)\right)_{T=T_0+\beta t} \sin\omega t\right.$$

$$\left. + \frac{df}{dx}(\langle x\rangle)HAe^{-E/R(T_0+\beta t)}\frac{x_c}{\omega}\cos\omega t\right) + \text{higher order terms ... the response to the modulation}$$

$$= C_p\beta + Hf(\langle x\rangle)Ae^{-E/R\langle T\rangle} + \text{higher order terms ... the underlying signal}$$

$$+ B\left(\omega C_p \cos\omega t + \omega C_{pK}\sin\omega t + \frac{D}{\omega}\cos\omega t\right) \quad \text{... the response to the modulation}$$

$$+ \text{higher harmonics and other higher order terms.} \tag{80}$$

The higher order terms are those in B^2, B^3, etc. which appear because of the non-linearity of the kinetic process.

Under usual operation, the temperature variation should be small enough and the frequency high enough for terms involving B^2 or $1/\omega$ to be negligible. The various signals can then be related to the heat capacity and kinetics according to

$$\text{Average heat capacity} = C_p + Hf(\langle x\rangle)e^{-E/R\langle T\rangle}/\beta \tag{81}$$

$$\text{Phase-corrected reversing heat capacity} = C_{pPCR} = C_p \tag{82}$$

$$\text{Non-reversing heat capacity} = \langle dQ/dt\rangle/\beta - C_{pR}$$
$$= Hf(\langle x\rangle)e^{-E/R\langle T\rangle}/\beta \tag{83}$$

$$\text{Kinetic heat capacity} = C_{pK} = \frac{f(\langle x\rangle)}{\omega}\frac{d\left(HAe^{-E/RT}\right)_{T=\langle T\rangle}}{dT} \tag{84}$$

Of course, with ω large, the last is small compared with the reversing heat capacity and the response of the sample is dominated by the heat capacity of the sample.

The above analysis of MTDSC data in terms if kinetics has been proposed and developed by Lacey and Reading and co-workers [2,5,9,12,14] and also by Toda and co-workers [42,43]. It is essentially a kinetic approach to MTDSC theory and this is the basis of the theory throughout this chapter.

4.4 FREQUENCY DEPENDENT HEAT CAPACITY AND THE GLASS TRANSITION

Heat content stored in molecular motions, such as vibration, is assumed to be rapid when compared with the modulation of the temperature; kinetic effects influence heat flow, but only through reaction rates. This contrasts with cases where the heat in molecular vibrations itself is not rapid and some sort of kinetics plays a role in the heat capacity, or at least in some of the heat taken up and released. A relaxation time appears in the rate of change of enthalpy.

Perhaps the simplest example of how a time-scale can be involved in heat flow to and from a sample, and thereby give an (apparent) dependence of heat capacity upon frequency, is where the thermal conductivity of the sample is in some sense poor (or, equivalently, the sample's specific heat is very large). For simplicity, we may look at the case of a homogeneous spherical sample inside a locally uniform part of the calorimeter. More realistic cases are less easy to analyse, but the qualitative effects are much the same.

Using complex notation, the cyclic part of the temperature can be written as

$$\tilde{T} = \mathrm{Re}\{\hat{T}(r)\mathrm{e}^{i\omega t}\} \tag{85}$$

where r = distance from the centre of the sample and $\hat{T}(R) = \hat{T}_\mathrm{s}$ on the sample's surface $r = R$. By T_s we now must be quite precise and here we mean the temperature on this surface. From consideration of heat flow inside the sample

$$\frac{\mathrm{d}^2\hat{T}}{\mathrm{d}r^2} + \frac{2}{r}\frac{\mathrm{d}\hat{T}}{\mathrm{d}r} = \frac{i\omega\rho c}{\kappa}\hat{T} \tag{86}$$

where ρ = density
c = specific heat
κ = thermal conductivity

(All assumed constant here.) This complex temperature $\hat{T}$ can then be found to be

$$\hat{T} = \hat{T}_s\, R \sinh\left(\sqrt{\frac{\omega\rho c}{2\kappa}}(1+i)r\right) \Big/ r \sinh\left(\sqrt{\frac{\omega\rho c}{2\kappa}}(1+i)R\right) \tag{87}$$

and the total power flow into the sample is

$$\begin{aligned}\frac{\mathrm{d}Q}{\mathrm{d}t} &= \mathrm{Re}\left\{4\pi R^2 \times \kappa \frac{\partial \tilde{T}}{\partial r}\right\} \text{(area} \times \text{heat flow/unit area)} \\ &= \mathrm{Re}\left\{4\pi R\kappa \left(\sqrt{\frac{\omega\rho c}{2\kappa}}(1+i)R \coth\left(\sqrt{\frac{\omega\rho c}{2\kappa}}(1+i)R\right) - 1\right) \hat{T}_s e^{i\omega t}\right\}\end{aligned} \tag{88}$$

This gives rise to a complex heat capacity of

$$\hat{C} = -4i\pi \frac{R\kappa}{\omega} \left(\sqrt{\frac{\omega\rho c}{2\kappa}}(1+i)R \coth\left(\sqrt{\frac{\omega\rho c}{2\kappa}}(1+i)R\right) - 1\right) \tag{89}$$

For sufficiently small samples, $R\sqrt{\frac{\omega\rho c}{\kappa}} \ll 1$, this expression simplifies to the true specific heat

$$\hat{C} \simeq C_\mathrm{p} = \frac{4\pi R^3}{3}\rho c \tag{90}$$

If the frequency is insufficiently small, $\hat{C}$ will deviate significantly from C_p, as indicated. Estimates on how large the frequency may be taken for specific sizes of samples, still with desired accuracy, can be found in papers by Hatta [24] and Toda [25]. These effects can be used to measure the specimen's thermal conductivity.

In the rest of the discussions on transitions throughout this chapter, such size-dependent effects are taken to be negligible. Interpretation of results is easiest if the sample is small enough for its temperature to be uniform, $T_\mathrm{s}(t)$.

A more interesting case of where heat capacities, cyclic, phase-corrected reversing, etc., vary with frequency ω is of a material undergoing a phase transition. This time, the size of sample is not so important – the phenomenon is an intrinsic property of the material – and the underlying temperature $\langle T \rangle = T_0 + \beta t$ plays a key role.

With a relaxation time that decreases rapidly as temperature increases, heat is gained and lost more easily at higher temperatures. In particular, for temperature $T > T_g$, the glass transition temperature, $\mathrm{d}Q/\mathrm{d}t = C_{pl}\mathrm{d}T/\mathrm{d}t$, where C_{pl} is the liquid heat capacity. For lower temperatures, relaxation is slow and the heat capacity is C_{pg}, the heat capacity of the glass, which is smaller than C_{pl}. This observed heat capacity depends on how the relaxation time compares with the time-scale of the changing temperature. Roughly speaking, time-scale > relaxation time ($T_g < T$) leads to C_{pl}, while time-scale < relaxation time ($T_g > T$), gives C_{pg}. As a consequence, even for standard DSC, T_g depends on the temperature ramp β. More specifically the glass transition temperature increases with β.

This carries over to the cyclic measurements. Provided that the relaxation time is small enough compared with the period of oscillation, $2\pi/\omega C^* \simeq C_{pl}$, while if the period of oscillation is too short, $C^* \simeq C_{pg}$. The change occurs at a glass transition temperature dependent on the frequency of modulation, ω. By the same reasoning as for standard DSC, and consequently the total signal in MTDSC, this cyclic-glass transition temperature increases with ω. Moreover, because the oscillations must in some sense be fast as noted above, in MTDSC the cyclic T_g is higher than the underlying T_g [32].

How this T_g varies with ramp β and frequency ω depends upon the nature of the relaxation process. A specific model with the enthalpy of a specimen changing according to a single ordinary differential equation and affected by both temperature and its rate of change is given in Eq. (42). Such models lead to predictions of not just the glass transition temperatures, but also the profiles of C^*, C_{pPCR} and C_{PK} in the transition regimes. MTDSC offers a way of determining key physical parameters related to the material, not just C_{pl} and C_{pg}, from comparing the variation of T_gs with β and ω and transition profiles with those predicted by such models.

A more general approach to materials exhibiting frequency-dependent heat capacities is that of Schawe and co-workers [18]. For some linear (or at least, for temperature not varying too much, approximated linear) process, the rate of heat intake and rate of change of temperature can be related through a convolution.

$$\frac{\mathrm{d}Q}{\mathrm{d}t} = \int_{-\infty}^{t} \psi(t-t')\frac{\mathrm{d}T}{\mathrm{d}t}(t')\mathrm{d}t' = \int_{-\infty}^{\infty} \psi(t-t')\frac{\mathrm{d}T}{\mathrm{d}t}(t')\mathrm{d}t' \tag{91}$$

taking $\psi(t') \equiv 0$ for $t' < 0$, for some kernel ψ which is fixed by the underlying physical process. For a simple material with $Q = C_pT$ and

$C_p = \text{constant}$, $\psi(t') = C_p\delta(t')$, where $\delta(t')$ is the 'Dirac delta function' ($\delta(t') = 0$, for $t' \neq 0$, $\int_{-a}^{a} \delta(t')\mathrm{d}t' = 1$ for $a > 0$.)

Taking the Fourier transform of this convolution leads to

$$F\left(\frac{\mathrm{d}Q}{\mathrm{d}t}\right)(\omega') = F(\psi)(\omega')F\left(\frac{\mathrm{d}T}{\mathrm{d}t}\right)(\omega') \tag{92}$$

where $F(f)(\omega') = \int_{-\infty}^{\infty} f(t)\mathrm{e}^{i\omega' t}\,\mathrm{d}t$.

Writing $\omega = -\omega'$ leads to

$$F\left(\frac{\mathrm{d}Q}{\mathrm{d}t}\right)(-\omega)\Big/F\left(\frac{\mathrm{d}T}{\mathrm{d}t}\right)(-\omega) = \hat{C}(\omega) = F(\psi)(-\omega) \tag{93}$$

The Fourier transform of the rates of change of enthalpy and temperature are related through that of the kernel function, which can be identified with the complex heat capacity. By carrying out a succession of experiments to determine $\hat{C}(\omega)$ and then doing a Fourier inversion it is then possible, in principle, to recover the function ψ, and hence gain information about the physical kinetics. However, to be useful this approach must refer to more specific models of realistic behaviour, which then brings us back to the kinds of results discussed in this chapter.

The basic model used for the glass transition is that of Hutchinson and Kovacs [41] (see also [31]).

$$\frac{\mathrm{d}\delta}{\mathrm{d}t} = -\Delta C_p\frac{\mathrm{d}T}{\mathrm{d}t} - \frac{\delta}{\tau_0}\mathrm{e}^{-\Delta h^*/RT} \tag{94}$$

Here, $\delta = \text{enthalpy} - \text{equilibrium enthalpy} = \text{enthalpy} - C_{pl}T$. This equation can be rewritten in terms of the difference between the enthalpy and that for the glass.

$$\eta = \text{enthalpy} - C_{pg}T = \delta + T\Delta C_p \tag{95}$$

Near the average glass transition, the Arrhenius term can be approximated as

$$\mathrm{e}^{-\Delta h^*/RT} = \mathrm{e}^{-\Delta h^*/RT_{g\beta}} \cdot \mathrm{e}^{\Delta h^*(T-T_{g\beta})/RT_{g\beta}^2} \tag{96}$$

and the equation for η becomes, on making use of this 'large activation energy' approximation ($\Delta h^* \gg RT_{g\beta}$, so terms in $RT_{g\beta}/\Delta h^*$ can be neglected)

$$\frac{d\eta}{dt} = \left(\frac{1}{\tau_0}\exp\left(\frac{-\Delta h^*}{RT_{g\beta}}\right)\right)\exp\left(\frac{\Delta h^*(T - T_{g\beta})}{RT_{g\beta}^2}\right)\left(T_{g\beta}\Delta C_p - \eta\right) \quad (97)$$

This equation indicates that η is the size of $T_{g\beta}\Delta C_p$, while T changes by an amount of size $RT_{g\beta}^2/\Delta h^*$ and does so, for the averaged measurements, at rate β. Balancing the terms in this equation leads to an expression for the glass transition temperature

$$\frac{\beta\Delta h^*}{RT_{g\beta}^2} = \frac{1}{\tau_0}\exp\left(\frac{-\Delta h^*}{RT_{g\beta}}\right) \quad (98)$$

The solution of the approximate equation for η leads, eventually, to an ageing term as well as an integral term which is independent of ageing. In section 3.5, an ad hoc model is used for the average signal in order to illustrate points more directly with regard to the phenomenology of this measurement.

Looking at the cyclic parts of η, $\mathrm{Re}\{\hat{\eta}e^{i\omega t}\}$ and of temperature, $\mathrm{Re}\{\hat{T}e^{i\omega t}\}$ it is seen that

$$i\omega\hat{\eta} = \left(\frac{1}{\tau_0}\exp\left(-\frac{\Delta h^*}{RT_{g\omega}}\right)\right)\exp\left(\frac{\Delta h^*(T - T_{g\omega})}{RT_{g\omega}^2}\right)\left(\frac{\Delta h^*\Delta C_p T\hat{T}}{RT_{g\omega}^2} + \hat{T}\Delta C_p - \frac{\bar{\eta}\Delta h^*\hat{T}}{RT_{g\omega}^2} - \hat{\eta}\right) \quad (99)$$

where the averaged temperature, written at T is within the order of magnitude $RT_{g\omega}^2/\Delta h^*$ of the cyclic transition temperature $T_{g\omega}$. For high frequencies, so that the cyclic transition temperature is significantly greater that the averaged one, the averaged enthalpy difference $\bar{\eta}$ is exponentially close to T. Thus,

$$\left(1 + i\omega\tau_0\exp\left(\frac{\Delta h^*}{RT_{g\omega}}\right)\cdot\exp\left(\frac{-\Delta h^*(T - T_{g\omega})}{RT_{g\omega}^2}\right)\right)\hat{\eta} = \hat{T}\Delta C_p \quad (100)$$

The location of the glass transition is then fixed by

$$\omega\tau_0 \exp\left(\frac{\Delta h^*}{RT_{g\omega}}\right) = 1 \qquad (101)$$

while the response to the modulation is

$$\frac{\hat{\eta}}{\hat{T}} = \frac{\Delta C_p}{1 + i\exp\left(-\Delta h^*(T - T_{g\omega})/RT_{g\omega}^2\right)}$$

Taking the real and imaginary part predicts according to this model

$$\Delta C_{pPCR} = \frac{\Delta C_p}{1 + \exp\left(-2\Delta h^*(T - T_{g\omega})/RT_{g\omega}^2\right)} \qquad (102)$$

and

$$C_{pK} = \frac{\Delta C_p \exp\left(-\Delta h^*(T - T_{g\omega})/RT_{g\omega}^2\right)}{1 + \exp\left(-2\Delta h^*(T - T_{g\omega})/RT_{g\omega}^2\right)}, \qquad (103)$$

respectively. To fix real behaviour (multiple relaxation times) it is necessary to include 'shape factors'. See equations 41 and 42.

A slightly more general formulation of this Hutchinson–Kovacs model [41] is given in Chapter 4. In that model, for C_{pg} and C_{pl} assumed constant, the equilibrium enthalpy is given by $C_{pl}T = C_{pg}T + \varepsilon_h N^*$ with $N^*(T)$ the equilibrium number of configurations of energy ε_h. More generally, the enthalpy is $Q = C_{pl}T + \delta = C_{pg}T + \varepsilon_h N$, with $N(t)$ the instantaneous number of configurations. Using the relation between N^* and T, the relaxation law $\mathrm{d}N/\mathrm{d}t = (N^* - N)/\tau$ (see Chapter 4) gives, on eliminating N in favour of δ, the above equation for excess enthalpy δ.

4.5 MELTING

One simple model for melting, which has shown good agreement with experimental results, is to represent it in the same way as an irreversible, endothermic chemical reaction (see above). For a polymer consisting (partly) of crystals with a range of melting temperatures, T_m, at any time t the crystalline mass distribution can be given in terms of a density function $m(t, T_m)$ (so the total mass fraction of crystals is $x(t) = \int m(t, T_m)\mathrm{d}T_m$). During melting (and ignoring possible recrystallisation), the crystalline density reduces according to a rate law of the form

$$\frac{\partial m}{\partial t} = F(T, T_m)m \qquad (104)$$

Cf. eqns (38) and (39) [23]. As in the above consideration of Arrhenius kinetics, the cyclic signal then takes the approximate form shown in Eq. (105).

$$\text{Cyclic signal} = \omega B(C_p \cos \omega t + C_{PK} \sin \omega t) \qquad (105)$$

Thus, the reversing heat capacity gives a good estimate of C_p. The kinetic heat capacity, which decreases as the reciprocal of modulation frequency, is determined by how the melting rate depends upon temperature.

$$C_{PK} = \frac{L}{\omega} \int \left(m(t, T_m) \frac{\partial F(T, T_m)}{\partial T} \right) dT_m \qquad (106)$$

where L is the latent heat for the transition $\times$ mass of sample.

A more accurate consideration of the reversing heat capacity (still following the Arrhenius analysis, or again see Toda *et al.* [23,25]) will give its variation with frequency (the difference $C_{pPCR} - C$ decreases as $1/\omega^2$ for 'large' frequency ω).

A very different approach to polymer melting stems from the detailed description by Wunderlich [31,44]. In this, for the time-scales involved in the calorimetry, melting is considered as instantaneous. The fraction $m(t, T_m)$ drops to zero as $T(t)$ increases through T_m. (Melting is not kinetically hindered; or if it is, its time-scale is very short compared with that of the MTDSC.) Nucleation of crystals (which take the form of lamellae) is still taken to be negligible (the time-scale for this is long compared with an experiment), but now existing crystals (with $T_m > T$) grow when either there is available melt or melt is being made available through the melting of smaller lamellae (with lower values of T_m). In this model, a simple version of which has been formulated and discussed in Lacey and Nikolopoulos [45], the number densities $n(t, T_m)$ of lamellae stays fixed until T exceeds T_m, while initially the mass fraction $m(t, T_m)$ increases in a well-determined way, again until T exceeds T_m, when both n and m fall to zero.

In the particular model considered by Lacey and Nikolopoulos [45], there were three distinct phases:

1) the small initial rate of melting of crystals allows instant recrystallisation of the melted polymer, so there is no net melting. Both the average and the phase-corrected reversing heat capacities remain equal to C_p;

2) there is now sufficient crystalline material being melted for net melting to occur for part of each cycle, but insufficient for recrystallisation not to take up the excess melt before the cycle is complete. Thus, there is no melting on average. So, the average signal only manifests the heat capacity while the cyclic heat capacity is now increased due to the melting and crystallisation in each cycle;

3) an increased crystalline fraction (and reduced number of surviving lamellae for melt to recrystallise onto) means that net melting occurs throughout each cycle and both average and phase-corrected heat capacities exceed C_p.

A more quantitative analysis of the model for this third phase shows that, if the amplitude of modulation is sufficiently small for the temperature to be always increasing:

$$\begin{aligned}\text{Average heat capacity} &= C_p + \text{latent heat} \times \text{mass} \times (m(t, T_0 + \beta t) \\ &\quad - \text{fractional recrystallization rate}/\beta) \\ \text{Phase-corrected heat capacity} &= \text{cyclic heat capacity} \\ &= C_p + \text{latent heat} \times \text{mass} \times m(T_0\beta t, T_0 + \beta t)\end{aligned}$$

(to leading order). So, the average heat capacity is less than the phase-corrected heat capacity. (With a larger amplitude, with temperature decreasing for a part of each cycle, the expression for the phase-corrected heat capacity is rather more complicated due to the melting rate not being sinusoidal – it is zero whenever the temperature falls. In this case, the phase-corrected heat capacity can drop below the average heat capacity towards the end of the phase.) The above relation for small amplitude agrees with the observation that the phase-corrected reversing heat capacity is higher than the average heat capacity. The model, as currently constituted, does not allow for the fact that real polymers almost never achieve 100% crystallinity. Thus, there is always a substantial amorphous fraction that, in effect, cannot crystallise. This could be accounted for by describing a 'background' amorphous fraction that participates to a lesser extent than the fraction closer to the lamellae and thus is more able to crystallise. The apparent frequency dependence of the melting and rearrangement peak in, for example, PET shown in Figure 1.21 has its origins, in our view, in thermal transport difficulties within the sample. This could be accounted for by allowing for either the sample temperature dropping below the programme $T_0 + \beta t + B \sin \omega t$ when melting occurs or for the sample to have an internal temperature that varies significantly during the phase change. Each of these effects will act to hinder the melting and produce the crossing of the underlying and phase-corrected heat capacities. Toda's concept of kinetic hindering of melting [23,25] can also be interpreted in the same way (i.e. as arising from temperature differences rather than true kinetics of melting), and so this approach might be combined with that of [45] particularly near the end of melting where the phase-corrected reversing heat capacity is below that of the average signal.

4.6 CALIBRATION

About the simplest model for an ideal calorimeter, i.e. one which is unbiased and has perfect measurements of both sample and reference (but which nevertheless allows for direct heat transfer between sample and reference), is due to Wunderlich *et al.* [46]. Heat flow to the sample is given by the temperature difference between the reference and the sample and that between block and sample.

$$(C_R + C_p)\frac{dT_s}{dt} = K_1(T_R - T_s) + K_2(T_F - T_s) \tag{107}$$

where C_R is the heat capacity of the actual pan and its environs. Similarly,

$$C_R\frac{dT_R}{dt} = K_1(T_s - T_R) + K_2(T_F - T_R) \tag{108}$$

The similarity between these two equations is associated with the lack of bias. This sort of ordinary differential equation model relies on the calorimeter being adequately represented by a finite number of parts (here two) each of which has a uniform temperature. The heat transfer coefficients will be independent of temperature for a truly linear system (but the device can be regarded satisfactorily as linear as long as their values do not change significantly over the temperature range inside the calorimeter at any instant or from the minimum to the maximum of a modulation). Eliminating T_F, the model reduces to

$$C_R\frac{d\Delta T}{dt} + K\Delta T = C_p\frac{dT_s}{dt} \tag{109}$$

where $\Delta T = T_R - T_s$ and $K = 2K_1 + K_2$.

For $T_s = T_0 + \beta t + B\sin\omega t = T_0 + \beta t + \mathrm{Re}\{-iBe^{i\omega t}\}$,

$$\begin{aligned}\Delta T &= \beta C_p/K + \mathrm{Re}\left\{\omega BC_p e^{i\omega t}/(K + i\omega C_R)\right\}\\ &= \beta C_p/K + \frac{\omega BC_p}{K^2 + \omega^2 C_R^2}(K\cos\omega t + \omega C_R \sin\omega t)\\ &= \frac{\beta C_p}{K} + \frac{\omega BC_p}{\sqrt{(K^2 + \omega^2 C_R^2)}}\cos(\omega t - \varphi)\end{aligned} \tag{110}$$

where 'phase lag' $\varphi = \tan^{-1}(C_R/K)$.

A single calibrating run with a specimen of known heat capacity in a sample pan then suffices to find:

the value of K used in the average signal, underlying heat capacity $= K\langle\Delta T\rangle/\beta$;

the value of C_R used, for instance, in obtaining the cyclic heat capacity $= \sqrt{(K^2 + \omega^2 C_R^2)} \times$ amplitude of $\Delta T/\omega B$.

It is apparent that there are three pieces of data available for finding just two device unknowns, K and C_R. The calibrating factor $\sqrt{(K^2 + \omega^2 C_R^2)}$ and phase lag φ are fixed by K, C_R and the frequency ω. So, even if the calorimeter is to be used with different modulation periods further calibration is unnecessary—according to this model.

More sophisticated models, with a greater number of calorimeter parts, their temperatures and interconnections lead to similar relationships between ΔT and C_p but involve more internal device parameters. It follows that if a more complicated model of the calorimeter is required, a greater number of calibrating runs are needed to fix the calorimeter constants before the calibrating factor for the cyclic heat capacity, and the phase lag will be known for any frequency.

Allowances for bias or for imperfect temperature measurement can be made with simple variations of the Wunderlich model [46].

Considering first imperfect measurements, the temperatures registered by the thermocouples are

$$T_{sM} = (1 - \eta)T_s + \eta T_F = T_0 + \beta t + B\sin\omega t \tag{111}$$

$$T_{RM} = (1 - \eta)T_R + \eta T_F \tag{112}$$

where T_s and T_R are the true sample and reference temperatures.

The parameter η, $0 < \eta < 1$, gives a measure of the imperfection of the device. Manipulations similar to those by the basic model lead, taking a very simple case of

$$K_1 = 0, K = K_2$$

to

$$\langle \Delta T \rangle = \beta(1 - \eta)C_p/K \tag{113}$$

and

$$\begin{aligned} \Delta\tilde{T} &= \text{cyclic part of } \Delta T \\ &= \omega K(1-\eta)C_p \text{Re}\left\{\frac{Be^{i\omega t}}{(K + i\omega C_R)(K + i\omega\eta(C_R + C_p))}\right\} \end{aligned} \tag{114}$$

where now $\Delta T = T_{RM} - T_{SM}$.

Writing $\Delta\tilde{T} = \text{Re}\{\Delta\hat{T}e^{i\omega t}\}$,

$$\Delta\hat{T} = \frac{K(1-\eta)\omega BC_p}{(K + i\omega C_R)(K + i\omega\eta(C_R + C_p))} \tag{115}$$

This indicates a non-linear relation between the modulation of the temperature difference and the heat capacity. However, a single calibration can again be sufficient. The average measurement fixes $k \equiv K/(1 - \eta)$. Use of the in-phase and the out-of-phase parts of the modulated measurements determines the real and imaginary parts of

$$\frac{\omega BC_p}{(K/(1-\eta))\Delta\hat{T}} \equiv a_1 + ia_2 = (1 + ib_1)(1 + ib_2) \tag{116}$$

where $b_1 = \omega C_R/K$ and $b_2 = \omega\eta(C_R + C_p)/K$

Then, this means that it is possible to find b_1 (hence C_R/K) by solving a quadratic equation and b_2. This leads to a second quadratic equation this time, for η.

All this determines η, K and C_R from the single calibration provided it is clear, e.g. from past experience, which roots of the quadratic equations are appropriate. If this were not the case, a second run would be necessary.

Although the relationship between C_p and $\Delta\hat{T}$ is more awkward than the simpler, ideal case, the evaluation of the (complex) cyclic heat capacity is not a particular problem (once η, K and C_R are known):

$$C_p = \hat{C} = \frac{(K + i\omega C_R)(K + i\omega\eta C_R)(\Delta\hat{T}/B)}{\omega[K(1-\eta) - (K + i\omega C_R)(i\omega\eta\Delta\hat{T}/B)]} \tag{117}$$

Bias can result either from an asymmetric distribution of heat capacity or from asymmetric thermal conductivities. The simple case of no direct thermal connection between sample and reference, symmetric heat capacity, but uneven heat transfer can be modelled by

$$(C_R + C_p)\frac{dT_s}{dt} = \lambda K(T_F - T_s), \quad C_R\frac{dT_r}{dt} = K(T_K - T_R) \tag{118}$$

where $\lambda \neq 1$. (Good temperature measurement is assumed in this model.)

In the present case

$$\langle \Delta\hat{T} \rangle = \frac{\beta}{K\lambda}(C_p + (1-\lambda)C_R) \tag{119}$$

$$\Delta\hat{T} = \frac{B\omega}{\lambda}\frac{(C_p + (1-\lambda)C_R)}{K + i\omega C_R} \tag{120}$$

The bias is apparent from the offset term $(1-\lambda)C_R$, so that $\Delta T \neq 0$ even when $C_p = 0$. Now two calibrating runs are needed. For instance, with an empty sample pan, a first run fixes $(1-\lambda)C_R/K\lambda$ from the average signal. In a second run, with $C_p > 0$, the average signal gives $K\lambda$ and the cyclic signal determines C_R/K. In a subsequent experiment, the (complex) cyclic heat capacity measurement is then

$$C_p = \frac{\lambda(K + i\omega C_R)\Delta\hat{T}}{B\omega} - (1-\lambda)C_R \tag{121}$$

Although these two departures from ideality are themselves somewhat specialised, much more general models lead to rather similar results as will be seen below.

The models based on ordinary differential equations, such as those above, vary in simplicity and accuracy. Some allow for bias and/or temperature measurements differing from true temperatures. All are linear. This is the key fact which can be exploited to get general results relating temperature measurements to heat capacities, without making detailed assumptions on heat transfer within the calorimeter. Once linearity holds, calibration, of some type, can be done. (The only drawback in such a general consideration is that how calibration factors depend on frequency is no longer clear. Calibration should be done at the particular frequency ω of the experiment or at least at similar frequencies so interpolation can be employed.)

The calorimeter (and its contents) can be considered as a body which contains heat (specific heat of its parts and heat capacity of the pans and sample) and which is capable of transferring heat, through forced convection and conduction, in some linear way. Flow of heat is proportional to temperature or temperature gradient. Significant non-linear heat transport, for instance due to natural convection or through thermal properties of the calorimeter varying noticeably over the range of temperatures found within the device during a few oscillations, would have a major effect on the ease of use of the method and interpretation of experimental results.

Assuming that the calorimeter does behave linearly, the temperature $T(\mathbf{x}, t)$ satisfies some linear heat equation. The underlying part $\bar{T}$ (or the actual temperature if conventional DSC is being done) then takes the form $\bar{T} = S(\mathbf{x}) + \beta t$ for a linear ramp β, provided that the heating of the calorimeter is controlled through its external underlying temperature being some $\bar{T}_{\mathrm{f}} = S_{\mathrm{f}} + \beta t$. The same goes for the underlying forcing temperature being position dependent, of the form $\bar{T}_{\mathrm{f}} = S_{\mathrm{f}}(\mathbf{x}) + \beta t$ on part of the calorimeter surface, with the remaining part being perfectly insulated. If the temperature on part of the surface were ambient, say $\bar{T}_{\mathrm{f}} = \bar{T}_0 =$ constant for some points $\mathbf{x}$, the temperature ramp in the calorimeter would be position dependent, $\bar{T} = S(\mathbf{x}) + b(\mathbf{x})t$, and the following discussion of calibration would need to be modified. Should this be the case, there would tend to be a steady linear drift between the temperatures at two points in the calorimeter, for example, the sample and reference temperatures.

Because of the linear equation satisfied by temperature inside the calorimeter and outside the sample and its pan, the temperature at any point, in the case of the steady rise, can be given as a linear combination of $\bar{T}_{\mathrm{f}}$ and the sample's temperature $\bar{T}_{\mathrm{s}}$. In particular, the underlying temperatures as measured for the sample and the reference pan are given by:

$$\bar{T}_{\mathrm{m}} = J_{\mathrm{ms}}\bar{T}_{\mathrm{s}} + J_{\mathrm{mf}}\bar{T}_{\mathrm{f}} \tag{122}$$

$$\bar{T}_{\mathrm{r}} = J_{\mathrm{rs}}\bar{T}_{\mathrm{s}} + J_{\mathrm{rf}}\bar{T}_{\mathrm{f}}, \tag{123}$$

respectively. All the coefficients J are independent of the ramp β (but might be weakly dependent on temperature so the characteristics of a calorimeter might be rather different at the finish of a run from what they are at start). The

Js also satisfy $J_{\rm ms} + J_{\rm mf} = J_{\rm rs} + J_{\rm rf} = 1$ (if $\bar{T}(\mathbf{x}) = S(\mathbf{x}) + \beta t$ is a possible temperature distribution then so is $S(\mathbf{x}) + \beta t + S_0$ for any constant S_0). Solving these, $\bar{T}_{\rm s}$ and $\bar{T}_{\rm f}$ can instead be determined in terms of $\bar{T}_{\rm m}$ and $\bar{T}_{\rm r}$.

$$\bar{T}_{\rm s} = J_{\rm sm}\bar{T}_{\rm m} + J_{\rm sr}\bar{T}_{\rm r} \tag{124}$$

$$\bar{T}_{\rm f} = J_{\rm fm}\bar{T}_{\rm m} + J_{\rm fr}\bar{T}_{\rm r} \tag{125}$$

where, again, $J_{\rm sm} + J_{\rm sr} = J_{\rm fm} + J_{\rm fr} = 1$ and, what is more, $\bar{T}_{\rm f} = (J_{\rm sm} - J_{\rm fm})\Delta\bar{T} + \bar{T}_{\rm s}$, where $\Delta\bar{T} = \bar{T}_{\rm r} - \bar{T}_{\rm m}$ is the underlying measured temperature difference.

The underlying rate of heating of the sample and its pan (and possibly of its environs) $\beta(C_R + C_{\rm p})$ is of course proportional to the temperature difference $\bar{T}_{\rm f} - \bar{T}_{\rm s}$, so

$$\beta(C_{\rm R} + C_{\rm p}) = K(\bar{T}_{\rm f} - \bar{T}_{\rm s}) \tag{126}$$

and

$$\beta C_{\rm p} = \left\langle \frac{{\rm d}Q}{{\rm d}t} \right\rangle = J_1\Delta\bar{T} + J_2\beta \tag{127}$$

for some constants (or weakly temperature-dependent functions) J_1 and J_2. (Equivalently, solving a (linear) heat equation throughout the interior of the calorimeter with the sample absent, $C_{\rm p} = 0$, gives $\Delta\bar{T} \propto \beta$. Now including the sample gives the same heat equation, but with a heat sink $\beta C_{\rm p}$ and so an extra contribution to the temperature difference, one proportional to $\beta C_{\rm p}$, must be included. This again gives:

$$\beta C_{\rm p} = \langle {\rm d}Q/{\rm d}t \rangle = J_1\Delta\bar{T} + J_2\beta.)$$

The calorimeter can now be calibrated by doing an empty run to find J_2/J_1, and then a run with a sample of known heat capacity, for example some sapphire, which will determine J_1 and hence J_2.

The cyclic signal can be looked at in a very similar way. Taking the cyclic parts of the temperature, for a purely sinusoidal modulation (or any harmonic for a less simple wave form) to be of the form $\tilde{T}(\mathbf{x}, t) = \tilde{T}_1(\mathbf{x})\cos\omega t + \tilde{T}_2(\mathbf{x})\sin\omega t = \operatorname{Re}\left\{\hat{T}(x){\rm e}^{iwt}\right\}$ for

$$\hat{T} = T_1 - iT_2 \tag{128}$$

The measured temperatures and those of the sample and in the exterior are related via

$$\hat{T}_{\mathrm{m}} = a_{\mathrm{ms}}\hat{T}_{\mathrm{s}} + a_{\mathrm{mf}}\hat{T}_{\mathrm{f}} \tag{129}$$

$$\hat{T}_{\mathrm{r}} = a_{\mathrm{rs}}\hat{T}_{\mathrm{s}} + a_{\mathrm{rf}}\hat{T}_{\mathrm{f}} \tag{130}$$

or equivalently

$$\hat{T}_{\mathrm{s}} = a_{\mathrm{sm}}\hat{T}_{\mathrm{m}} + a_{\mathrm{sr}}\hat{T}_{\mathrm{r}} \tag{131}$$

$$\hat{T}_{\mathrm{f}} = a_{\mathrm{fm}}\hat{T}_{\mathrm{m}} + a_{\mathrm{fr}}\hat{T}_{\mathrm{r}} \tag{132}$$

(For the modulation, it is not too important that no part of the calorimeter's surface be fixed at ambient.) As with the Js, the as should be at most weakly dependent upon temperature, but they will depend upon frequency and no simple relation between pairs should be expected. (For very high frequency, a_{rs} and a_{rf}, for example, will both be small).

The rate of intake of heat by the sample and its surroundings, $\frac{\mathrm{d}\tilde{Q}}{\mathrm{d}t} = \mathrm{Re}\{\hat{Q}_{\mathrm{T}}\mathrm{e}^{i\omega t}\} = \mathrm{Re}\{\frac{\mathrm{d}}{\mathrm{d}t}(C_{\mathrm{R}} + C_{\mathrm{p}})\hat{T}_{\mathrm{s}}\mathrm{e}^{i\omega t}\} = \mathrm{Re}\{i\omega(C_{\mathrm{R}} + C_{\mathrm{p}})\hat{T}_{\mathrm{s}}\mathrm{e}^{i\omega t}\}$, is again going to be proportional to the complex cyclic temperature difference $\hat{T}_{\mathrm{f}} - \hat{T}_{\mathrm{s}}$, but now with a factor of proportionality which varies with angular frequency:

$$\hat{Q}_{\mathrm{T}} = \hat{Q} + i\omega C_{\mathrm{r}}\hat{T}_{\mathrm{s}} = i\omega(C_{\mathrm{p}} + C_{\mathrm{r}})\hat{T}_{\mathrm{s}} = k(\omega)(\hat{T}_{\mathrm{f}} - \hat{T}_{\mathrm{s}}) \tag{133}$$

Here, $\hat{Q} = i\omega C_{\mathrm{p}}\hat{T}_{\mathrm{s}}$ gives the rate of heat intake by the sample and $i\omega C_{\mathrm{r}}\hat{T}_{\mathrm{s}}$ by its environs.

Because the as do not satisfy the same identities as the Js and because the different temperatures have rather different rates of change (amplitude and phase vary with position) the expression relating to $\hat{Q}$ to $\hat{T}_{\mathrm{m}}$ and $\hat{T}_{\mathrm{r}}$ is not quite as simple as the formula for $\langle \mathrm{d}Q/\mathrm{d}t \rangle$:

$$\begin{aligned}\hat{Q} &= i\omega C_{\mathrm{p}}\hat{T}_{\mathrm{s}} = i\omega C_{\mathrm{p}}\big(a_{\mathrm{sm}}\hat{T}_{\mathrm{m}} + a_{\mathrm{sr}}\hat{T}_{\mathrm{r}}\big)\\ &= k\big((a_{\mathrm{fm}} - a_{\mathrm{sm}})\hat{T}_{\mathrm{m}} + (a_{\mathrm{fr}} - a_{\mathrm{sr}})\hat{T}_{\mathrm{r}}\big) - i\omega C_{\mathrm{r}}\big(a_{\mathrm{sm}}\hat{T}_{\mathrm{f}} + a_{\mathrm{sr}}\hat{T}_{\mathrm{s}}\big)\end{aligned} \tag{134}$$

This leads, after a little manipulation, to:

$$\Delta\hat{T} = \frac{K_1 + K_2 C_p}{1 + K_3 C_p} \hat{T}_m \tag{135}$$

where the Ks vary with angular frequency ω (and possibly depend weakly upon temperature).

To do the calibration for a required frequency in finding the three Ks, it is now necessary to carry out three runs: one with an empty sample pan to fix K_1, and then two more with different heat capacities for the sample to determine K_2 and K_3. Once these have been established, the calorimeter can be used to determine the C_p (or rather the complex heat capacity $\hat{C}$) for a sample by:

$$\hat{C} = \frac{\hat{Q}}{\hat{T}_s} = \frac{K_1 \hat{T}_m - \Delta\hat{T}}{K_3 \Delta\hat{T} - K_2 \hat{T}_m} = \frac{K_1 - (\Delta\hat{T}/\hat{T}_m)}{K_3(\Delta\hat{T}/\hat{T}_m) - K_2} \tag{136}$$

(Note that K_1, K_2 and K_3 are complex quantities, and therefore this relation contains information about the phase lag ϕ as well as the cyclic heat capacity C^*.)

For a standard modulation $\tilde{T}_m = B \sin \omega t$. So $\hat{T}_m = -iB$, and

$$\hat{C} = \frac{i(\Delta\hat{T}/B) - K_1}{K_2 - iK_3(\Delta\hat{T}/B)} \tag{137}$$

With a biased calorimeter, one which indicates non-zero ΔT even without a sample, both K_1 and J_2 are non-zero. Without bias, only the calibrations with known samples need to be done. Rather more of a problem with devices deviating from the ideal is the inaccuracy of the measurement of the sample's temperature: $T_m \neq T_s$. For the ideal case, $T_m = T_s$, in the above $J_{ms} = J_{sm} = a_{ms} = a_{sm} = 1$ and $J_{mf} = J_{sr} = a_{mf} = a_{sr} = 0$. The form of the underlying measurement remains unchanged:

$$C_a = \text{average heat capacity} = J_1(\Delta\bar{T}/\beta) + J_2 \tag{138}$$

but the cyclic measurement simplifies to

$$\hat{C} = \text{complex heat capacity} = B_1(\hat{T}/B) + B_2 \tag{139}$$

That is, $K_3 = 0$, $B_1 = i/K_2$ and $B_2 = -K_1/K_2$. Regarding calibration, in such a case, one of the runs for the cyclic calibration could be dispensed with. (For a truly ideal calorimeter, it will also be unbiased, so $J_2 = B_2 = 0$. Then, only a single run with known $C_p \neq 0$ is needed to find J_1 and B_1.)

The lack of accuracy for the measurement of the temperature of the sample has three possible drawbacks.

- The cyclic heat flow, i.e. $\hat{Q}$, is given by $\hat{C}\hat{T}_s(= C_p\hat{T}_s$ for the standard inert case) and is then only known if $\hat{T}_s$ is known. This is a minor difficulty as it tends to be $\hat{C} = \hat{Q}/\hat{T}_s$ which is of interest, as this gives information about heat capacities and temperature dependencies of kinetic processes.

  The relation between $\hat{C}$ and $\Delta\hat{T}$ is non-linear, because it can be expected that K_3 is non-zero. Again, this should cause no real problems because K_3 is determined through the extra calibration run and a cyclic measurement such as

$$\hat{C} = \frac{(\Delta\hat{T}/B) + (iK_1)}{(-iK_2) - K_3(\Delta\hat{T}/B)} \tag{140}$$

  can then be used.
- For measurements of temperature-dependent properties of a sample, it is how things vary with $T_s \simeq \bar{T}_s$, not with T_m or $\bar{T}_m$, which are really important. It can generally be expected that the difference is small enough for events that occur when $T_m = T^*$ to be interpreted as happening when $T_s = T^*$ but in the case of a phase transition this is not so clear. (In these more extreme processes, it is obviously important that no significant variation of the thermal properties of the calorimeter occur over the range of temperatures during the events.) To allow for this, how the actual underlying temperature of the sample relates to $\bar{T}_m = T_0 + \beta t$ and the measurement $\Delta\bar{T}$ needs to be found. With $\bar{T}_s = \bar{T}_m + J_3\Delta\bar{T}$, some form of calibration is needed to determine J_3. One possibility is to use a sample with known transition, for instance a glass transition: $J_3 = (T_g - T_{gm})/\Delta\bar{T}$ if a glass transition, which is known to occur at T_g appears to take place at T_{gm}. A variant of this might be to use a sample which, although free of transition, does have significant but known variation of heat capacity with temperature (the properties of the calorimeter should vary less significantly). Now,

$$\beta C_p(\bar{T}_s) = J_1\Delta\bar{T} + J_2\beta \quad \text{with} \quad \bar{T}_s = \bar{T}_m + J_3\Delta\bar{T} \tag{141}$$

  (neglecting any modulation) and for a 'slow' ramp (so temperature variations in the device are small enough for C_p to be assumed to have locally linear dependence upon temperature),

$$\beta C_{\rm p}(\bar{T}_{\rm m}) + \beta \frac{{\rm d}C}{{\rm d}T}(\bar{T}_{\rm m}) J_3 \bar{T} \cong J_1 \Delta \bar{T} + J_2 \beta \tag{142}$$

The bias J_2/J_1 can be found, as before, by an empty run, and then from two further runs, with different sized samples, J_1 and J_3 may be determined. It might be noted that should the first derivative, ${\rm d}C_{\rm p}/{\rm d}T$, be noticeable, improved accuracy in a cyclic calibration should be given by replacing $C_{\rm p}$ by $\hat{C} = C_{\rm p} - (i\beta/\omega)({\rm d}C_{\rm p}/{\rm d}T)$, since

$$\begin{aligned} \frac{{\rm d}Q}{{\rm d}t} &= (\beta + B\omega \cos \omega t) C_{\rm p}(T_0 + \beta t + B \sin \omega t) \simeq \beta C_{\rm p} \\ &\quad + B \left(\omega C_{\rm p} \cos \omega t + \beta \frac{{\rm d}C_{\rm p}}{{\rm d}T} \sin \omega t \right) \end{aligned} \tag{143}$$

All the as appearing in the relationships between heat flows and temperatures depend, as noted earlier, upon frequency in some generally unknown way which means that ideally calibrations should be done at the frequency of an experiment. One thing, however, is clear, as ω is reduced towards zero (so the modulation gets more like a ramp), the coefficients in the equations for $\hat{T}$ approach those in the earlier equations involving $\hat{T}$ (scaled by an appropriate power of ω). This means that the calibration factors and constants tend to limiting values as ω is reduced to zero: extrapolation of uncorrected results can lead to measurements for $C_{\rm p}$ improved over the basic (uncorrected) values.

One further difficulty is the variation in thermal contract between calorimeter and pan (and/or between pan and sample) from run to run. Hatta and co-workers [21,22] produced a method to account for varying thermal resistance taking an inert sample ($C_{\rm p}$ real and positive) and a simple model for a calorimeter. The same method can be extended and combined with the above general model to account for an uncertain heat transfer coefficient between the sample and its pan, but assuming good thermal contact between the pan and the calorimeter. (If heat transfer coefficients between the pan and its contents and between the pan and its environs are both unknown—and finite—correction will be significantly harder.)

Replacing $T_{\rm s}$ by the pan's temperature $T_{\rm p}$, the above procedure using cyclic temperature measurements can be used to obtain an overall heat capacity measurement for the sample *and* its pan:

$$C_{ps} = \frac{(\Delta\hat{T}/B) + iK_1}{(-iK_2) - K_3(\Delta\hat{T}/B)} \tag{144}$$

Now focusing on the pan and sample,

$$C_p\frac{dT_s}{dt} = K_I(T_p - T_s), \quad \dot{Q} = C_{sp}\frac{dT_p}{dt} + C_p\frac{dT_s}{dt}, \tag{145}$$

where $\dot{Q}$ is the heat flow into the pan from the calorimeter, C_{sp} is the heat capacity of the pan and K_I is the heat transfer coefficient between sample and pan. The modulated parts are then related through:

$$(K_I + i\omega C_p)\hat{T}_s = K_I\hat{T}_p, \quad i\omega C_{ps}\hat{T}_p = i\omega C_{sp}\hat{T}_p + i\omega C_p\hat{T}_s$$

So,

$$\frac{1}{C_{ps} - C_{sp}} = \frac{1}{C_p} + \frac{i\omega}{K_I} \tag{146}$$

where the sample is inert (before, after and between transitions), C_p is real and positive, as is C_{sp}, the heat capacity of the actual pan (also assumed known). The real and imaginary parts can then be used to determine C_p and K_I. During a transition C_p is no longer real, but could still be found if K_I were known. One approach is then to interpolate for K_I between its values before and after the transition.

Of course, during calibration the K_I might again vary (and be finite). This complicates the initial determination of K_1, K_2, K_3. It seems likely that now 6 runs could be needed: giving a total of 12 (real) pieces of data (from the real and imaginary parts) to fix the three complex constants K_1, K_2 and K_3 (6 real bits of information) and 6 different K_Is, all real—although these are not really wanted.

The above discussion outlines how it is possible, in principle, to deal with almost all of the non-idealities of real world DSC cells with sufficient ingenuity and effort. It is probable that commercial instruments will gradually implement these procedures so that they will become available to many scientists. However, not considered above are the problems that arise from poor thermal conductivity within the sample. Once calibrated, the different thermal resistances within the calorimeter should behave in a predictable way. Of course, this is not true of the sample. We are not talking about measuring the thermal conductivity of the sample while no transition is occurring, which has already been done [47] by MTDSC, but doing so while a transition is occurring and, preferably, also measuring the reversing and non-reversing signals as well. This remains a challenge for the future. For the present, most workers are best advised to use relatively long periods, of the order of 60 s, as this reduces the effects of thermal resistances.

References

[1] S.L .Simon, Thermochim. Acta, 374 (2001) 55.
[2] M. Reading, D. Elliott and V.L. Hill, Proc. NATAS, 20 (1992) 145.
[3] P.S. Gill, M. Reading, I.M. Salin and J.C. Seferis, Proc. Greek Acad. Sci., 67 (1992) 311.
[4] M. Reading, B.K. Hahn and B.S. Crowe, US patent 5,224,775 (1993).
[5] M. Reading, Trends Polym. Sci., 1 (1993) 8.
[6] M. Reading, D Elliott and V.L.Hill, J. Thermal Anal., 40 (1993) 949.
[7] P.S. Gill, S.R. Sauerbrunn and M. Reading,J. Thermal Anal., 40 (1993) 939.
[8] M. Reading, R. Wilson and H.M. Pollock, Proc. NATAS, 22 (1994) 2.
[9] M. Reading, A. Luget and R. Wilson, Thermochim. Acta, 238 (1994) 295.
[10] M. Reading, US Patent 5,474,385 (1995).
[11] M. Song, A. Hammiche, H.M. Pollock, D.J. Hourston and M. Reading, Polymer, 36 (1995) 3313.
[12] A.A. Lacey, C. Nikolopoulos and M. Reading, J. Thermal Anal., 50(1–2) (1997) 279.
[13] D.J. Hourston, M. Song, A. Hammiche, H.M. Pollock and M. Reading, Polymer, 38 (1997) 1.
[14] K.J. Jones, I. Kinshott, M. Reading, A.A. Lacey, C. Nikolopoulos and H.M. Pollock, Thermochim. Acta, 305 (1997) 187.
[15] M. Reading, Thermochim. Acta, 292 (1997) 179.
[16] M. Reading and R. Luyt, J. Thermal Anal. Cal., 54 (1998) 535.
[17] M. Reading, J. Thermal Anal. Cal., 54 (1998) 411.
[18] J.E.K. Schawe, Thermochim. Acta, 261 (1995) 183.
[19] P.J. Haines, M. Reading, F.W. Wilburn, In Handbook of Thermal Analysis and Calorimetry, Volume 1: Principles and Practice (Chapter 5), M.E. Brown, Ed., Elsevier Science B.V., Amsterdam (1998).
[20] R. Androsch and B. Wunderlich, Thermochim. Acta, 333 (1999) 27.
[21] I. Hatta and S. Muramatsu, Jpn. J. Appl. Phys., 35 (1996) L858.
[22] I. Hatta and N. Katayama, J. Thermal Anal., 4 (1998) 577.
[23] A. Toda, C. Tomita, M. Hikosaka and Y. Sarayuma, Polymer, 39 (1998) 5093.
[24] I. Hatta and A.A. Minakov, Thermochim. Acta, 330 (1999) 39.
[25] A. Toda, T. Arita, C. Tomita and M. Hikosaka, Polymer, 41 (2000) 8941.
[26] I. Hatta and S. Muramatsu, Jpn. J. Appl. Phys., 35 (1996) L858.
[27] I. Hatta and N. Katayama, J. Thermal Anal., 54 (1998) 577.
[28] A.K. Galwey and M.E. Brown, In: Handbook of Thermal Analysis and Calorimetry, Volume 1: Principles and Practice (Chapter 3), M.E. Brown, Ed., Elsevier Science B.V., Amsterdam (1998).

[29] A. Toda, T. Arita and M. Hikoska, J. Thermal Anal. Cal., 60 (2000) 821.
[30] M. Reading, K.J. Jones and R. Wilson, Netsu Sokutie, 22 (1995) 83.
[31] B. Wunderlich, In Thermal Characterization of Polymeric Materials, Vol. 1 (Chapter 2), E.A. Turi Ed., Academic Press, San Diego (1997).
[32] J.M. Hutchinson and S. Montserrat, Thermochim. Acta, 377 (2001) 63.
[33] J.M. Hutchinson, A. Boon Tong and Z. Jiang, Thermochim. Acta, 335 (1999) 27.
[34] S. Montserrat, J. Polym. Sci. B Polym. Phys., 38 (2000) 2272.
[35] D.J. Hourston, M. Song, A. Hammiche, H.M. Pollock and M. Reading, Polymer, 37 (1996) 243.
[36] N.A. Bailey, J.N. Hay and D.M. Price, Thermochim. Acta, 367–368 (2001) 425.
[37] M. Reading, D.M. Price and H. Orliac, In Material Characterization by Dynamic and Modulated Thermal Analytical Techniques, ASTM *STP 1402*, A.T. Riga and L.H. Judovits Eds., American Society for Testing and Materials, West Conshohocken, PA (2001) pp. 17–31.
[38] S.M. Sarge, E. Gmelin, G.W.H. Höhne, H.K. Cammenga, W. Hemminger and W. Eysel, Thermochim. Acta, 247 (1994) 129–168.
[39] S. Weyer, A. Hensel and C. Schick, Thermochim. Acta, 305 (1997) 267.
[40] W.P. Brennan, B. Miller and J.C. Whitwell, Ind. Eng. Chem. Fundam. 8 (1969) 314.
[41] J.M. Hutchinson and A.J. Kovacs, J. Polym. Sci. B Polym. Phys., 14 (1976) 1575.
[42] A. Toda, T. Oda, M. Hikosaka and Y. Saruyama, Thermochim Acta, 293 (1997) 47.
[43] A. Toda, T. Oda, M. Hikosaka and Y. Surayama, Polymer, 38 (1997) 231.
[44] B. Wunderlich, Macromolecular Physics, Volume 3: Crystal Melting, Academic Press, New York (1980).
[45] A.A. Lacey and C. Nikolopoulos, IMA J. Appl. Math., 66 (2001) 449.
[46] B. Wunderlich, Y.M. Jin and A. Boller, Thermochim. Acta, 238 (1994) 277.
[47] S.M. Marcus and M. Reading, US Patent 5,335,993 (1994).

Chapter 2

THE APPLICATION OF MODULATED TEMPERATURE DIFFERENTIAL SCANNING CALORIMETRY FOR THE CHARACTERISATION OF CURING SYSTEMS

Bruno Van Mele, Hubert Rahier, Guy Van Assche, Steven Swier
Department of Physical Chemistry and Polymer Science, Vrije Universiteit Brussel, Pleinlaan 2, 1050 Brussels, Belgium

1 Introduction

Modulated temperature differential scanning calorimetry (MTDSC, also called temperature modulated DSC or TMDSC) is an extension of conventional DSC in which a modulated temperature input signal is used. This modern technique has proven to be very beneficial for the thermal characterisation of many materials, especially polymers [1–5].

The simultaneous measurement of the amplitude (modulus) of the complex heat capacity, the heat flow and the phase angle between heat flow and heating rate (termed heat flow phase) enables a more detailed study of complicated material systems, both in quasi-isothermal and non-isothermal conditions. The extraction of the signals is briefly summarised below. More details on theory and applications are given in dedicated special issues of *Thermochimica Acta* [6,7] and *Journal of Thermal Analysis and Calorimetry* [8] and in other chapters of this book.

In MTDSC, a sample is subjected to a modulated temperature programme that is obtained by superimposing a sine wave on the conventional isothermal or linearly changing temperature:

$$T = T_0 + \frac{\beta}{60}t + A_T \sin(\omega t) \quad (1)$$

where T is the temperature, T_0 is the initial temperature, β is the (linear) heating rate (in K min^{-1}), A_T is the temperature modulation amplitude, ω

M. Reading and D. J. Hourston (eds.), Theory and Practice of Modulated Temperature Differential Scanning Calorimetry, 83–160.

is the modulation angular frequency and t is the time (in s). This modulated temperature input gives rise to a modulated heat flow response, which consists of an underlying and a cyclic heat flow signal. Assuming that the temperature modulation and the heating rate are sufficiently small so that over the temperature interval of one modulation the response of the rate of the kinetic processes to the temperature can be approximated as linear, the heat flow response can be written as [1,3]:

$$\phi = \frac{\mathrm{d}Q}{\mathrm{d}t} = C_\mathrm{p}\left(\frac{\beta}{60} + A_\mathrm{T}\omega\cos(\omega t)\right) + f(t,T) + A_\mathrm{K}\sin(\omega t) \tag{2}$$

where ϕ is the modulated heat flow, which equals the amount of heat transferred to the sample Q per unit of time, $f(t,T)$ is the average response of a kinetic phenomenon to the underlying temperature programme and A_K is the amplitude of the kinetic response to the temperature modulation.

The extraction of the signals from the modulated heat flow and temperature is done by a continuous integration, averaging and smoothing over more than one modulation period. The underlying signals for both temperature and heat flow are calculated by an averaging process that subtracts the effects of the perturbation. The resulting underlying or 'total' heat flow, ϕ_tot, and underlying temperature reconstitute quantitatively the thermoanalytical curve measured by conventional DSC. Using a discrete Fourier transform algorithm, the amplitude and the phase of the cyclic component of both temperature input and heat flow response is extracted. The ratio of the amplitudes of the cyclic heat flow output, A_HF, and the heating rate input, $A_T\omega$, results in an additional signal: the modulus of the complex heat capacity, $|C_\mathrm{p}|$, also termed 'cyclic' heat capacity (in J K^{-1}):

$$|C_\mathrm{p}| = \left(\frac{A_\mathrm{HF}}{A_T\omega}\right) \tag{3}$$

In other thermal analysis techniques that use an oscillating excitation of the material, for example dynamic mechanical analysis, torsional braid analysis, dynamic rheometry and dielectric thermal analysis, a phase angle is defined between the modulated input and the resulting output signal. Since MTDSC is a dynamic technique as well, a heat flow phase is defined as the phase angle of the modulated heat flow output (with the convention of a negative heat flow for exothermic events) with respect to the modulated heating rate input. One of the major differences with other dynamic analysis techniques, however, is the large instrument contribution to the heat flow phase. It can be corrected by shifting the heat flow phase curve to the zero level for reference points where no transition occurs [3,5,9].

Using the corrected heat flow phase (material contribution) and the modulus of the complex heat capacity, $|C_\mathrm{p}|$, the components in-phase (C'_p) and out-of-phase (C''_p) with the modulated heating rate can be calculated using Eq. (4) (for more detailed approaches see Refs. [3,9,10]):

$$C'_\mathrm{p} = |C_\mathrm{p}| \cos \varphi$$
$$C''_\mathrm{p} = -|C_\mathrm{p}| \sin \varphi \tag{4}$$

Multiplying C'_p by the (measured) underlying heating rate gives the 'reversing' heat flow, ϕ_R in W; the 'non-reversing' heat flow, ϕ_NR in W, is the difference between the total heat flow and the reversing heat flow:

$$\phi_\mathrm{R} = C'_\mathrm{p}\beta \cong |C_\mathrm{p}|\beta$$
$$\phi_\mathrm{NR} = \phi_\mathrm{tot} - \phi_\mathrm{R} \tag{5}$$

It should be noted that the corrected heat flow phase is very small in most cases, so that the difference in value between C'_p and $|C_\mathrm{p}|$ is negligible. For isothermal experiments, the reversing heat flow equals zero because of a zero underlying heating rate and consequently the non-reversing heat flow equals the total heat flow.

For quantitative MTDSC measurements, it is necessary to calibrate temperature and heat flow as in a conventional DSC. The heat capacity signal is calibrated at a single temperature or as a function of temperature using a reference material.

In this chapter, the major benefits of MTDSC to characterise *reacting* polymer systems are highlighted, with a special focus on polymer network formation. All MTDSC experiments shown are performed on TA Instruments 2920 DSC equipment with the MDSC$^\mathrm{TM}$ accessory. Dynamic rheometry measurements were made with a TA Instruments AR1000-N rheometer in parallel plates mode using disposable aluminium plates.

2 Polymer Networks and General Nature of Curing

2.1 GENERAL ASPECTS OF POLYMER NETWORK FORMATION

Polymer networks, such as thermosets and elastomers, constitute a major class of polymeric materials. Their properties differ in many aspects from those of linear and branched thermoplastic polymers. Thermosets are usually very rigid and insoluble three-dimensional network structures in which chain motion is greatly restricted by a high degree of crosslinking. Thermosetting materials are produced by polymerisation in situ since they cannot flow and be reshaped upon the application of heat after formation. Their insolubility, together with the amorphous nature of the network, restricts the applicability of diffraction and other morphological techniques to characterise the final structure.

Thermosets can be divided into several classes depending on the chemical composition of the monomers or pre-polymers (resins). Important thermosetting resins in current commercial applications are the condensation products of formaldehyde with phenol (phenolic resins), urea or melamine (amino resins). Other major classes are epoxy resins, unsaturated polyester resins, allyl resins and isocyanate resins.

The in situ polymerisation or cure process implies an irreversible change under influence of heat and/or pressure from a low molecular weight and soluble material (the resin) into one which is insoluble through the formation of a covalently crosslinked, thermally stable network. During this chemical curing process, many physical properties change remarkably. One of the most obvious changes concerns the ‘chemorheology’ of the system evolving from viscous flow for the resin to energy elasticity for the cured thermoset. Simultaneously with the change in mechanical properties, such as the elastic modulus increasing from almost zero to more than 10^9 Pa, the glass transition temperature, T_g, usually increases from a value below room temperature for the resin formulation up to a final value far beyond room temperature for the cured thermoset.

A basic requirement for the overall transformation from a liquid reaction mixture to a solid amorphous three-dimensional network is the involvement of monomers of which at least one type possesses a functionality greater than two. Initially, somewhat larger molecules or ‘oligomers’ are produced. Subsequently, even larger and branched molecules are formed. As cure further progresses, highly branched structures develop and when these structures extend throughout the whole sample ‘gelation’ is attained. The point of gelation, corresponding to the incipient formation of an infinite network, marks the end of macroscopic flow of a curing resin. From this point on, the mixture is divided into gel and sol fractions. The gel fraction is insoluble and increases at the expense of the sol content as cure proceeds beyond gelation. The sol fraction remains soluble and can be extracted from the gel. The moment that the weight average molecular weight diverges to infinity is referred to as chemical gelation. According to Flory’s theory of gelation for step-growth polymerisation [11], the conversion at chemical gelation is constant. As cure continues, the sol is further incorporated into the three-dimensional network causing an increase of the crosslink density.

As the cure progresses, T_g increases with the increasing molecular weight and/or crosslink density. While curing isothermally, T_g will often rise up to or beyond the cure temperature, T_{cure}. When T_g becomes equal to T_{cure}, the material gradually transforms from a liquid or rubbery state to a glassy state. This process is called 'vitrification', solidification or hardening. Vitrification determines the ultimate degree of reaction conversion of the thermosetting system because it dramatically affects the progress of the cure reaction. The reaction rate slows down due to mobility restrictions and eventually is almost quenched [12,13]. The slow diffusion-controlled curing at T_{cure} lower than T_g is accompanied by a structural relaxation or physical ageing process due to the fact that the glassy state is not an equilibrium thermodynamic state. Due to structural relaxation, volume, entropy and enthalpy decrease with reaction time. In addition to the decrease in the enthalpy [14,15], the mechanical [16,17] and the dielectric properties [18,19] of thermosetting systems are also affected.

To obtain a thermosetting material with optimum properties, it is necessary to understand the chemorheology of cure. The above-mentioned rheological events, i.e. gelation and vitrification, are conveniently assembled in a time–temperature–transformation (TTT) or a continuous heating transformation (CHT) cure diagram, both proposed by Gillham [12,13,20–29]. Since gelation impedes macroscopic flow and vitrification retards chemical reaction, many applications can be devised where gelation and vitrification are exploited to reach optimum properties [22,25]. TTT and CHT diagrams provide for every specific combination of processing time and temperature the state of the reactive mixture, allowing to design time/temperature cure paths for optimum processing and final material properties. Cure paths can be isothermal, non-isothermal or a combination of both. To construct these important diagrams in an accurate and quantitative manner, the cure mechanism and reaction kinetics of the reactive mixture need to be known. A rate law taking into account chemical and diffusion effects has to be modelled according to a mechanistic or an empirical approach.

2.2 CURE REACTION MECHANISM

The important classes of polymerisation mechanisms are also used in thermosetting systems. The cure reaction mechanism will be one determining factor for the evolution of the system. Results for several polymerisation systems will be discussed in this chapter, and some background information about the reaction mechanisms is given in this section.

2.2.1 Step-Growth Polymerisation

Addition and condensation polymerisation belong to this class of polymerisation mechanisms. The specific feature is that all species with reactive groups can take part in the curing process (also called 'polymer coupling').

Epoxy–amine systems follow an addition step-growth polymerisation mechanism. The two principal reactions of primary and secondary amines with epoxy oligomers are shown in Reaction scheme 1 [30]. These reactions are catalysed by acids, phenols and alcohols (e.g. impurities in commercial epoxy resins). The presence of water causes a tremendous acceleration, but does not alter the network structure. The hydroxyl groups formed by the amine–epoxy addition steps are also active catalysts, so that the curing reaction usually shows an accelerating effect in its early stage (autocatalysis).

Reaction scheme 1

Primary amine–epoxy reaction:

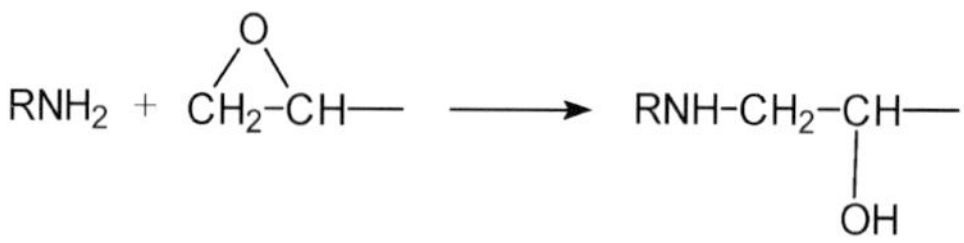

Secondary amine–epoxy reaction:

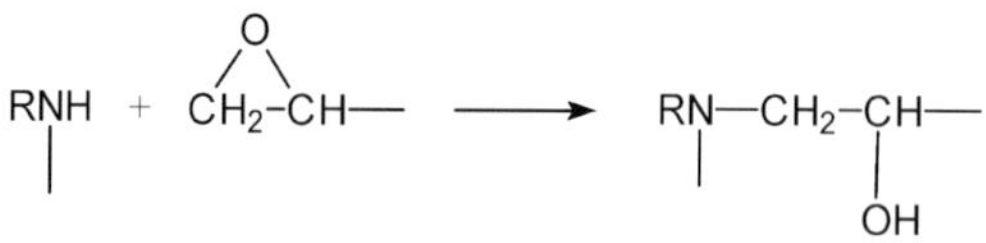

Melamine–formaldehyde resins cure according to a condensation polymerisation. Starting from a fresh mixture of melamine and formaldehyde, methylol groups are initially formed. By subsequent condensation reactions of the methylol groups, methylene bridges (MB) and methylene–ether bridges, in short termed ether bridges (EB) are formed (Reaction scheme 2). Water is released in each of these reactions. These thermosetting systems are widely applied in current commercial applications such as decorative laminates and moulding powders.

Reaction scheme 2

Methylene bridge formation:

$$\mathrm{M{-}NRH + M{-}NR'{-}CH_2OH \longrightarrow M{-}NR{-}CH_2{-}NR'{-}M + H_2O}$$

Ether bridge formation:

$$M\text{–}NR\text{–}CH_2OH + M\text{–}NR'\text{–}CH_2OH$$
$$\longrightarrow M\text{–}NR\text{–}CH_2O\text{–}CH_2\text{–}NR'\text{–}M + H_2O$$

with R, R′=H, CH_2OH or bridge; M = triazine moiety.

2.2.2 Chain-Growth Polymerisation without Termination

Typical examples are living anionic polymerisations. In general terms, the complicated cure of catalysed epoxy–anhydride systems is interpreted according to this mechanism (Reaction scheme 3). When no free hydroxyl groups are present, the reaction can only take place in the presence of an initiator, which opens the epoxy ring. The initiation by tertiary amines proceeds through formation of a zwitterion with the epoxide group, creating a hydroxylate. This hydroxylate reacts with an anhydride to form a carboxylate (acylation), which in turn reacts with an epoxy to generate a hydroxylate again (esterification). The tertiary amine seems to be irreversibly bound to the epoxide [31]. Etherification, only interfering at elevated temperatures or when epoxy is present in excess with aliphatic or aromatic amine curing systems, is more readily observed in catalysed epoxy–anhydride systems [32,33].

In contrast to epoxy–amines, water really participates in the cure chemistry of the epoxy–anhydride system and alters the living polymerisation mechanism because of interfering termination reactions [34]. A less dense network structure with altered properties results [35].

Reaction scheme 3

Initiation:

Acylation:

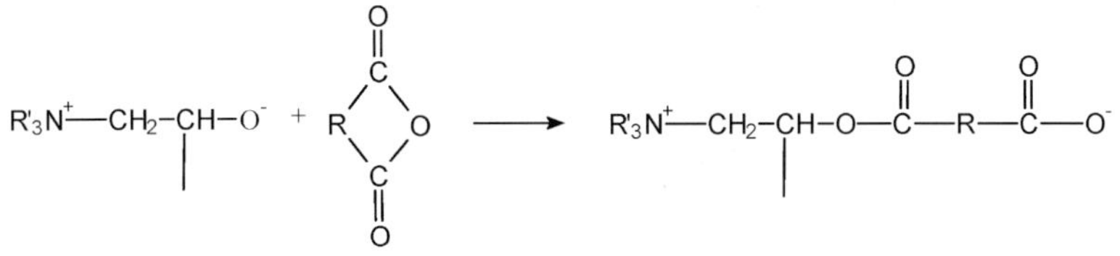

Esterification:

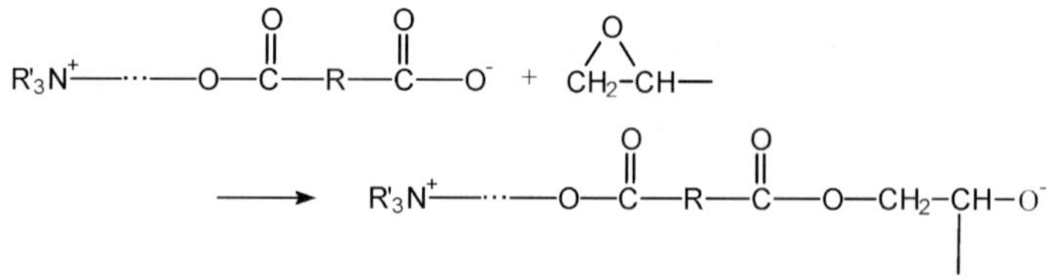

2.2.3 Chain-Growth Polymerisation with Termination

A major exponent of this class of polymerisations are free radical polymerisations in the presence of a radical initiator. A classic example is the crosslinking (co)polymerisation of unsaturated polyester resins with styrene, initiated by the decomposition of a peroxide initiator. Some important reaction steps involved in free radical polymerisations are sketched in Reaction scheme 4.

Reaction scheme 4

Dissociation:

$$\mathrm{ROOR'} \xrightarrow{\Delta} \mathrm{RO^{\cdot}} + {}^{\cdot}\mathrm{OR'}$$

Chain Initiation:

$$\mathrm{RO^{\cdot}} + \mathrm{CH_2{=}CXY} \longrightarrow \mathrm{RO} - \mathrm{CH_2{-}\dot{C}XY}$$

Chain propagation:

$$\mathrm{RO{-}(CH_2{-}CXY{-})_{\mathit{i}-1}{-}CH_2{-}{}^{\cdot}CXY} + \mathrm{CH_2{=}CXY}$$
$$\longrightarrow \mathrm{RO{-}(CH_2{-}CXY{-})_{\mathit{i}}{-}CH_2{-}{}^{\cdot}CXY}$$

Chain termination:

$$\mathrm{R_{\mathit{i}}{-}CH_2{-}{}^{\cdot}CXY} + {}^{\cdot}\mathrm{CXY{-}CH_2{-}R_{\mathit{j}}}$$
$$\longrightarrow \mathrm{R_{\mathit{i}}{-}CH_2{-}CXY{-}CXY{-}CH_2{-}R_{\mathit{j}}} + \mathrm{R_{\mathit{i}}{-}CH{=}CXY} + \mathrm{CHXY{-}CH_2{-}R_{\mathit{j}}}$$

While for step-growth polymerisations (section 2.2.1) and chain-growth polymerisations without termination (section 2.2.2) an overall distribution of reacting species or one type of reacting species (the monomer) is mechanistically characterising the observed reaction rate, the balance between two distinct and mechanistically different species (the monomer and a macromolecular radical) is determining the observed rate of chain-growth polymerisations with termination.

A different evolution of molar mass distribution is noticed, depending on whether the underlying reaction mechanism is step-growth or chain-growth polymerisation. The average molecular size in step-growth polymerisation remains small even if a large part of reactive groups has reacted. In chain-growth polymerisation with termination, high molar mass species are already produced in an early stage of the reaction. Chain-growth polymerisation without termination gives rise to an intermediate situation.

2.2.4 Heterogeneous Polymerisation

Next to homogeneous reaction conditions, multi-phase or heterogeneous polymerisation conditions frequently occur. Suspension and emulsion polymerisation are examples, but also condensation polymerisation with phase separation of water during cure. The low-temperature production of inorganic polymer glasses (IPGs) is a special case of suspension polymerisation involving clay particles in a reactive silicate solution.

2.3 MECHANISTIC VERSUS EMPIRICAL CURE RATE LAW

Starting from the cure reaction mechanism, a proper cure rate law, describing the evolution of the system from initial to final state, can be proposed. In the case of a mechanistic approach, in which the reaction model consists of a set of chemical reaction steps, a set of (stiff) coupled differential equations has to be solved to describe the evolution of the important reacting species of the system. In this case, effects of the composition of the fresh reaction mixture (such as a stoichiometric unbalance of resin and hardener, the concentration of accelerator, initiator or inhibitor) and the influence of additives (such as moisture and fibres in composites) can be studied. Because this set of equations may be rather complex and/or even partly unknown, various simplifications have to be made.

A simplified mechanistic model for the epoxy–anhydride cure of Reaction scheme 3 is given in Reaction scheme 5.

Reaction scheme 5

Initiation reaction:

$$R_3N + Ep \xrightarrow{k_i} Ep^*$$

Propagation reactions:

Acylation reaction:

$$Ep^* + An \xrightarrow{k_{p1}} An^*$$

Esterification reaction:

$$An^* + Ep \xrightarrow{k_{p2}} Ep^*$$

Etherification reaction:

$$Ep^* + Ep \xrightarrow{k_e} Ep^*$$

where R_3N is the tertiary amine initiator, Ep is the epoxy group and Ep^* is the hydroxylate, An is the anhydride molecule and An^* is the carboxylate.

In this reaction mechanism, the initiation step is slow, whereas the acylation step is much faster than the esterification step [36,37]. Based on this reaction scheme, the concentration dependence of initiator, anhydride and epoxy are described.

For the free radical copolymerisation of unsaturated polyester resins containing an inhibitor, the following simplified mechanism could be used.

Reaction scheme 6

Initiation reaction:

$$I_2 \xrightarrow{k_i} 2fR^{\cdot}$$

Propagation reactions:

$$R^{\cdot} + St \xrightarrow{k_{p1}} R^{\cdot}$$

$$R^{\cdot} + UP \xrightarrow{k_{p2}} R^{\cdot}$$

Inhibition reaction:

$$R^{\cdot} + Z \xrightarrow{k_Z} P$$

Termination reaction:

$$2R^{\cdot} \xrightarrow{k_t} P$$

where I_2 is the free radical initiator, f is the chain initiation efficiency, $R^{\cdot}$ is the growing radical, St and UP represent reactive C=C units from the polyester macromonomer and styrene, Z is an inhibitor and P is polymer formed.

For many cure conditions, a detailed mechanistic approach is unrealistic, so that an empirical approach might be preferred for modelling the in situ conversion process of a thermosetting material in practical conditions. An empirical rate law only describes the time/temperature evolution of the overall reaction conversion, x, ranging from $x = 0$ for the unreacted stoichiometric mixture ($T_g = T_{g0}$), to $x = 1$ for the fully-cured network ($T_g = T_{g\infty}$). Only one differential equation has to be solved. An empirical rate equation of the type proposed by Kamal [38] can be applied for the conversion rate of any epoxy resin cure showing auto-catalytic behaviour:

$$\frac{dx}{dt}(x,T) = (k_1 + k_2 x^m)(1 - x)^n \tag{6}$$

with x the conversion[1] of epoxy groups, k_1 and k_2 rate constants, m and n reaction orders.

2.4 SPECIFIC VERSUS OVERALL DIFFUSION CONTROL [39]

As chemical, mechanical and electrical properties attain their ultimate values during the last stages of cure, the diffusion-controlled regime appears to be a very important part of the curing process. An accurate quantitative description of the impact of mobility restrictions on cure is therefore essential. Diffusion control can be specific or non-specific (overall).

Specific diffusion control concerns translational diffusion of individual species (or reactive groups) and is characteristic for dilute solutions (e.g. free radical (co)polymerisation with the monomer as diluent). The diffusion rate of the active centres then depends on the molecular weight of the molecules to which they are attached, on their symmetry, on the composition and on the interactions. For example, in free radical polymerisations, a large difference between the size of polymer and monomer molecules (which remains the case along the entire cure path) is found and so a large difference in their diffusion coefficients is observed.

For step-growth reactions (or in dense systems), the diffusion becomes rather *non-specific* because of the continuous size distribution of reactive species with a gradual transition from monomer to polymer during the cure process. Experimental evidence of overall diffusion control is found in properties correlated to structure growth, like gel point conversion or equilibrium elastic modulus, which are independent whether or not the crosslinking reactions partially occur in the glassy state. These observations suggest that the reactivity of species is uninfluenced by the size of the molecules to which they are attached. Gel point conversion would otherwise be shifted to higher values when partially cured in the glassy state since the reaction

[1] Instead of x, the symbol α is also often used for denoting the reaction conversion.

between small oligomers would be favoured over reaction between larger molecules [40]. Non-specific or overall diffusion control is governed by segmental mobility and is associated with the glass transition region. A tremendous deceleration of the reaction rate near vitrification is expected because of the reduced segmental mobility.

Note that some ionic reactions involving smaller scale mobilities might remain unaffected below the main (glass) transition. An example is the production of low-temperature IPGs [41,42].

2.5 GLASS TRANSITION–CONVERSION RELATIONSHIP

To develop a cure rate law suitable in both chemically and diffusion-controlled conditions, the glass transition–conversion relationship is also important. Many $T_g - x$ models have been reported in the literature [23,43–48]. Those with a sound physical background take into account that the increase in T_g of a polymeric system undergoing cure is the result of several processes: (i) a reduction in the concentration of chain ends by chemical reactions to form linear polymers and by crosslinking reactions consuming end groups and sites along the linear polymer; (ii) the formation of branch points and crosslinks restricting the movement of the chains at these points; (iii) a further decrease in the conformational entropy due to the departure from Gaussian behaviour, affecting systems with relatively short chains between crosslinks.

3 Experimental Procedures to Monitor the Cure Process

3.1 GELATION

Macroscopic gelation is characterised by the strong increase in viscosity, the beginning of elasticity and the formation of an insoluble gel fraction.

With dynamic rheometry, the measurement of the dynamic moduli G' and G'' in small amplitude oscillatory shear is exploited. The gelation point is reported to be the intersection point of the curves of storage and loss moduli, i.e. the moment at which tan δ equals one [49]. However, the crossover is observed to correspond to the gel point only for stoichiometrically balanced network polymers and networks with excess crosslinking agent at temperatures much above T_g [50].

Alternative approaches are based on (dc)conductivity measurements with dielectric thermal analysis (DETA) [51], the cessation of flow measured by rotational viscometry [52], an iso-viscosity method and the determination of the insoluble fraction by extraction and weighing [20].

3.2 VITRIFICATION AND THE GLASS TRANSITION

Techniques measuring the (thermo)mechanical properties as cure proceeds are very appropriate for the assessment of vitrification. One of the most important changes upon vitrification is the increase in modulus by two or three orders of magnitude (from 10^6 Pa in the rubbery state to 10^9 Pa in the glassy state), together with a change in cure shrinkage.

With thermomechanical analysis (TMA), the expansion or shrinkage of the sample under constant stress is monitored against time or temperature. It is also possible to measure dilatometric changes [53]. Near vitrification, the change in volume contraction versus isothermal reaction time will be reduced due to diffusion control.

Dynamic mechanical thermal analysis (DMTA, also called dynamic mechanical analysis or DMA) enables the measurement of the storage and loss modulus of the sample under an oscillating load against time, temperature or frequency of oscillation. The instrument design may allow fixed frequency or resonant frequency operation. A convenient technique operating in the latter mode is torsional braid analysis (TBA) [25,26,54]. Passing through the region of vitrification, the storage modulus increases, whereas tan δ and the loss modulus show a maximum. The vitrification time or temperature can be evaluated as the midpoint or inflection point of the transition in the storage modulus or at the peak position in tan δ or the loss modulus [53,55].

DETA is a complementary dynamic method for the evaluation of vitrification. The advantage of DETA is that the dynamic range is large, e.g. from about 0.1 Hz to 100 kHz. Vitrification can be assessed either as the midpoint or as the inflection point in the stepwise decrease of the permittivity, ε', or as the peak in the loss factor, ε'' [56,57].

For the calculation of T_g, measured in cooling or heating experiments by DMTA, TMA or DETA, similar calculation procedures as for the vitrification point are valid [58].

One of the easiest and most widely practised methods of determining T_g is to follow the change in heat capacity, C_p, as a function of temperature by differential scanning calorimetry (DSC). This calorimetric method has the advantage of requiring the least amount of sample and can usually be carried out in much less time than most of the other methods. A plot of C_p or heat flow against temperature shows a more or less abrupt stepwise increase in the glass transition region and remains almost constant below and above the transition (Figure 2.1).

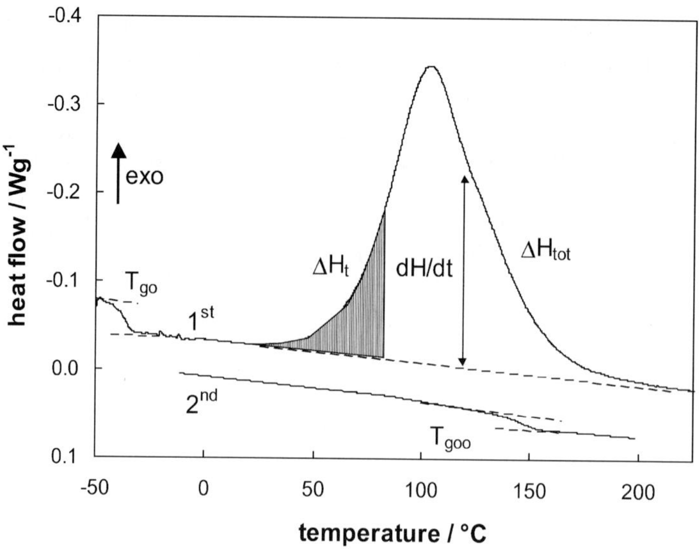

Figure 2.1. DSC heat flow signal as a function of temperature representing the non-isothermal cure of a thermosetting system: 1st and 2nd heating.

3.3 CONVERSION AND CONVERSION RATE

The changes in the concentration of the reactive groups or the products formed can be followed during cure either by measuring the individual components with spectroscopic techniques or chromatographic methods (mechanistic approach) [59–61] or by measuring an overall extent of conversion (empirical approach). DSC is the most widely applied and useful technique for the latter approach. It is a very elegant and extensively used technique and it is not as time-consuming as the above-mentioned direct analysis methods [55,62,63].

If the exothermic heat evolved during chemical reaction is proportional to the extent of consumption of reactive groups, the *overall conversion*, x, and the *overall rate of conversion*, $\mathrm{d}x/\mathrm{d}t$, can be calculated as:

$$x = \frac{\Delta H_t}{\Delta H_{\mathrm{tot}}} \tag{7}$$

and

$$\frac{\mathrm{d}x}{\mathrm{d}t} = \frac{1}{\Delta H_{\mathrm{tot}}} \frac{\mathrm{d}H}{\mathrm{d}t} \tag{8}$$

with ΔH_t the overall reaction enthalpy evolved up to time t and ΔH_{tot} the overall reaction enthalpy for full reaction ($x = 1$) or total reaction enthalpy.

Figure 2.1 displays a typical DSC heat flow signal of an initially unreacted sample submitted to a non-isothermal curing. The first heating is recorded from −70 to 250°C. It shows first a stepwise shift of the baseline around −40°C, which is assigned to the glass transition temperature of the fresh reaction mixture, T_{g0}. The exothermic cure reaction starts near 25°C and ends at about 225–250°C. The theoretical overall conversion attained at this point is one. The area underneath the exothermic peak is a measure of the total heat released. The shift in the second heating curve around 145°C represents the glass transition temperature of the fully-cured system, $T_{g\infty}$. Due to the polymerisation, T_g increased more than 150°C.

An alternative evaluation of the progress of the cure process by DSC can be achieved by performing partial and residual cure experiments. In a partial cure experiment, at a predetermined time or temperature (for isothermal or non-isothermal cure conditions) the sample is quenched to a much lower temperature in order to stop the chemical reactions. In a subsequent heating experiment, termed the *residual cure* experiment, T_g and the residual reaction enthalpy, ΔH_{res}, are measured. The overall conversion at the predetermined time or temperature is then calculated as:

$$x = \frac{\Delta H_{\text{tot}} - \Delta H_{\text{res}}}{\Delta H_{\text{tot}}} \tag{9}$$

Because of the existence of a one-to-one relationship between T_g and x [13], it is often more adequate to use T_g instead of ΔH_{res}, especially towards complete conversion as the ΔH_{res} is small and difficult to quantify in these conditions. It should be noted that any method that involves increasing the temperature of a partially cured sample may allow cure reactions to proceed. Thus, T_g may increase while it is being measured.

3.4 DIFFUSION EFFECTS DURING CURE

There are few reports on the direct measurement of diffusion during cure. Although conventional DSC has been extensively used to study chemically controlled cure kinetics, the opposite is true for diffusion-controlled cure kinetics [64]. Cure reactions under mobility-restricted conditions are largely reduced in rate. The use of heat flow for the estimation of diffusion-controlled cure kinetics is therefore often unsuccessful because of the small size of the heat flow signal. Moreover, the effect of diffusion is generally neglected under non-isothermal conditions. Wisanrakkit and Gillham [47] suggested the use of T_g to analyse the diffusion-controlled regime. The evolution of T_g with reaction temperature or time is then determined from a residual cure of a number of partially reacted samples. Although T_g is easily measured, and although a considerable increase in T_g accompanies a small change in chemical conversion in the later stages of cure, the interference of enthalpy recovery might obstruct an accurate T_g determination [14]. Moreover, it is not always possible to determine T_g throughout the entire range of cure for high-T_g thermosetting systems. Besides, to use this procedure to estimate the effect of diffusion control on cure kinetics is very time-consuming.

An extensive study of the effect of molecular weight, branching and free volume on long-range self-diffusion in epoxies during cure was performed using a pulsed-gradient spin-echo NMR method [65]. The decrease in the diffusion coefficient of the reactive species as cure proceeds was shown to be dependent on the decrease in free volume by measuring the diffusion of a plasticiser taken as a tracer. The free volume-corrected diffusion coefficient thereby obtained was then correlated to molecular weight. No effect of gelation on the diffusivity could be observed.

Recently, an approach was used based on dielectric measurements of the mobility-related properties, including ion conductivity and relaxation time, to estimate the diffusion coefficient [66,67]. The effect of diffusion on cure kinetics appeared to be insignificant in the pre-gel stage, but close to vitrification, the curing reaction became significantly slower.

Note that although the results reported by Yu and von Meerwall [65] and by Deng and Martin [66,67] showed a dependency of diffusion coefficient on molecular weight, the diffusion of reactive groups toward each other concerned with epoxy cure reactions (step-growth polymerisation reactions) is only controlled by segmental diffusion irrespective of the size of molecules to which they are attached. Neither molecular weight nor symmetry or composition play any role within overall diffusion control.

4 Procedures for (MT)DSC Cure Measurements [68–70]

4.1 GENERAL CONSIDERATIONS FOR ACCURATE KINETIC ANALYSIS

4.1.1 Sample Preparation, Sample Size and Storage

During sample preparation, precautions have to be taken that no, or very little, reaction occurs prior to the (MT)DSC experiment. One procedure might be to mix all components at room temperature and to store this mixture at low temperature (e.g. −20°C) before analysis. The recommended procedure, however, is to use freshly prepared reaction mixtures for which the initial conversion approaches zero.

An adequate sample size has to be chosen, e.g. a flat sample of 5–10 mg, as a compromise between the thermal detection limit and the existence of thermal gradients in the sample.

4.1.2 Case of Volatile Reaction Products

A cure reaction yielding volatile products, such as water in step-growth condensation reactions, should always be examined in sealed pans withstanding high internal pressure to obtain accurate kinetic data. In pans with loose fitting lids, the endothermic evaporation of volatile products is sometimes sufficiently large to compensate for the exothermic heat of reaction. Moreover, it is impossible to obtain quantitative data when sample mass is continuously changing by loss of volatile products.

4.1.3 (Quasi-)isothermal (MT)DSC Cure Experiments

The interval of appropriate isothermal cure temperatures is limited. At too low cure temperatures, the reaction time is too long and the corresponding heat flow data may not exceed baseline noise. At too high cure temperatures, the reaction time is too short, so that a significant degree of conversion is unrecorded in the equilibration period at the start of the experiment.

The sample can be immediately inserted into the DSC furnace, which is previously equilibrated at the desired curing temperature, or the sample can be placed into the DSC furnace at room temperature and then heated to the required temperature at a rapid but controlled rate. With both methods, however, some heat of reaction might remain unrecorded in the initial equilibration period and an extrapolation to zero time is needed for accurate kinetic data.

4.1.4 Isothermal or Non-Isothermal Data for Kinetic Analysis?

Fundamental kinetic studies are by preference performed in isothermal rather than in non-isothermal reaction conditions because frequently, as cure proceeds, parallel reactions with different activation energies occur, changing the relative rates of reactions with temperature. In theory, one non-isothermal experiment comprises all the kinetic information normally enclosed in a series of isothermal experiments, which makes the kinetic analysis of non-isothermal DSC data very attractive. The criteria for judging the kinetic parameters derived from non-isothermal experiments must be its ability to describe the isothermal experiments as well. Yet, in practice, the parameters derived from the kinetic analysis of one non-isothermal experiment are inaccurate; the activation energy and the pre-exponential factor are generally overestimated. If several non-isothermal experiments are considered, more accurate values for the activation energy and the pre-exponential factor are obtained [71,72]. A lot of discrepancies among the parameters derived from isothermal and non-isothermal experiments are ascribed to the thermal lag of the instrument when a heating rate is applied [73]. However, a lot of the unreliability comes from attempting to fit measured data to a kinetic equation that does not truly describe the real course of the reaction.

To enlarge the experimental temperature window, a combination of isothermal and non-isothermal experiments is preferred in this work.

4.1.5 *Baseline for Isothermal and Non-Isothermal (MT)DSC Cure*

The calculation of conversion [Eq. (7)] and conversion rate [Eq. (8)] requires the numerical integration of the (partial) areas of exothermic reaction peaks (Figure 2.1) and therefore the need to draw a baseline.

For isothermal DSC data, the heat flow signal goes asymptotically to a plateau value when the reaction is completed. Because it is difficult to separate residual reaction from instrumental instability at this final stage, it is useful to let any reaction that appears to be completed at time t, continue for a time $2t$. The baseline is a horizontal at this final steady state signal.

For non-isothermal DSC data, the baseline can be approximated as linear. However, if there is a significant change in heat capacity between reactants and products, $\Delta C_{p,react}$, a better approach might be to assume a baseline proportional to the reaction conversion [62]. In this context, MTDSC is extremely useful because both heat capacity and heat flow are measured. Therefore, one can consider the change in heat flow due to chemical reaction only (associated with the non-reversing heat flow signal). A re-run or second heating curve (for the fully-cured resin) provides additional indications for the beginning and the end of the baseline.

4.1.6 *Total Reaction Enthalpy*

The overall reaction enthalpy at full conversion or total reaction enthalpy is an important parameter in Eqs. (7–9). Because during an isothermal cure well below $T_{g\infty}$, reactivity is frozen out and full conversion is never reached, an accurate value of ΔH_{tot} is calculated from non-isothermal experiments and preferably at more than one heating rate (e.g. between 5 and 20°C min^{-1}). For too low heating rates, some initial and final reaction may remain unrecorded because of insufficient instrument sensitivity. For too high heating rates, the later stages of cure may interfere with thermal decomposition processes.

At advanced conversions, functional groups can no longer meet and react because of the progressively increasing topological complexity of a reacting system. Therefore, the complete conversion of functional groups, i.e. full overall chemical conversion, may never be attained. Even in the complete absence of any diffusion hindrance, this topological constraint cannot be removed by a simple temperature rise. This is in contrast to reactions controlled by mobility restrictions, i.e. an isothermal cure at temperatures well below $T_{g\infty}$, where heating the sample above $T_{g\infty}$ removes the mobility restrictions. In the case of a topological limit, the experimental value of ΔH_{tot} is always lower than the theoretical limit, as calculated from reactions of model compounds.

It should be noted that each reaction enthalpy, ΔH, depends on the average temperature of the experiment. If the change in heat capacity between reactants and products, ΔC_{pr}, is significant, a temperature-dependent value of ΔH might be necessary.

4.2 MTDSC PARAMETERS

In an MTDSC experiment, a repeated temperature modulation is superimposed on the normal linear temperature programme [1–5,74]. The modulation amplitude and frequency, and the underlying heating rate can be chosen independently.

4.2.1 Modulation Amplitude

For non-isothermal experiments, 'heating only' conditions, with the modulation amplitude chosen so that no cooling occurs over one complete cycle, are of no use in cure studies. On the contrary, for experiments with very low underlying heating rate, or when C_p should be measured as accurately as possible, it is advisable to use a larger modulation amplitude. Of course, the amplitude of the temperature modulation has to be limited, since its effect on the cure kinetics has to be negligible. Typical amplitudes are between 0.1 and 1°C.

4.2.2 Modulation Period

The range of frequencies that can be used in practice is limited to about one decade, so that no strong frequency effects are expected, as opposed to the conditions of DETA where frequencies can easily be changed from 1 to 10^5 Hz. The modulation frequency in all MTDSC experiments shown is always fixed at 0.017 or 0.01 Hz (corresponding to a modulation period of 60 or 100 s, respectively). The frequency dependence of the heat capacity during cure is discussed in section 6.4.

4.2.3 Temperature-Dependent Heat Capacity Calibration

Commonly, the heat capacity signal is calibrated at a single temperature. However, the experimental error on the heat capacity can further be reduced by a 'dynamic' calibration over the entire temperature range instead of at a single temperature. The heat capacity calibration constant, K_{C_p}, shows a gradual evolution over the entire temperature range, with a total variation of 4% between −50 and 300°C. Below −50°C, the deviation increases.

For the cure studies in this work, this deviation is not so important. Firstly, because most of the MTDSC experiments are performed above −50°C, and secondly, because for quantitative analyses a mobility factor is calculated by normalising the heat capacity between reference heat capacities determined at the same temperature. Thus, changes in K_{C_p} with temperature have no effect on this result (section 5.8).

5 MTDSC Characterisation of Cure: Experimental Observations

5.1 EXPERIMENTAL SYSTEMS

5.1.1 Epoxy Systems

Different types of epoxies, such as monofunctional phenyl glycidyl ether (PGE), bifunctional epoxy (DGEBA LY 564 or 556, Ciba-Geigy) and tetrafunctional epoxy (MY 720, Ciba-Geigy) have been studied. The curing agents are anhydride hardener (HY 917, Ciba-Geigy) using 1 wt% of an accelerator (1-methyl imidazole DY 070, Ciba-Geigy), or different types of amines, such as bifunctional aniline or N,N'-dimethylethylenediamine, and tetrafunctional methylenedianiline (MDA) or 3,3′-dimethyl-4,4′-diaminodicyclohexylmethane (Araldite HY 2954, Ciba-Geigy). The molar ratios of amine/epoxy functional groups, r, are varied. Stoichiometric mixtures ($r = 1.0$) and mixtures with an excess of amine ($r > 1.0$) or epoxy ($r < 1.0$) are used.

5.1.2 Polyester–Styrene Systems

A mixture of unsaturated polyester resin with ca. 45 wt% of styrene (Polylite P51383, Reichhold), methyl ethyl ketone peroxide as initiator

(Butanox M-60, AKZO) and 1% cobalt 2-ethylhexanoate in styrene as accelerator (NL 49S, AKZO) in a w/w mixing ratio of 100/2/1 was used. A weight fraction of 200 ppm 2-methylhydroquinone inhibitor was added to delay the reaction for cure experiments above room temperature.

5.1.3 Melamine–Formaldehyde Resins

Melamine–formaldehyde (MF) resins of a molar ratio F/M = 1.70 were prepared at 95°C by dissolving 505 g melamine in 592 g formalin (34.5 wt% aqueous formaldehyde with a pH of 9.2). The reaction was stopped when the reaction mixture reached the cloud point [75]. At 25°C, the pH of the MF resin was adjusted to 7.5 and 9.5. These resins were spray-dried using a Büchi spray dryer and further dried for half an hour in a vacuum oven at 60°C before each MTDSC experiment. Liquid ^{13}C-NMR spectra showed that more methylene bridges and ether bridges and fewer residual methylol groups (see section 2.2.1 were present in MF pH 7.5 compared to MF pH 9.5.

5.1.4 Inorganic Polymer Glasses

The IPG in this study was an amorphous aluminosilicate formed by the low-temperature reaction of an alkaline sodium silicate solution (Sil) and a dehydroxylated clay (metakaolinite, Mk) [41,76]. The molar ratio Sil/Mk is one.

5.2 REMARKS CONCERNING MTDSC SIGNALS

5.2.1 Non-Reversing MTDSC Heat Flow Versus Conventional DSC Heat Flow

The total heat flow obtained in quasi-isothermal MTDSC experiments agrees very well with the heat flow evolution obtained in a conventional DSC experiment, performed under the same conditions without of the modulation (Figure 2.2a). Neither changing the modulation amplitude nor the period had an effect on the reaction exotherm seen in the non-reversing heat flow. This illustrates the negligible effect of the perturbation on the cure reaction.

For a correct calculation of the non-reversing heat flow in non-isothermal experiments, attention has to be paid to the heat capacity and heating rate used to obtain the reversing heat flow. Firstly, the *dynamic heat capacity calibration* should be used in order to correct for the temperature dependence of K_{C_p} (section 4.2.3). However, for the curing experiments the gradual change in K_{C_p} (when using a dynamic calibration instead of a single point calibration) results only in minor baseline variations for the reaction exotherm in the non-reversing heat flow. Secondly, one should use the *measured* underlying

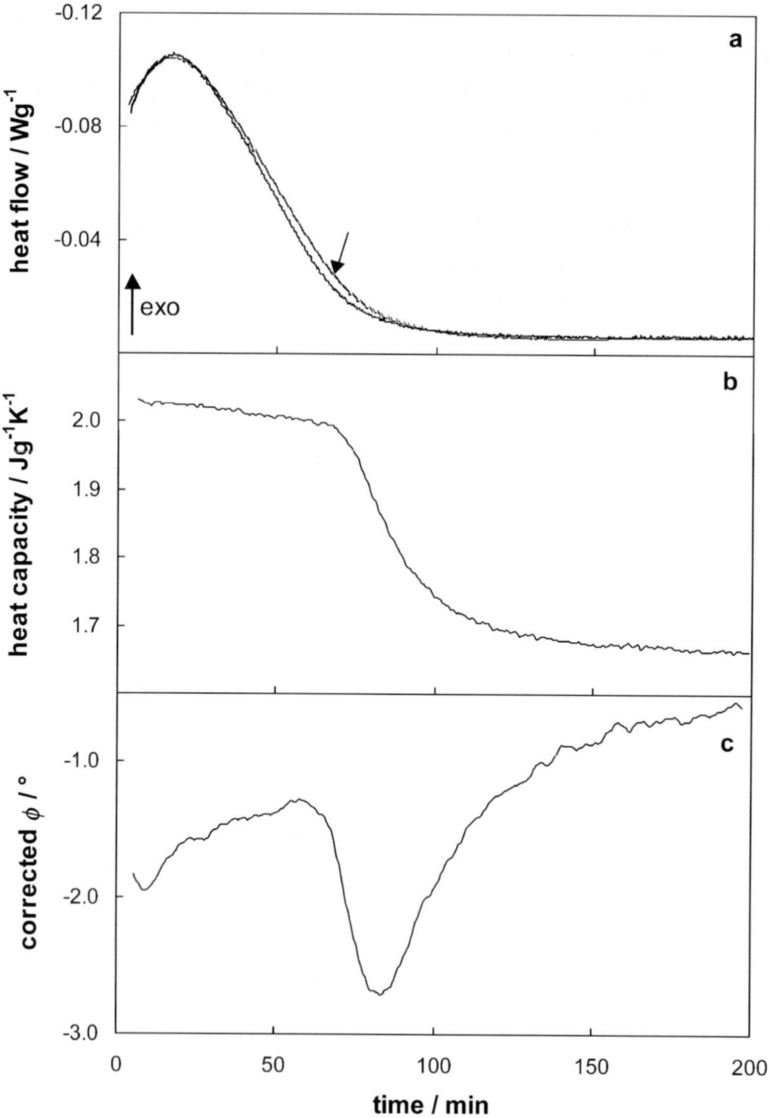

Figure 2.2. Quasi-isothermal cure of an epoxy–anhydride at 100°C: (a) comparison of the non-reversing heat flow obtained in MTDSC to the heat flow obtained in conventional DSC (arrow), (b) heat capacity and (c) corrected heat flow phase.

heating rate to subtract correctly the heat capacity contribution (reversing heat flow) from the total heat flow. In general, the corrections are much smaller than the experimental error, which is due to small differences in the mixing ratios (a fresh epoxy–hardener mixture is made for each experiment) and the influence of small (irreproducible) amounts of moisture on the reaction kinetics. Because the effect of the dynamic heat capacity calibration and the heating rate corrections on the non-reversing heat flow is small, the chemical rate of reaction can be obtained quantitatively from the non-reversing heat flow.

5.2.2 Heat Flow Phase

Figure 2.2c shows the corrected heat flow phase for the epoxy–anhydride system cured at 100°C; the heat flow phase measured quasi-isothermally at 100°C for the fully cured resin was used as a reference point. The evolutions will be discussed in detail in the following section. In this paragraph, the magnitude of the signal is considered.

For the systems and conditions studied in this work, the corrected heat flow phase always remains small: the maximum difference with respect to the fully-cured reference state (in the absence of any transition) is 10° and in most cases much smaller. The cosine and sine for this angle equals 0.98 and 0.17, respectively. Thus, for all experiments shown, C_p' and $|C_p^*|$ coincide within 2% and C_p'' is always close to zero. This implies that the reversing and non-reversing heat flows calculated using C_p' (according to the complete deconvolution) or $|C_p^*|$ (cyclic C_p according to the simple deconvolution) are virtually identical. Hence, the use of the phase angle is not necessary for the quantitative interpretation of heat capacity and non-reversing heat flow for the cure experiments discussed. Nevertheless, the evolution of the phase angle over the course of the reaction contains valuable information on the reaction and the mechanism (see further discussion).

Recently, a (semi-)quantitative use of the heat flow phase was discussed. From the contribution of the temperature dependence of the reaction rate to the heat flow phase signal, an overall activation energy was estimated for an epoxy cure without vitrification [77]. If vitrification occurs, one obtains the phase shift due to thermal relaxation only by removing the contributions of the temperature dependence of the reaction rate and of the heat transfer conditions from the heat flow phase signal [78].

5.3 ISOTHERMAL CURE WITH VITRIFICATION

5.3.1 Epoxy Systems

The experiment of Figure 2.2 will now be considered in more detail as a typical example of isothermal cure with vitrification. It shows the non-reversing heat flow (Figure 2.2a), the heat capacity (Figure 2.2b) and the heat flow phase (Figure 2.2c) as a function of reaction time for the quasi-isothermal cure of an epoxy–anhydride resin at 100°C for 200 min. The reaction exotherm obeys an auto-catalytic behaviour: the heat flow increases at first and passes through a maximum. The heat capacity, C_p, first decreases slightly. Subsequently, a stepwise decrease in C_p is observed, simultaneously with a sharp decrease in heat flow. The heat capacity change, ΔC_p, equals 0.33 J g^{-1} K^{-1}. The time at half of the change in heat capacity, $t_{1/2\Delta C_p}$, equals 86 min. The glass transition temperature at this instant, measured in a partial cure experiment, amounts to ca. 100°C. Therefore, the stepwise decrease in C_p corresponds to the transition of the polymerising system from the liquid or rubbery state to the glassy state. The time $t_{1/2\Delta C_p}$ obtained in an MTDSC experiment can be used to quantify the time of vitrification. This time can be interpreted as the time that half of the material has transformed to the glassy state (on the time-scale of the modulation) [42,68].

Figure 2.2c shows the evolution of the corrected heat flow phase, φ. The fully-cured glass state is always used as a reference (zero value) for the instrument correction [68, 69]. The phase angle corrected in this way has a small negative value, tending to more positive values due to the chemical reactions. Indeed, in Figure 2.2c the corrected heat flow phase, φ, initially amounts to $-2.0°$ and then slowly evolves toward zero as the reaction proceeds. Relaxation phenomena are superimposed as local (downward) extremes. Thus, the (downward) local extreme in φ observed at 83 min confirms the vitrification process observed in C_p in Figure 2.2b. At the end of the quasi-isothermal experiment, φ equals $-0.6°$.

5.3.2 Unsaturated Polyesters

Although the auto-acceleration or 'gel effect' in free radical polymerisation is well understood in general terms of a decrease in the mobility of growing chains, quantitative results on detailed aspects to develop more predictive models are still scarce. One of the major reasons is the difficulty of maintaining isothermal conditions in bulky samples, and the lack of truly isothermal conversion–time data and reliable rate constants [79]. The benefits of MTDSC in this respect are illustrated with the isothermal free radical copolymerisation of an unsaturated polyester–styrene system [80–82]. The combined use of MTDSC and dynamic rheometry is very beneficial for an isothermal study of gelation, vitrification and auto-acceleration (gel effect) in terms of reaction conversion. In Figure 2.3a, the non-reversing heat flow at 30°C is compared with rheological information. The typical auto-acceleration, observed as a second maximum in the heat flow signal, coincides with the ultimate increase of the complex viscosity, η^* (beyond 10^5 Pa s), and occurs just before the onset of vitrification observed in MTDSC by the start of the final decrease in heat capacity and the decrease in the heat flow phase signal (Figure 2.3b). The onset of vitrification occurs at a

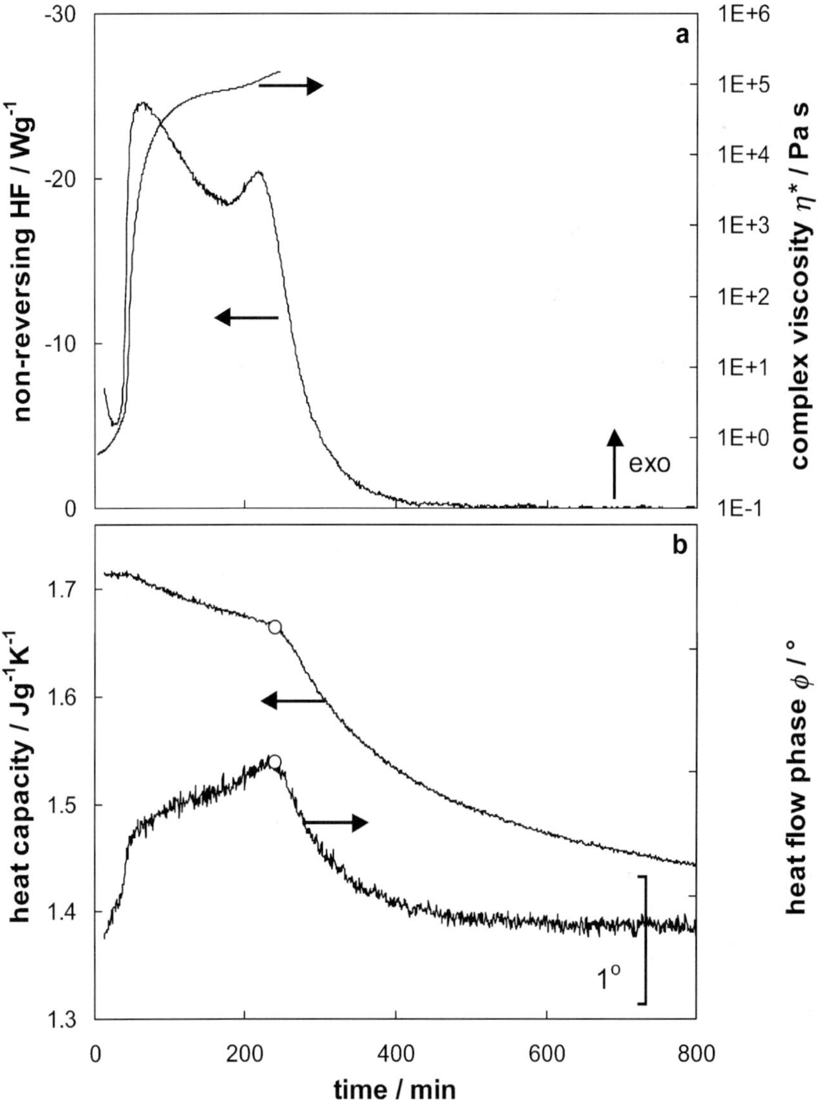

Figure 2.3. Quasi-isothermal cure of an unsaturated polyester at 30°C: (a) non-reversing heat flow and complex viscosity (logarithmic scale); (b) heat capacity and heat flow phase; the symbol (o) denotes the point at maximum auto-acceleration in the non-reversing heat flow.

conversion close to 80% and coincides with the point of maximum auto-acceleration (if the decreasing monomer concentration is taken into account [81]).

The conversion at the onset of auto-acceleration (the heat flow minimum) is close to 60%. Because it occurs at a much more advanced conversion than gelation, the term gel effect for indicating the auto-acceleration is somewhat misleading and should be avoided in this case. Since the phenomenon occurs closely before the onset of vitrification, it is not likely it is due to the sharp increase in bulk viscosity at gelation, but rather it is caused by a change in molecular mobilities at higher conversion. This effect of changing (decreasing) molecular mobilities on the (increasing) reaction rate of the unsaturated polyester system is caused by the specific features of a free radical chain-growth polymerisation mechanism and is in contrast with thermosetting systems, such as epoxy–amines, obeying an addition step-growth polymerisation mechanism.

It is interesting to point out that the observed evolution of the complex viscosity is quite different for both polymerisation types (Figure 2.4). In a step-growth (bifunctional) epoxy–(tetrafunctional) amine system, η^* rises continuously with extent of polymerisation: gelation occurs near 60% conversion [83]. In the polyester system, η^* rises more sharply and at a much lower conversion.

The different conversion-dependence of η^* is related to the molecular weight evolution and network development. For addition step-growth polymerisation systems, the molecular weight of the polymer chains gradually increases, while for (linear) free radical chain-growth polymerisations the

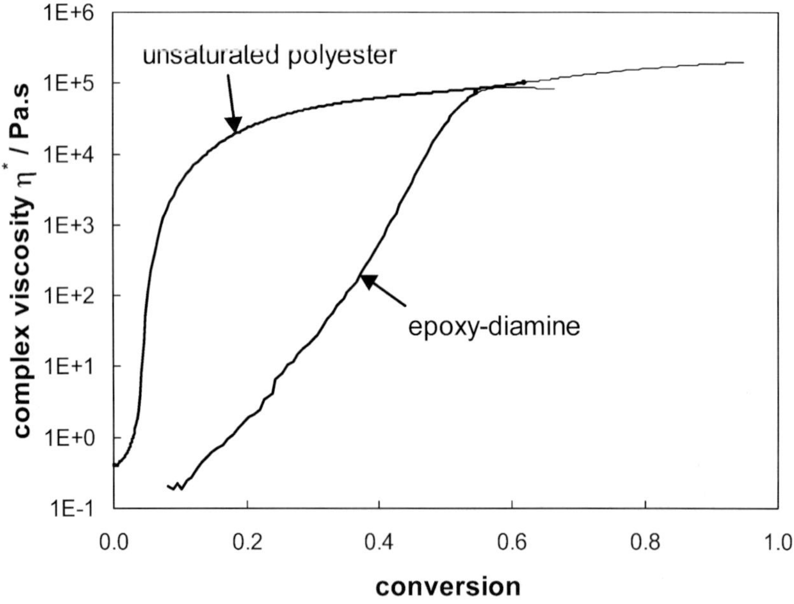

Figure 2.4. Comparison of the evolution of complex viscosity with conversion for the isothermal cure of an epoxy($f = 2$)–amine($f = 4$) system at 60°C and of an unsaturated polyester at 40°C (logarithmic scale).

highest average degree of polymerisation of the polymer chains is attained at the start of the reaction (section 2.2).

It also should be pointed out that the auto-acceleration at high conversion closely before the onset of vitrification is not observed for step-growth polymerisation thermosetting systems [42,68,80,83–85].

A model including initiation, propagation, transfer, inhibition and termination steps should be used to describe the auto-acceleration. Such a mechanistic model should deal with termination and propagation rates controlled in a different way by the decreasing mobility or diffusion. In fact, at a high monomer conversion, almost all elementary reactions, including the termination of growing radical chains, propagation of the growing chain and even chemical initiation reactions, can become diffusion-controlled and cause several effects: the gel effect (an auto-acceleration), a glass effect (slow down) and a cage effect, respectively [86–90]. Since entities of very different sizes are involved, from single styrene units to reactive groups that are part of the network, these free radical polymerisations are typical examples of systems where a specific diffusion control should be taken into account.

5.3.3 Melamine–Formaldehyde Resins

Step-growth condensation copolymerisations give rise to additional experimental difficulties, with respect to the former reactions studied, due to the continuous release of e.g. water. Indeed, the evaporation of water produced by the reaction may obscure the detection of the cure process and prohibit a reliable quantification of the reaction heat and the reaction conversion. To illustrate how condensation polymerisations can be studied by MTDSC, the post-cure condensation reactions of melamine–formaldehyde (MF) resins will be discussed [91].

Two main steps are to be distinguished in the production and further conversion of melamine–formaldehyde resins: (i) the addition reaction of formaldehyde and melamine to form methylolated melamine compounds, and (ii) methylene and ether bridge formation via condensation reactions releasing water (section 2.2.1). Since spray-dried methylolated melamine resins (MF resins) are studied (see experimental), the cure of MF resins is confined to the condensation reactions of the latter step.

Experimental Requirements and Reproducibility of Condensation Reactions

In order to study the kinetics of the condensation reactions with (MT)DSC, evaporation of water must be suppressed, because this strongly endothermic process (2257 J g^{-1} at 100°C [92]) masks the exothermic cure reaction completely. Pressurisation or encapsulation of the sample can solve this problem. However, high-pressure DSC is not available in the modulated DSC mode. For this reason, reusable high-pressure stainless steel pans (HPS pans, volume 30 μl), withstanding an internal pressure up to 150 bar are more convenient. In order to minimise the heat of evaporation, the HPS pans are (nearly) entirely filled with sample. The remaining free volume is small (10–15 μl), so the heat required for evaporation of water in this volume can be neglected and does not disturb the exothermic effect of the ongoing condensation reactions.

Figure 2.5a shows the MDSC results on MF pH 9.5 cured isothermally at 119°C in closed and open reaction conditions. In open conditions, due to the evaporation of more than 9 wt% of water and the continuous loss of mass, a large decrease in C_p is observed, going from the beginning till the end of cure. The heat flow signal shows an irregular and mainly endothermic effect. Initially, the exothermic cure effect dominates the heat flow, but at the maximum cure rate with a maximum release of water around 50 min, the evaporation of water takes over and sharp endothermic peaks appear. In closed conditions, on the other hand, exothermic curing can be observed during the entire experiment. Instead of a continuous decrease in C_p, the heat capacity displays a slight increase up to about 150 min, followed by a more pronounced stepwise decrease.

The results of Figure 2.5a clearly demonstrate the requirement of closed reaction conditions to study these condensation reactions in a reproducible way. The reusable high-pressure steel pan guarantees the indispensable closed reaction conditions and enables a reliable study of the characteristic features of MF condensation reactions. The general trends in heat flow and heat capacity can be related to the effects of pH and vitrification on MF cure (see discussion below).

Vitrification During MF Cure

The final stepwise decrease in C_p is related to a vitrification process along the isothermal cure path. Simultaneously with the stepwise change in C_p, the heat flow signal further decreases toward zero. During vitrification, the mobility of the reactive functional groups becomes more and more restricted, and cure finally ceases, even in the presence of remaining functionalities. Similar to the results obtained for the cure of epoxy thermosetting systems, a slight increase in the phase angle, φ, is seen due to the proceeding reaction and the heat capacity change during vitrification (Figure 2.5b). Moreover, a relaxation peak associated with vitrification is superimposed [68,69,83].

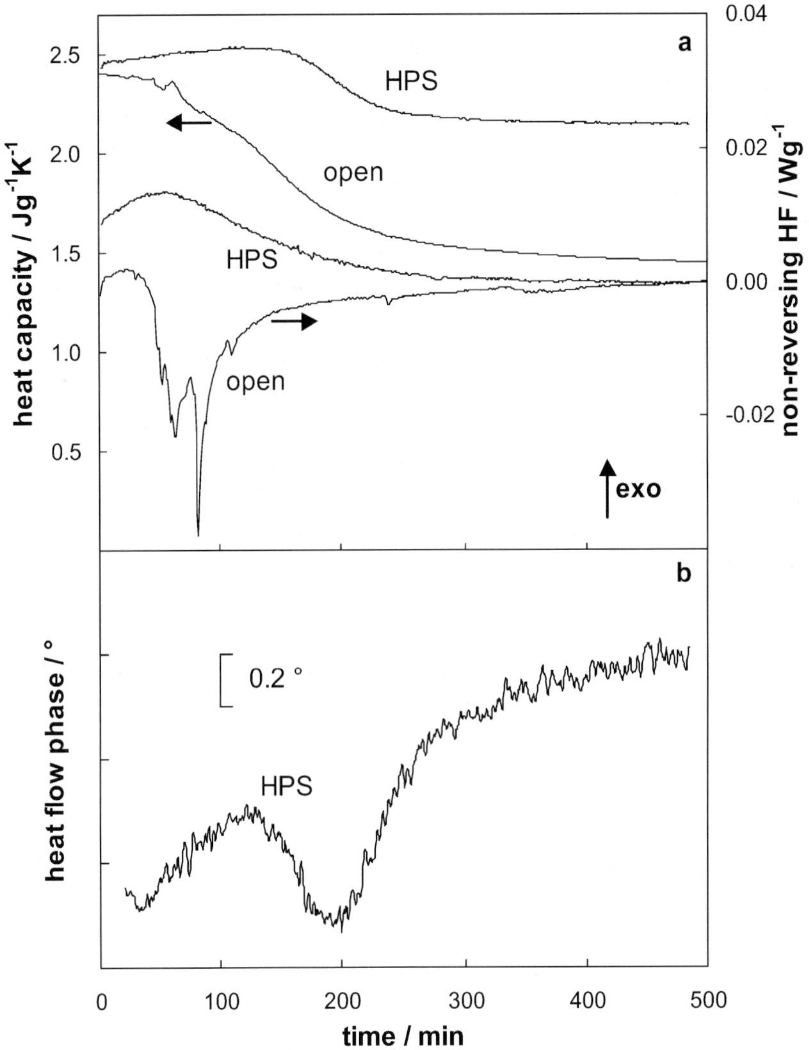

Figure 2.5. Quasi-isothermal cure of a melamine–formaldehyde (MF) resin (pH 9.5 F/M = 1.7) at 119°C in closed high-pressure steel (HPS) and open Al pans: (a) non-reversing heat flow and heat capacity; (b) heat flow phase.

The local minimum of ϕ, at a time of 196 min, coincides with the inflection point of the stepwise decrease in C_p.

5.3.4 *Inorganic Polymer Glasses*

The formation of IPG will be discussed to illustrate that different chemical systems are influenced differently by the proceeding reaction. These IPGs are formed by the heterogeneous reaction of an alkaline Sil with dehydroxylated clays (metakaolinite). The properties of the final amorphous aluminosilicates depend on the low-temperature reaction conditions [76,83,93].

The heat flow, heat capacity and heat flow phase signals during production of IPG at 35°C starting from 1.8 μm Mk particles are shown in Figure 2.6. The reaction exotherm shows a typical behaviour with the maximum rate of reaction at the start of the experiment followed by a shoulder of more or less constant rate of reaction. The heat capacity remains nearly constant up to ca. 20% conversion, then a gradual decrease is observed. The rate of reaction stays more or less the same over the first half of the change in C_p and decreases slowly over the second half. The concurrent hardening process is independently measured by dynamic mechanical analysis. The onset of the steep increase in storage modulus (DMA) coincides with the onset of the heat capacity decrease (MTDSC), and a local minimum in the heat flow phase relaxation peak is observed at this early stage of the reaction. Whereas the DMA set-up employed is limiting the window of moduli to follow experimentally the hardening process (both the low and high modulus values are not uniquely related to the IPG properties [41]), the MTDSC heat capacity curve enables one to investigate the entire process. This is important for studying these low-temperature IPG reactions, since vitrification is not slowing down the reaction rate and the largest part of the reaction enthalpy (more than 50%) is set free in the solidifying state.

5.4 NON-ISOTHERMAL CURE WITH VITRIFICATION AND DEVITRIFICATION

5.4.1 Epoxy–Anhydride [68,84]

Figure 2.7 shows the non-reversing heat flow and heat capacity as a function of temperature for the cure of the epoxy–anhydride at a heating rate of 0.2–0.7°C min^{-1} (curves 1–3). The heat capacity evolution for the completely cured resin is also shown (curve 4). First the experiment at 0.2°C min^{-1} will be considered. The reaction exotherm shows a maximum around 90°C, and is followed by a shoulder of more or less constant heat flow. The shoulder's height is 7% of the peak height or ca. 50 μW, and it spans a temperature interval of 40°C. In the heat capacity curve of the first heating, three transitions are observed. The first transition is T_{g0} of the uncured resin at −37°C (not shown). The second transition, a decrease in C_p, occurs at the instant the heat flow decreases. Near the end of the heat flow shoulder, C_p increases again. In between the transitions, C_p rises slowly with temperature. In the second heating (curve 4), $T_{g\infty}$ of the fully cured network amounts to 135°C.

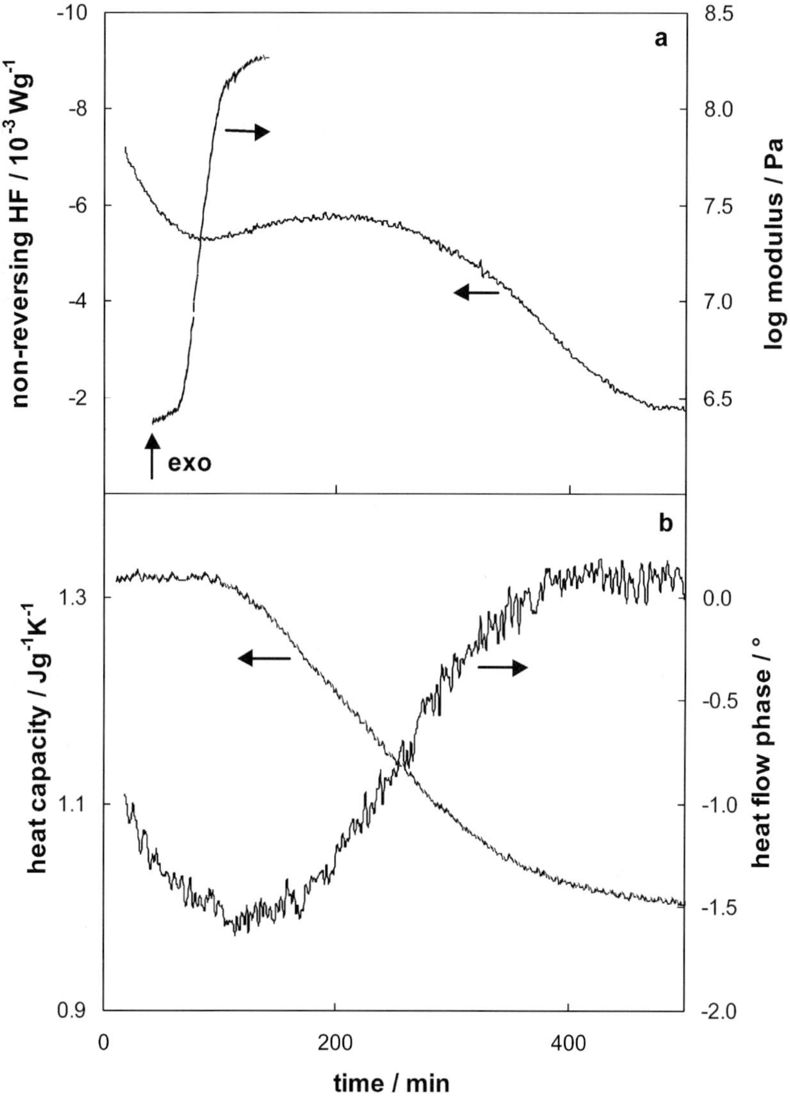

Figure 2.6. Production of an inorganic polymer glass (IPG) for a metakaolinite (Mk) particle size of 1.8 μm at 35°C; (a) non-reversing heat flow and storage modulus; (b) heat capacity and heat flow phase.

Measured against this second heating reference line, the temperatures at half of the heat capacity difference, $T_{\frac{1}{2}\Delta C_p}$, equal -37 (T_{g0}), 104 and 131°C, respectively.

Comparison of the evolutions of C_p and T_g with reaction temperature clearly shows that the three transitions in heat capacity, characterised by

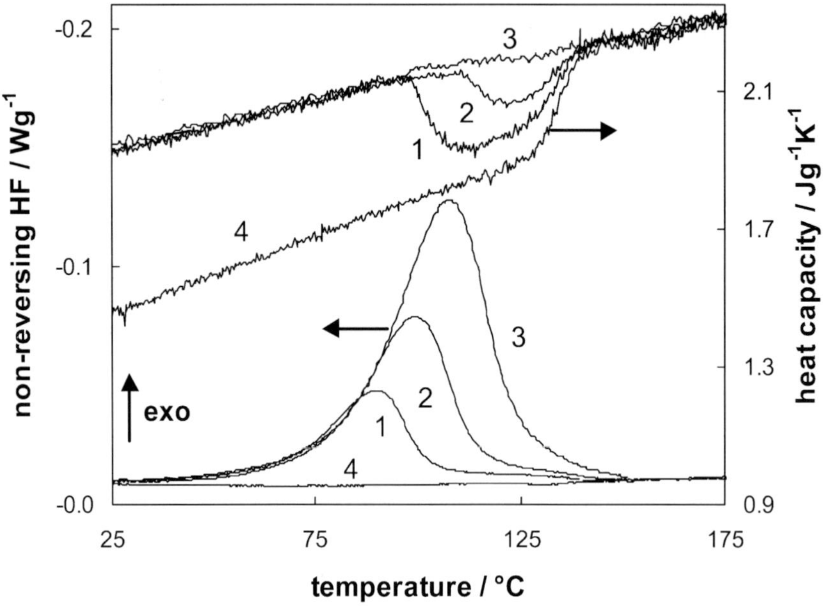

Figure 2.7. Non-isothermal cure of an epoxy–anhydride at 0.2 (1), 0.4 (2), and 0.7°C min^{-1} (3) and for the fully-cured material (4): non-reversing heat flow and heat capacity.

the temperatures $T_{1/2\Delta C_p}$, subsequently correspond to (i) devitrification at T_{g0}, (ii) vitrification due to a more rapid increase of T_g during cure (up to 10 times faster than the applied heating rate of 0.2°C min^{-1}) and (iii) a final devitrification process near the end of reaction because the conversion of reactive groups nears completion and T_g ceases to increase. In between the vitrification and the final devitrification process, the rate of reaction is diffusion-controlled and the T_g evolution runs more or less parallel with T, with a maximum difference ($T_g - T$) of 7°C.

5.4.2 Epoxy–Amine [68,84]

The results for the amine-cured epoxy for the same heating rate of 0.2°C min^{-1} are shown in Figure 2.8 (curve 1). The overall picture is similar to the result for the epoxy–anhydride (Figure 2.7). T_{g0} equals −25°C (not shown). The non-reversing heat flow passes through a maximum at 70°C, and then decreases sharply. Simultaneously, a sharp decrease in heat capacity occurs. The cure of another sample was stopped at this point. The glass transition temperature measured in a second heating is 81°C. Considering the reactivity of this resin, these data confirm the occurrence of vitrification at the instant that heat flow and heat capacity decrease. Between 85

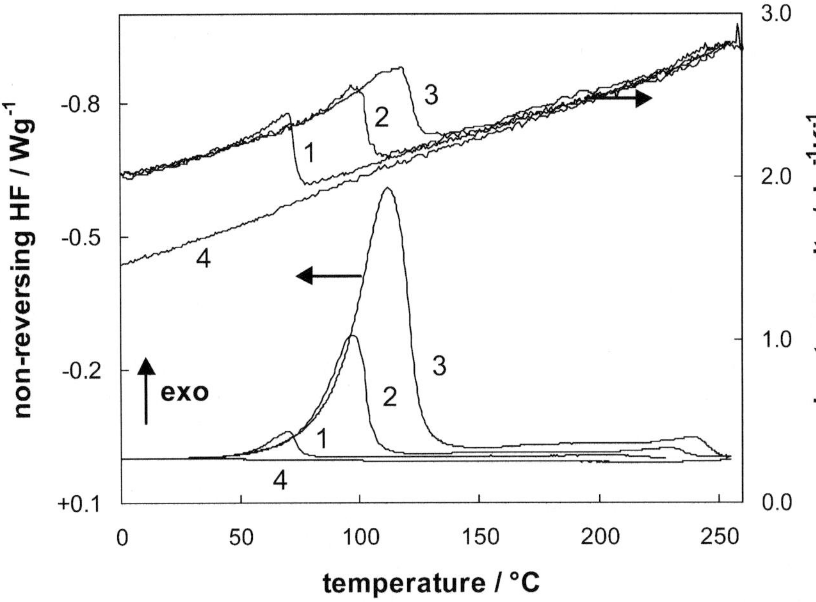

Figure 2.8. Non-isothermal cure of an epoxy($f = 4$)-amine($f = 4$) at 0.2 (1), 1 (2), and 2.5°C min^{-1} (3), and for the fully-cured material (4): non-reversing heat flow and heat capacity.

and 235°C, a temperature interval of 150°C, a small heat flow is still observed (25–60 μW). A second (low) maximum is attained around 200°C, then the heat flow decreases to reach the baseline level near 235°C. Over the 150°C wide interval, the C_p of the curing network (curve 1) and the C_p of the fully cured, vitrified resin (curve 4) slowly converge. The first and second heating both show a small increase in C_p ending at ca. 255°C. Since the reaction is completed near 235°C, this change in C_p corresponds most probably to devitrification. Both this transition and $T_{g\infty}$ cannot unambiguously be determined because the step change in heat capacity is small, and because degradation of the polymer network becomes prominent at 275°C. The smallness of $\Delta C_p(T_{g\infty})$ can be attributed to the high crosslink density of the fully cured resin: little mobility will be freed beyond $T_{g\infty}$ due to the restrictions of the tight network. For the tetrafunctional epoxy–amine, the final crosslink density is higher than for the bifunctional epoxy–anhydride, causing a higher $T_{g\infty}$ (ca. 255°C) and a smaller $\Delta C_p(T_{g\infty})$.

5.4.3 Effect of Heating Rate [68]

Figures 2.7 and 2.8 show the influence of the heating rate on the vitrification–devitrification behaviour of the anhydride and amine-cured epoxies, respectively. For the epoxy–anhydride system, no vitrification is observed at heating rates of 0.7°C min^{-1} or higher (curve 3). In contrast, for the epoxy–amine system even at a heating rate of 2.5°C min^{-1} (curve 3) a strong vitrification can be seen. The heating rate has to be at least 20°C min^{-1} to avoid vitrification (section 8). This indicates that attention needs to be paid when using non-isothermal experiments to study the reaction kinetics: the reaction is not de facto occurring in chemically controlled conditions, even at higher heating rates.

At a higher heating rate, it takes less time to reach a certain temperature. Consequently, a lower conversion and T_g are attained at that temperature. As a result vitrification is shifted to higher temperatures. This shift is obviously limited by $T_{g\infty}$, the devitrification temperature of the fully-cured resin. For the epoxy–anhydride, when the heating rate is increased, the minimum level of C_p in between T_{vit} and T_{devit} is closer to C_p of the liquid state (Figure 2.7). This implies that less mobility is frozen out and that a smaller fraction of the material transforms to the glassy state. This is a condition of non-isothermal 'partial vitrification'. For the amine-cured epoxy, even at 2.5°C min^{-1} the interval between vitrification and devitrification is at least 100°C wide. In this interval, C_p converges to C_p of the fully-cured (glassy) network (Figure 2.8), indicating that the material vitrifies almost completely during non-isothermal diffusion-controlled cure.

5.5 COMBINED CURE PATHS [80]

Examples of non-isothermal post-cure experiments after a preceding partial cure with mobility restrictions are given for the epoxy–anhydride system in Figure 2.9, and for the epoxy–amine system in Figure 2.10. The combinations of partial cure time and temperature are chosen in a way that both systems (partially) vitrify before the final non-isothermal cure stage. For the epoxy–anhydride system, up to the isothermal onset of vitrification after 165 min of reaction at 85°C, no enthalpy relaxation is observed in the non-reversing heat flow signal of the subsequent heating (Figure 2.9). As vitrification proceeds, due to the decreasing reaction rate (controlled to an increasing extent by diffusion) and the slow variation of T_g in these conditions, the structural relaxation effect is getting more pronounced [15,16]. Figure 2.9 also shows that the residual reaction exotherms become more asymmetrical and are delayed to higher temperatures. The (increasing) T_g of the system is acting as a physical barrier to the final cure: the residual reaction only starts when the necessary mobility is regained upon going through the glass transition. Nevertheless, at a heating rate of 2.5°C min^{-1}, this final cure always proceeds with a chemically controlled rate (once $T > T_g$). Note that T_g is still increasing considerably, even in mobility-restricted isothermal partial cure conditions, whereas the small residual reaction

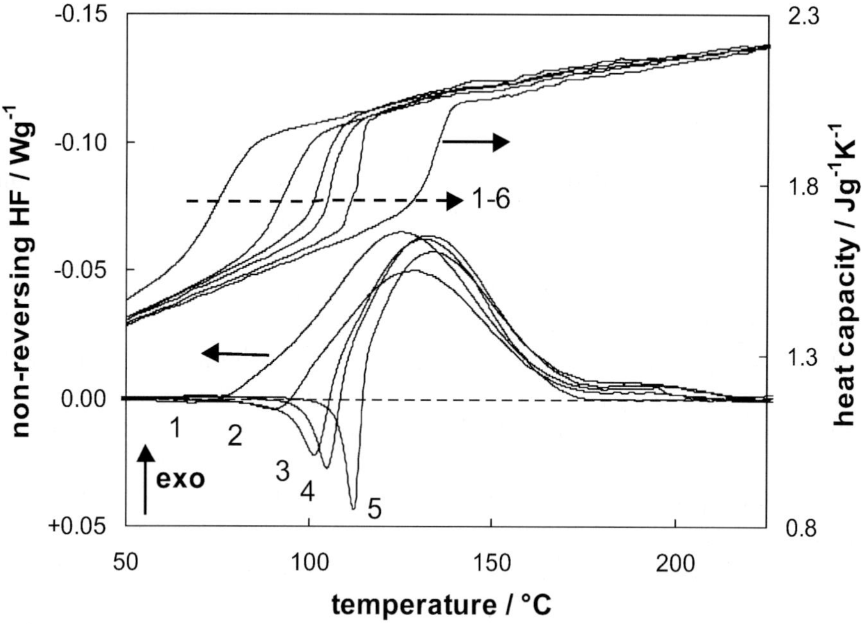

Figure 2.9. Post-cure of an epoxy–anhydride at 2.5°C min^{-1}: 1st heating after partial cure at 85°C for 165 min (1), 230 min (2), 500 min (3), 800 min (4), 3300 min (5) and 2nd heating in same conditions (6): non-reversing heat flow and heat capacity.

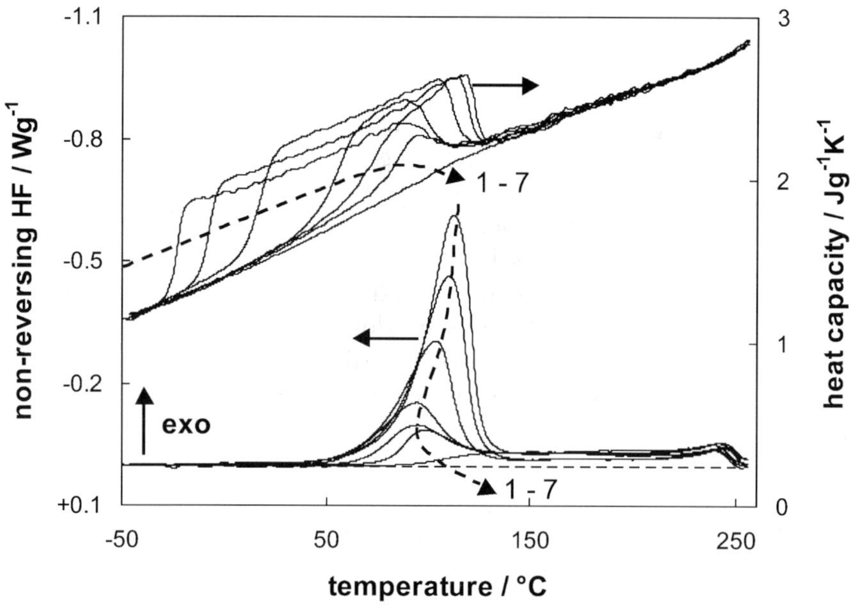

Figure 2.10. Post-cure of an epoxy($f = 4$)-amine($f = 4$) at 2.5°C min^{-1} after partial cure at 70°C for 0 min (1), 25 min (2), 50 min (3), 75 min (4), 90 min (5), 110 min (6) and partial cure at 1.0°C min^{-1} to 116°C (= T_{vit}) (7): non-reversing heat flow and heat capacity.

enthalpy is only slightly decreasing. Curve 6 shows $T_{g\infty}$ measured in a (second) heating after post-cure.

For the epoxy–amine system of Figure 2.10, a different behaviour is observed. No relaxation effects are noticed in the non-reversing signal of the post-cure. In this case, even in diffusion-controlled isothermal conditions, the variation of T_g with reaction time is high enough to avoid this effect. Accordingly, at a heating rate of 2.5°C min^{-1}, only the initial part of the post-cure is chemically controlled, as indicated by the heat capacity signals and the broad tails in the non-reversing heat flow signals of Figure 2.10. The start of mobility-restricted non-isothermal cure is interfering sooner with increasing conversion (reaction time) of the preceding isothermal cure. When the conversion after partial cure is low enough, the residual cure reaction first proceeds under chemically controlled conditions, followed by vitrification and a diffusion-controlled reaction regime. The higher the conversion at the start of the residual cure experiments, the shorter the part of the chemically controlled residual cure reaction. From a certain conversion, the heat capacity remains below the full mobility level related to that conversion (no T_g is observed), and the post-cure occurs under at least partially diffusion-controlled conditions over the entire range of the reaction exotherm (curves 4–7).

5.6 SLOW ISOTHERMAL CURE

5.6.1 Unsaturated Polyester

If the polyester–styrene resin, studied above room temperature in Figure 2.3, is cured at lower temperatures, e.g. at 0°C, the rate of reaction is too slow for a quantitative analysis of the heat flow. Figure 2.11 shows that the evolution of heat capacity and heat flow phase, however, can still be measured accurately. This illustrates an extra benefit of MTDSC for studying quasi-isothermally slowly reacting systems (notice the extended time-scale of Figure 2.11). During the slow reaction-induced vitrification process, the heat capacity shows a stepwise decrease and the heat flow phase passes through a minimum and increases again to its final, more glassy level [81].

5.6.2 Inorganic Polymer Glasses: Influence of Particle Size

Since the hardening process of a metakaolinite (Mk)–silicate is a heterogeneous reaction, the particle size of the solid reaction component, Mk, might also influence the reaction kinetics. The formation of the highly crosslinked aluminosilicate glass (IPG) was followed at 35°C. Figure 2.12 demonstrates the influence of the Mk particle size on the production rate of IPG [80].

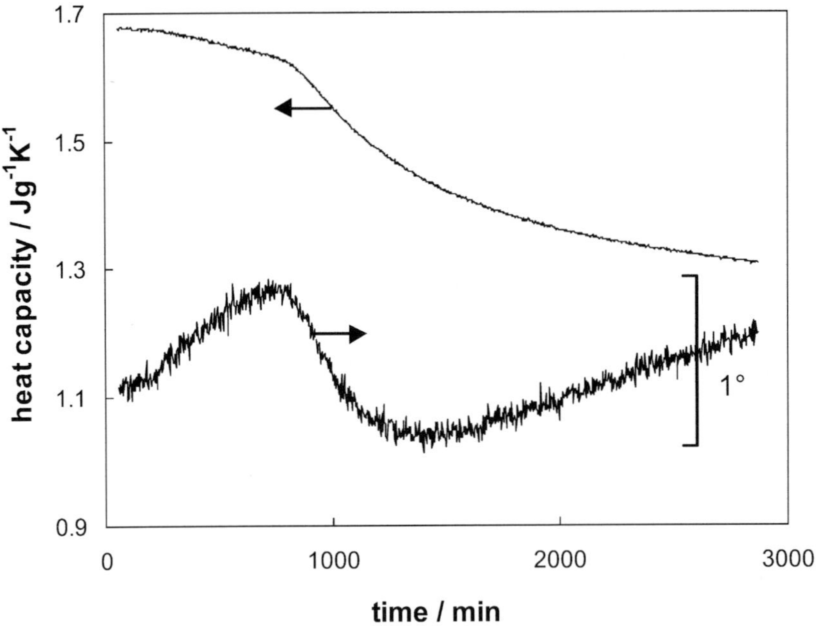

Figure 2.11. Quasi-isothermal cure of an unsaturated polyester at 0°C: heat capacity and heat flow phase.

The heat flow signal is weak, especially for the largest grains, and therefore baseline instabilities are causing the signal to become less accurate. However, it is clear that the smaller the particle size, the larger the heat flow at the beginning of the reaction and the faster the reaction tends to completeness. For the smallest 1.5 μm particles, the heat flow drops to zero before 500 min of reaction.

In the conditions studied, small heat capacity changes remain accurate and again allow the reaction to be followed long after the onset of vitrification. For the smallest particles, the heat capacity is still decreasing after 3000 min of reaction (not shown in Figure 2.12). The effect of particle size on the reaction rate can be quantified based on the reaction time to reach a certain decrease in heat capacity. In Figure 2.12b, the times for a drop in heat capacity of 10% and 50%, respectively, $t(10\%)$ and $t(50\%)$, are indicated on the heat capacity curves. The trends depicted by the interconnecting dashed lines sketch the effect of the initial reciprocal diameter, d^{-1} (plotted on the right Y-axis), which is proportional to the initial specific surface of the particles. It can be deduced that the reaction rate increases with decreasing particle size or increasing specific surface. For the largest specific surfaces, however, this tendency seems to level off. These conclusions are important

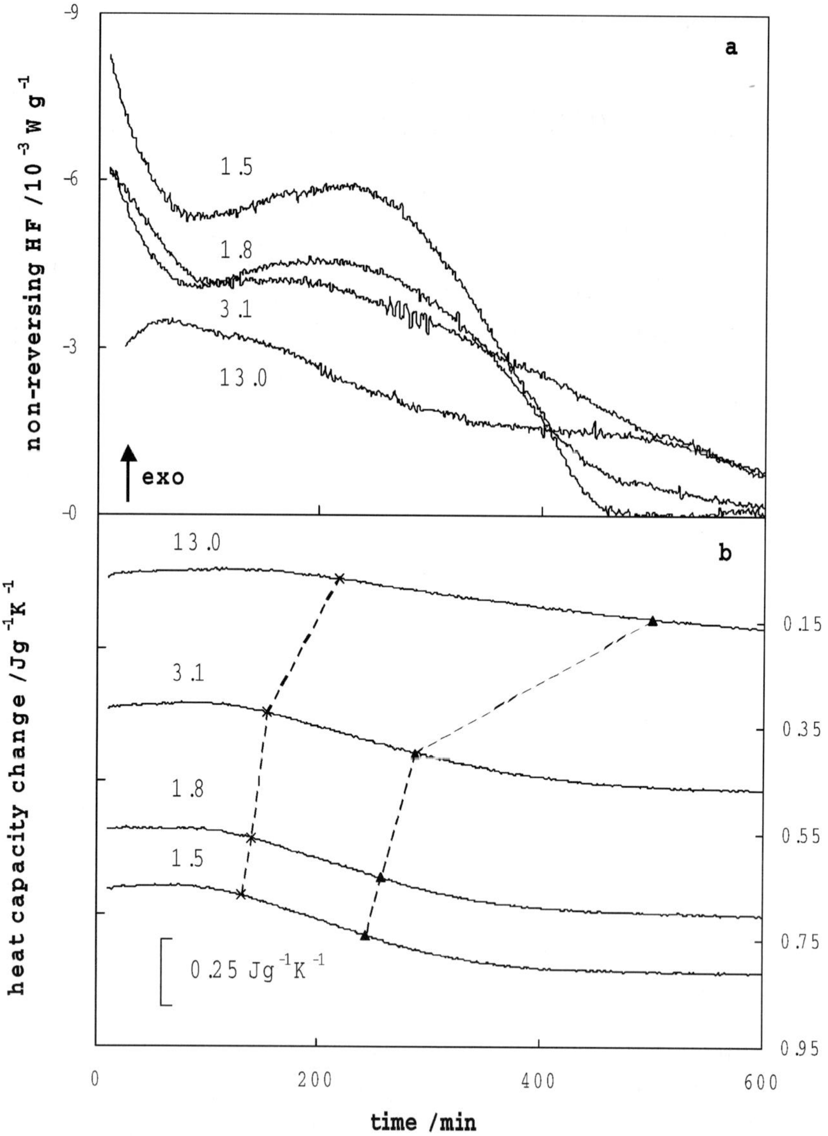

Figure 2.12. Production of inorganic polymer glasses for different Mk particle sizes (indicated values in μm) at 35°C: (a) non-reversing heat flow; (b) heat capacity (shifted according to initial reciprocal particle diameter, d^{-1} (right Y-axis), with t(10%) (×) and t(50%) (▲); dotted lines are a guide to the eye.

for a more elaborated model of the heterogeneous reaction kinetics of IPG.

5.7 PARTIAL VITRIFICATION

As observed in the previous sections, relaxation phenomena are superimposed as local (downward) extremes in the heat flow phase. Thus, the heat flow phase gives an indication of a vitrification or devitrification process during the thermal treatment. In non-isothermal experiments, conditions of partial vitrification—a zone where the material is in between the liquid/rubbery and the glassy state—can occur depending on the heating rate and the reactivity of the curing system (section 5.4).

In (quasi-)isothermal cure, partial vitrification is also seen if the cure temperature is chosen close to the glass transition temperature of the fully reacted polymer network, $T_{g\infty}$. An illustration is given in Figure 2.13 for the polyester–styrene cure at 30, 40 and 50°C [80,81]. With increasing temperature, the auto-acceleration becomes relatively less important. The maximum auto-acceleration always occurs just before the onset of vitrification (Figure 2.13a). Typical for these polyester systems is a broad fully-cured glass transition domain, in this case of at least 50°C around a low value of $T_{g\infty}$ of ca. 25–28°C. Because of the width of this broad transition domain, partial vitrification occurs in all isothermal cure conditions shown, even at 50°C. Although the degree of vitrification is decreasing with increasing cure temperature, a stepwise decrease in C_p is still clearly visible at 50°C (Figure 2.13b). The heat flow phase signal also indicates that the system is remaining in the relaxation regime for all conditions, even at the end of a cure at 30°C, and the phase angle is not evolving to less negative values (Figure 2.13b). In contrast, if cured at 0°C, an almost completely vitrified polyester system is obtained (Figure 2.11). In this case, the heat flow phase evolves towards zero after reaching the relaxation minimum.

Partial vitrification is also observed in isothermally cured epoxy systems. However, the effect is less pronounced since the glass transition domain at $T_{g\infty}$ is narrower for these networks [80]. An example is given in Figure 2.14 for the system DGEBA–MDA ($T_{g\infty} = 102$°C). At 80°C, a stepwise decrease in C_p and a relaxation peak are observed. At 100°C, the system is partially vitrifying and the phase angle remains in the relaxation regime at the end of cure. At 120°C, no vitrification effect is noticed any more, neither in C_p, nor in heat flow phase.

More recently, Montserrat and Cima [85] presented similar epoxy cure experiments, but their interpretation is somewhat misleading because they did not account for the partial vitrification phenomenon. The way the mobility factor, DF^*, was calculated in [85] is in contradiction with the

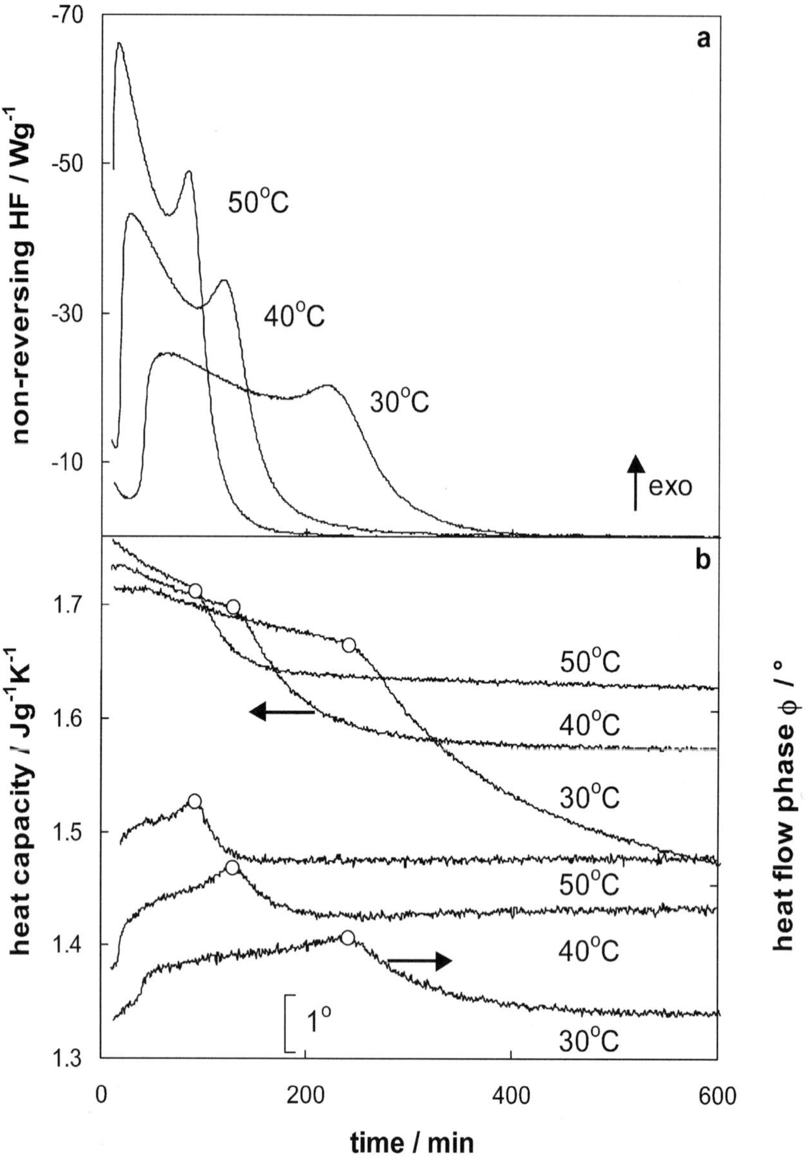

Figure 2.13. Quasi-isothermal cure of an unsaturated polyester at 30, 40, and 50°C: (a) non-reversing heat flow; (b) heat capacity and heat flow phase; the heat flow phase curves were shifted vertically to avoid overlap. The symbols (o) denote the points at maximum auto-acceleration in the non-reversing heat flow.

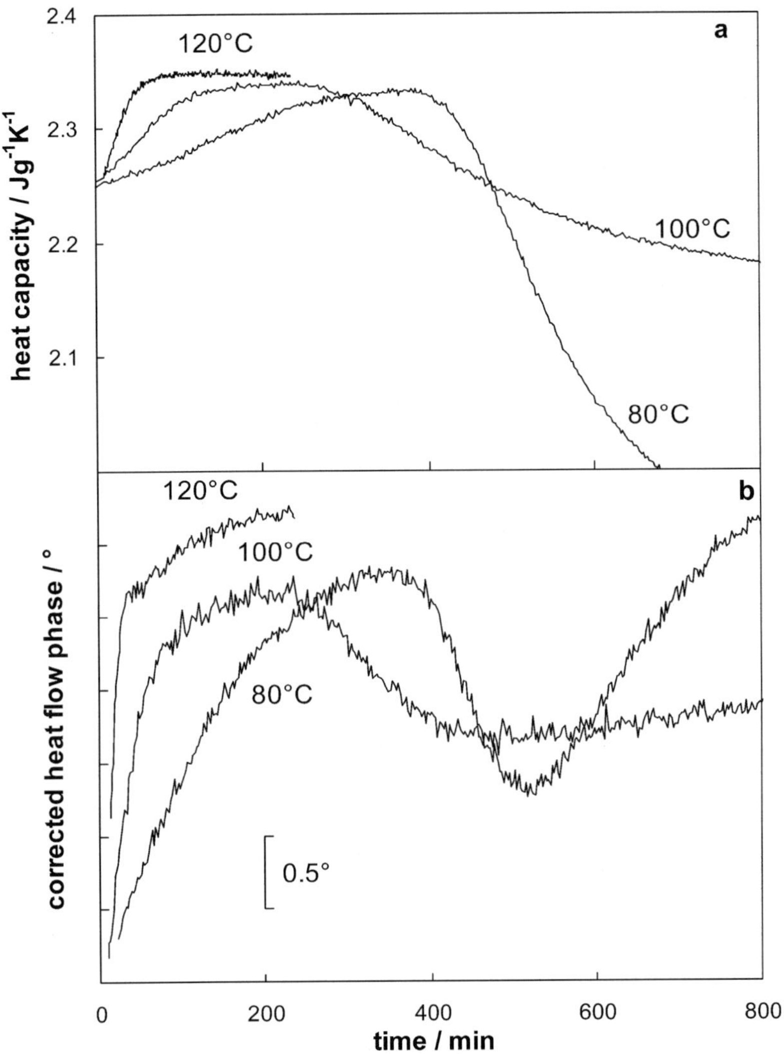

Figure 2.14. Quasi-isothermal cure of an epoxy($f = 2$)-amine($f = 4$) at 80, 100, and 120°C: (a) heat capacity; (b) heat flow phase (shifted for clarity).

fact that the final state is only partially glassy for cure temperatures that are within the glass transition of the fully-cured resin.

5.8 MOBILITY FACTOR TO QUANTIFY DEGREE OF VITRIFICATION

To quantify the degree of vitrification for conditions for which the reacting network is in between the glassy and the liquid or rubbery state, a mobility factor can be defined, based on the experimental heat capacity curves. This mobility factor, DF^*, is calculated by normalising the heat capacity variation according to [42,68,84]:

$$DF^*(x,T) = \frac{C_p(x,T) - C_{pg}(x,T)}{C_{pl}(x,T) - C_{pg}(x,T)} \tag{10}$$

The equation states that variations in heat capacity, C_p, are normalised between unity for the liquid or unrestricted state (with heat capacity C_{pl}), and zero for a frozen glassy state (with heat capacity C_{pg}). The evolution of this factor should mirror the reduction of mobility due to vitrification only, and not the changes in heat capacity due to changes in temperature or to the chemical changes themselves. Therefore, the influence of both temperature and conversion on the reference states, C_{pl} and C_{pg}, needs to be taken into account to obtain quantitative results.

The evolution of the mobility factor based on experimental heat capacity curves shown in sections 5.4 and 5.5 will be used for modelling the cure process and the results will be given in section 6.2.

5.9 HEAT CAPACITY CHANGE AS A RESULT OF REACTION BEFORE VITRIFICATION

If the (specific) heat capacity of the products formed differs from the heat capacity of the reagents consumed, the heat capacity will gradually change with reaction conversion.

5.9.1 Mechanistic Information

In Figure 2.15, the quasi-isothermal heat capacity change as a function of conversion is compared for the different thermosetting systems discussed in previous paragraphs. The largest (negative) change in C_p is always observed during the reaction-induced vitrification. The conversion at vitrification is dependent on the isothermal cure temperature chosen and on the $T_g - x$ relationship of the system.

Before vitrification, a heat capacity change as a result of chemical reaction, $\Delta C_{p,\mathrm{react}}$, is noticed. For the anhydride-cured epoxy and the polyester–styrene resin a minor, but reproducible, and almost linear decrease of C_p with conversion is observed. The former system is supposed to be an anionic chain-growth living polymerisation (without termination), the latter is a chain-growth copolymerisation with termination.

On the contrary, a rather important positive $\Delta C_{p,\mathrm{react}}$ is seen for the amine-cured epoxy and for the melamine–formaldehyde resin. These

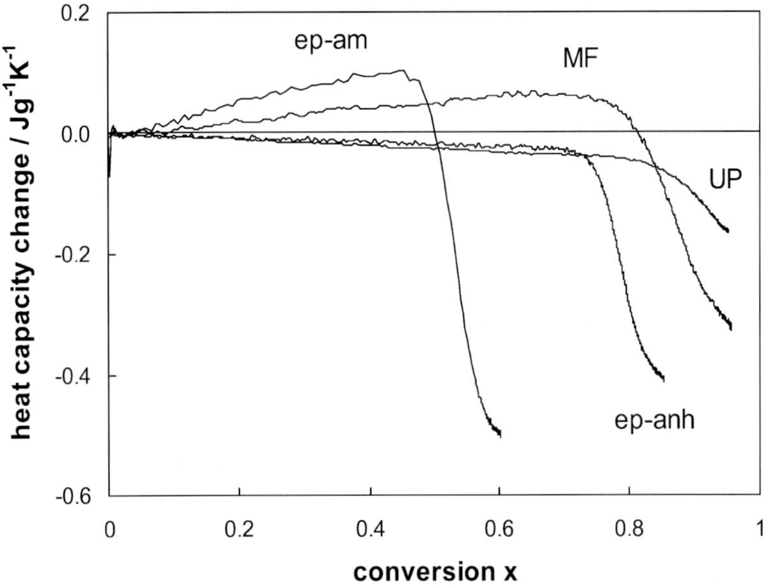

Figure 2.15. Comparison of the heat capacity change as a function of reaction conversion for an epoxy($f = 2$)-amine($f = 4$), epoxy–anhydride, unsaturated polyester and melamine–formaldehyde system.

systems cure according to an addition and a condensation step-growth polymerisation mechanism, respectively.

It is clear that $\Delta C_{p,react}$ provides information on the reaction mechanism. Instead of normalising the heat capacity signal in terms of a mobility factor ($DF^* = 1$ if no mobility restrictions before vitrification), the information available in $\Delta C_{p,react}$ can be exploited for mechanistic modelling.

5.9.2 Step-Growth Epoxy–Amine Polymerisation: Primary and Secondary Reactions

The importance for mechanistic investigations of $\Delta C_{p,react}$ in quasi-isothermal reaction conditions is further demonstrated in Figure 2.16 for the addition step-growth polymerisations of epoxy–amine systems with a different functionality of the reactive components. In stoichiometric mixing conditions, this leads to small molecules (PGE–aniline), linear macromolecules (DGEBA–aniline) and a polymer network (DGEBA–MDA), respectively. The differences in the rate of cure, caused by the nature of the epoxy and the mixing ratio, r, are visible. It is obvious that the heat capacity increases as a result of reaction to a maximum value (plateau) in curves 1–4. A similar evolution can also be observed in other experiments of previous sections (Figures 2.15 and 2.16), and analogous observations were made in literature

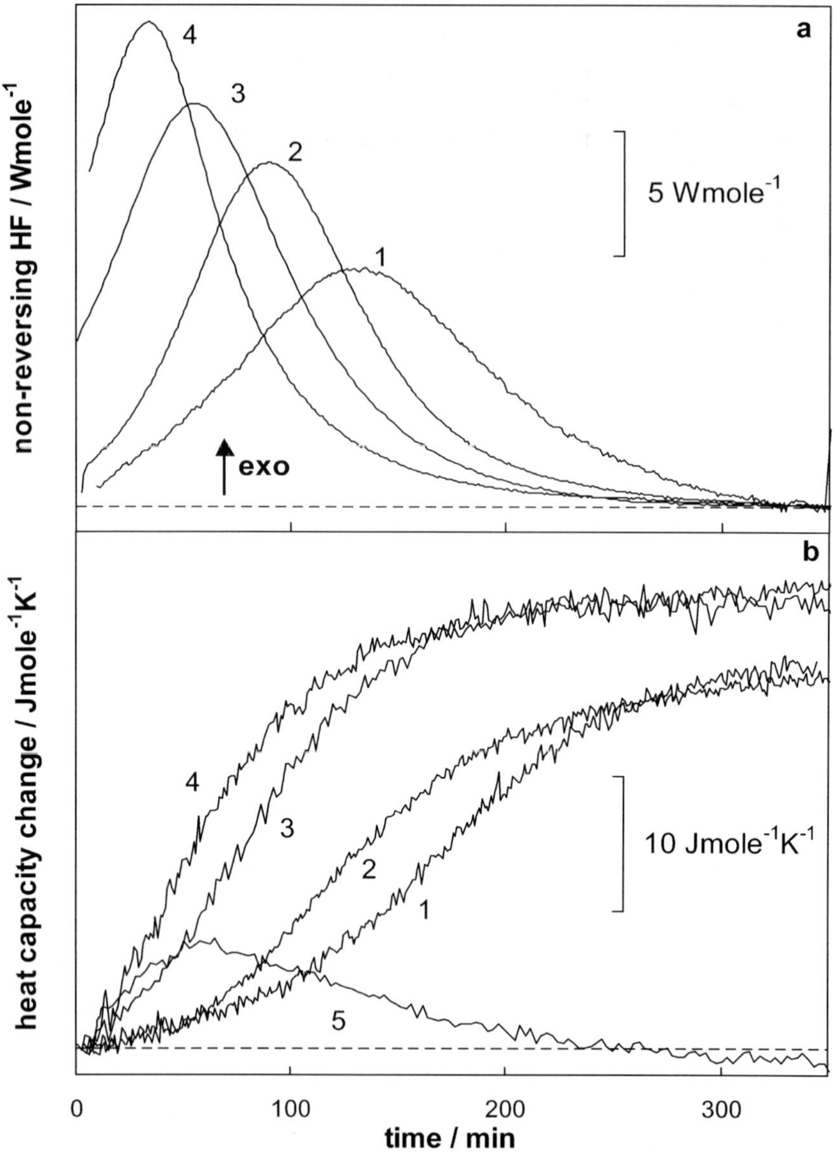

Figure 2.16. Cure of different epoxy–amine systems at 100°C; PGE–aniline in molar ratios of amine/epoxy functional groups $r = 0.6$ (1) and $r = 1.0$ (2); DGEBA-aniline with $r = 0.7$ (3) and $r = 1.0$ (4); PGE/N,N'-dimethylethylenediamine at 30°C is given for comparison (5): (a) non-reversing heat flow per mole of reacted (epoxy–NH) functional groups; (b) heat capacity change per mole of reacted (epoxy–NH) functional groups.

[94–96]. It should be noted that in all these experiments, *primary* amine–epoxy in combination with *secondary* amine–epoxy addition reactions are predominant in the reaction mechanism [30,97–99].

The heat capacity evolution is strikingly different if only *secondary* amine functionalities can react, as shown in curve 5 of Figure 2.16b (PGE–secondary amine system). The ability of MTDSC to make distinction between primary and secondary amine–epoxy reactions by using the C_p evolution can be explained in terms of a group additivity estimation method of thermodynamic properties of organic compounds at 298.15K in the liquid and solid phase [100]. According to this method, primary amine–epoxy reactions always give rise to positive values of $\Delta C_{p,react}$, whereas the predicted values of $\Delta C_{p,react}$ for secondary amine–epoxy reactions are almost zero or even negative (depending on the aliphatic or aromatic nature of the secondary amine). The experimental values of $\Delta C_{p,react}$ are in agreement with these predictions. As expected, the heat capacity curves 1–2 and 3–4 of Figure 2.16b reach a comparable limiting value, both for stoichiometric mixtures and for mixtures with an excess of epoxy ($r \leq 1.0$).

Note that the measurements of $\Delta C_{p,react}$ at full conversion can be disturbed by vitrification, as already illustrated with the heat capacity and the heat flow phase signals of the system DGEBA–MDA in Figure 2.14.

5.10 REACTION-INDUCED PHASE SEPARATION

Amorphous high-T_g engineering thermoplastics, introduced as an additive to the two-component reactive mixture prior to cure, can be expected to reduce the brittleness of a thermoset, without affecting the other properties significantly [101]. Starting from a homogeneous mixture, the system transforms into a phase-separated structure due to the increasing molecular weight or network formation of the reacting matrix [102]. The competition between the rates of cure and of phase separation determines the mechanism of this reaction-induced phase separation, i.e. nucleation and growth or spinodal demixing [101,103].

The potential of MTDSC for the real-time monitoring of reaction-induced phase separation is demonstrated with the cure of an epoxy—aniline–polyethersulphone (PES) mixture [80,104]. The epoxy–aniline system allows following the isothermal cure accurately above and below $T_{g\infty}$ (94°C). Choosing an isothermal cure temperature below $T_{g\infty}$ will provoke a combination of phase separation of a PES-rich phase and vitrification of the epoxy–aniline matrix. Figure 2.17 shows the quasi-isothermal cure at 80°C for both modified and unmodified epoxy–aniline systems. The effect of primary and secondary amine reactions is seen as a positive $\Delta C_{p,react}$. In the unmodified system, vitrification is seen after 91% conversion as a stepwise

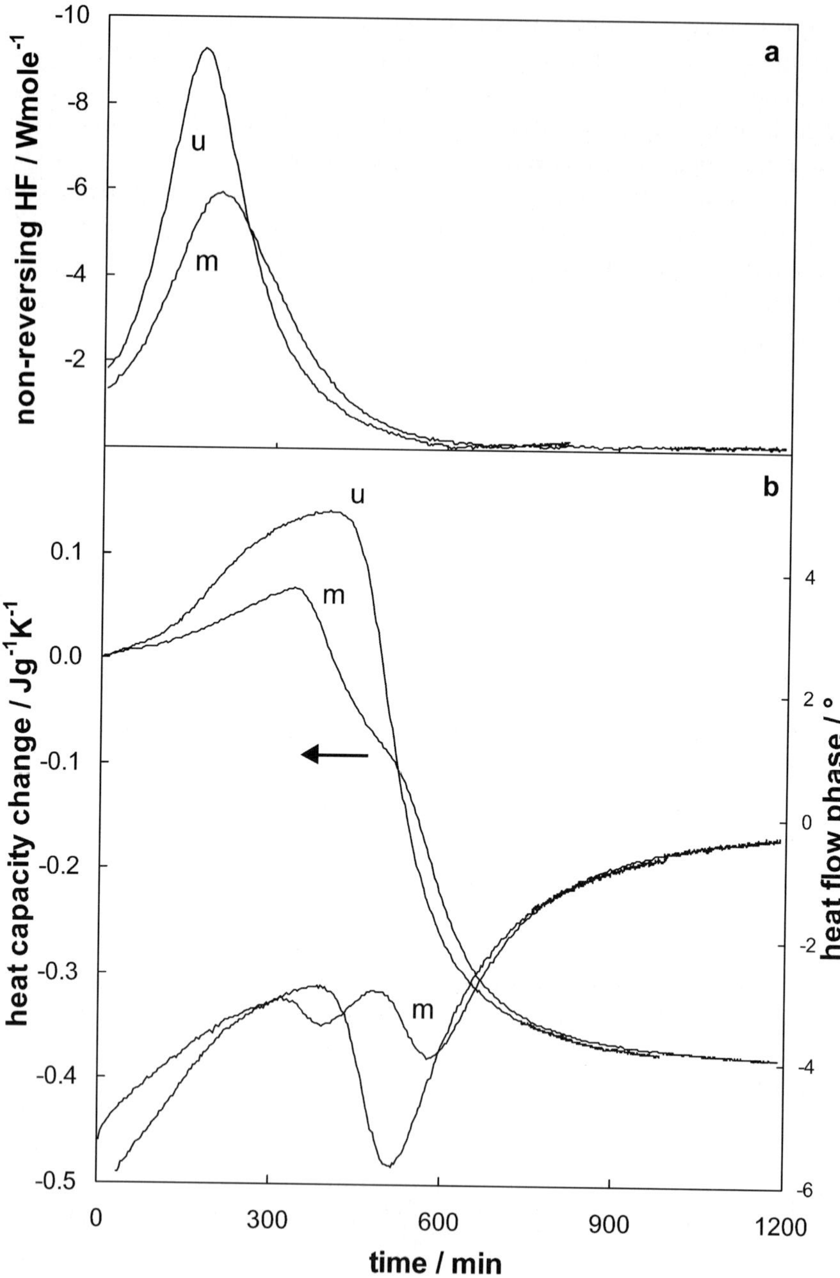

Figure 2.17. Quasi-isothermal cure at 80°C of an epoxy($f = 2$)-amine($f = 2$) (unmodified: u) and epoxy($f = 2$)-amine($f = 2$)/20% PES (modified: m): (a) non-reversing heat flow; (b) change in heat capacity and heat flow phase.

decrease in heat capacity, accompanied by a negative relaxation peak in the heat flow phase. The PES-modified epoxy–aniline exhibits two subsequent C_p decreases, which can be associated with the vitrification of a segregating dispersed PES-rich phase and the vitrification of the epoxy-rich matrix, respectively. Due to the higher T_g of PES, the PES-rich phase segregates and vitrifies first, while the epoxy matrix is still mobile. Vitrification of the epoxy-rich phase is occurring afterwards due to the progressing reaction. Two clearly separated relaxation peaks can be observed in the heat flow phase, which is a valuable asset in the analysis of phase separation during network formation [104].

6 Modelling the Diffusion-Controlled Overall Kinetics and Cure Rate Law of Epoxy Systems

The experimental MTDSC observations on anhydride-cured and amine-cured epoxies, described in the previous section, will now be modelled to illustrate the benefits of the technique to obtain a quantitative law of cure kinetics for such thermosetting systems. Because cure kinetics are often complicated by diffusion limitations and/or mobility restrictions, the effect of diffusion has to be incorporated into the overall reaction rate law. For this purpose, both heat capacity and non-reversing heat flow signals for quasi-isothermal and non-isothermal cure experiments are used.

6.1 PROPOSED MODEL

One valuable approach to quantify the effects of diffusion (or mobility restrictions) on the cure kinetics is via direct estimation of a diffusion factor, *DF*. The latter is defined as the ratio of the experimentally measured conversion rate $(\mathrm{d}x/\mathrm{d}t)_{\mathrm{obs}}$ over the predicted conversion rate at the *same* reaction conversion x in the absence of mobility restrictions $(\mathrm{d}x/\mathrm{d}t)_{\mathrm{kin}}$:

$$\left(\frac{\mathrm{d}x}{\mathrm{d}t}(x,T)\right)_{\mathrm{obs}} = \left(\frac{\mathrm{d}x}{\mathrm{d}t}(x,T)\right)_{\mathrm{kin}} DF(x,T) \tag{11}$$

The left-hand-side part of this equation $(\mathrm{d}x/\mathrm{d}t)_{\mathrm{obs}}$ is proportional to the non-reversing heat flow according to Eq. (8). For step-growth polymerisation reactions, diffusion control is governed by the vitrification process (section 5.1.1). Therefore, as a first approximation, the normalised heat capacity signal or mobility factor, DF^*, [Eq. (10)] is proposed to be an in situ measurement (simultaneous and independent of the heat flow) of the diffusion factor of Eq. (11). This approach will be justified a posteriori. Thus, the experimental data are:

$$\left(\frac{\mathrm{d}x}{\mathrm{d}t}(x,T)\right)_{\mathrm{obs}} = \frac{1}{\Delta H_{\mathrm{tot}}}\frac{\mathrm{d}H}{\mathrm{d}t}\Big|_{\mathrm{NRheatflow,obs.}} \tag{12}$$

$$DF(x,T) \cong DF^{*}(x,T) = \frac{C_{\mathrm{p}}(x,T) - C_{\mathrm{pg}}(x,T)}{C_{\mathrm{pl}}(x,T) - C_{\mathrm{pg}}(x,T)} \tag{13}$$

The empirical rate equation proposed by Kamal [38] is applied for the chemically controlled reaction rate of any epoxy resin cure showing auto-catalytic behaviour:

$$\left(\frac{\mathrm{d}x}{\mathrm{d}t}(x,T)\right)_{\mathrm{kin}} = (k_1 + k_2 x^m)(1-x)^n = k_{\mathrm{kin}}(1-x)^n \tag{14}$$

with x the conversion of epoxy groups, k_1 and k_2 the rate constants and m and n the reaction orders.

The temperature dependence of any rate constant is given by an Arrhenius relationship:

$$k = A\exp\left(-\frac{E}{RT}\right) \tag{15}$$

with A the pre-exponential factor, E the activation energy, T the absolute temperature and R the universal gas constant.

The phenomenological rate constant $k_{\mathrm{kin}}(x,T)$ describes the overall chemically controlled polymerisation reaction and is incorporating an increase in value due to auto-catalysis. Combining Eq. (11) and Eq. (14) results in

$$\left(\frac{\mathrm{d}x}{\mathrm{d}t}\right)_{\mathrm{obs}} = DF(x,T)k_{\mathrm{kin}}(1-x)^n = k_{\mathrm{app}}(1-x)^n \tag{16}$$

$$k_{\mathrm{app}} = k_{\mathrm{kin}}DF(x,T) \tag{17}$$

with $k_{\mathrm{app}}(x,T)$ an apparent rate constant describing the effect of diffusion on $k_{\mathrm{kin}}(x,T)$.

A quantification of *DF* to describe the transition from chemically-controlled to diffusion-controlled kinetics is based on the Rabinowitch equation, which is derived from the activated complex theory [39,105–107].

Whether a chemical reaction is controlled by diffusion depends on the relative time to diffuse and the time needed for the intrinsic chemical reaction resulting in bond formation:

$$t_{\text{reaction}} = t_{\text{chem.kinetics}} + t_{\text{diffusion}} \tag{18}$$

Two reactive groups diffuse toward each other to form an activated complex with possible creation of a stable bond. As long as diffusion in and out of the complex occurs more frequently than the process of bond formation, the reaction is controlled by chemical kinetics.

Consider the formation of a stable bond A–B via the formation of an activated complex, (AB)*, for an (overall) diffusion-controlled reaction

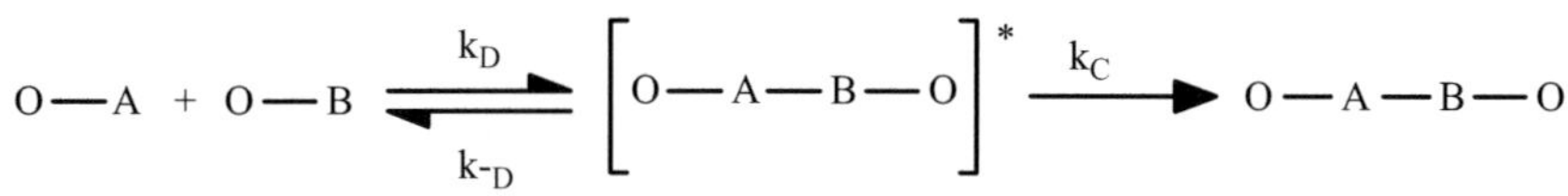

(19)

where –o means an undefined structure, k_D and k_{-D} rate constants for diffusion-controlled formation and dissociation of the activated complex, k_C the rate constant for transformation of the activated complex into a stable product.

When a steady state approximation for the activated complex is used, the rate of formation of A–B is given by:

$$\frac{d[AB]}{dt} = \frac{k_C k_D}{(k_C + k_{-D})}[A][B] = k_{\text{app}}[A][B] \tag{20}$$

which gives the Rabinowitch equation for the apparent rate constant of a bimolecular reaction:

$$\frac{1}{k_{\text{app}}} = \frac{1}{k_D} + \frac{1}{k_{\text{kin}}} \quad \text{and} \quad k_{\text{kin}} = k_C \frac{k_D}{k_{-D}} \tag{21}$$

with k_{kin} the experimentally measured rate constant under chemically controlled conditions. The limiting cases are $k_{\text{app}} = k_{\text{kin}}$ for $k_D \gg k_{\text{kin}}$ ($t_{\text{reaction}} = t_{\text{chem.kinetics}}$) and $k_{\text{app}} = k_D$ for $k_D \ll k_{\text{kin}}$ ($t_{\text{reaction}} = t_{\text{diffusion}}$).

In the case of overall diffusion control, the condition $k_D \gg k_{\text{kin}}$ is usually fulfilled prior to vitrification. Near vitrification, k_D becomes smaller than k_{kin} and under these conditions, k_{app} is temperature and conversion-dependent

(k_{kin} has generally an Arrhenius-type temperature dependency, k_D changes with temperature and conversion).

The following equation for the diffusion factor is derived from Eqs. (17) and (21):

$$DF(x,T) = \frac{k_D(x,T)}{k_D(x,T) + k_{kin}(x,T)} \tag{22}$$

The diffusion rate constant, k_D, can be expressed in terms of the overall diffusion coefficient, D [66]:

$$k_D = c(T)D(x,T) \tag{23}$$

with c a constant related to local conditions for creation of the chemical bond.

Since overall diffusion is governed by the diffusion of chain segments, the overall diffusion coefficient, D, is expected to be inversely proportional to the relaxation time of polymer segments [108], which enables a model based on the free volume concept and a description similar to the Williams–Landel–Ferry (WLF) equation [109–112]:

$$D = D_0 \exp\left(\frac{C_1'(T - T_g)}{C_2 + T - T_g}\right) \tag{24}$$

with C_1' a new constant, while C_2 is the universal WLF constant.

In the case of a curing system, the value of T_g is not constant but a function of reaction conversion. The diffusion rate constant, k_D, is finally given by:

$$k_D = k_{D0}(T) \exp\left(\frac{C_1'(T - T_g(x))}{C_2 + T - T_g}\right) \tag{25}$$

An Arrhenius temperature dependency of k_{D0} is considered and the following equation for the diffusion rate constant is deduced:

$$\ln k_D(x,T) = \ln A_D - \frac{E_D}{RT} + \frac{C_1'(T - T_g(x))}{C_2 + T - T_g(x)} \tag{26}$$

with A_D the pre-exponential factor and E_D the activation energy for the diffusion rate constant k_{D0}. A similar Arrhenius dependency is proposed in the literature, but instead of the WLF free volume contribution a more empirical approach was used [64].

The $T_g - x$ model used in Eq. (26) is [44,113]:

$$T_g(x) = \frac{T_{gu}}{1 - K_C X_C} = \frac{T_{g0} + ax}{1 - K_C X_C} \tag{27}$$

with a and K_C optimisation parameters, T_{g0} the glass transition temperature of the unreacted mixture and X_C the crosslink density. X_C equals 0 as long as x is below x_{gel} and rises up to 1 from x_{gel} to the final conversion of 1. K_C in the denominator describes the effect of crosslinking on T_g. T_{gu} is the glass transition temperature of an uncrosslinked system identical in every respect to the crosslinked system except that the crosslinks are missing. The parameter a in T_{gu} describes the effect of the decrease of the concentration of chain ends on T_g.

6.2 OPTIMISED CURE RATE LAW

To optimise the model parameters, the fitting strategy involves the *simultaneous* treatment of *all* isothermal and non-isothermal cure experiments with and/or without mobility restrictions of a thermosetting system of a *fixed* initial composition. The temperature range is largely extended to obtain an accurate estimate for all fitting parameters of the cure rate model.

The modelling was performed using FITME, a version of OPTKIN [114], a program for the mechanistic modelling of reaction kinetics, modified to enable the calculation and fitting of conversion, rate of conversion and reduced rate of conversion profiles for isothermal, non-isothermal and user-defined temperature programs. The optimum set of parameters derived corresponds to the least sum of squares of the differences between experimental and calculated values. The optimisation strategy is using an algorithm based on a combination of the methods of Newton–Raphson, Steepest Descent and Marquardt [115,116]. The method requires no assumptions (apart from the proposed cure rate law with kinetic and diffusion effects) and can therefore be applied to many kinds of cure reactions. The model parameters are summarised below.

Chemical kinetics [Eq. (14)]	$A_1, E_1\ (k_1), Z_2, E_2\ (k_2), m, n$
Diffusion model [Eq. (26)]	$A_D, E_D\ (k_{D0}), C_1', C_2$
$T_g - x$ relation [Eq. (27)]	T_{g0}, a, K_c, x_{gel}

Modelling results for the two epoxy systems will be discussed below.

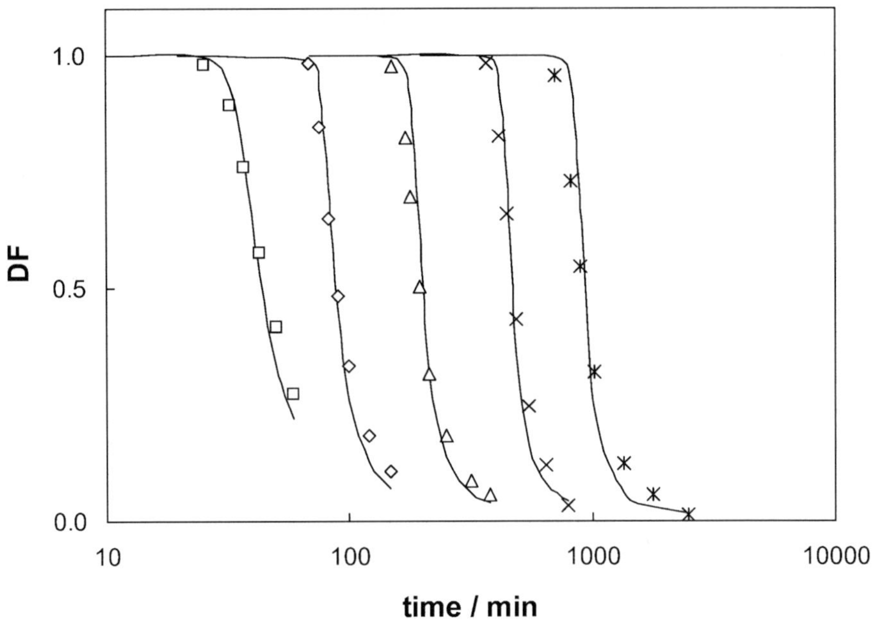

Figure 2.18. Diffusion factor for the quasi-isothermal cure of an epoxy–anhydride system at 120°C ( ), 100°C (◇), 85°C (△), 70°C (×), 60°C (✱); Data points from MTDSC (DF^*) and optimised model (——).

6.2.1 Epoxy–Anhydride System

After optimisation, the agreement between experiment and model is very satisfactory (Figures 2.18–2.21), considering the wide range of experimental conditions and the fact that *all* heat capacity and non-reversing heat flow profiles are fitted with *one* parameter set. The sudden decrease in diffusion factor near vitrification is well described for both quasi-isothermal and non-isothermal conditions (Figures 2.18 and 2.19). In the latter case, a similar level is reached in between vitrification and devitrification; even the increase due to devitrification is properly predicted. Moreover, Figures 2.20 and 2.21 illustrate well the effect of mobility limitations on the cure reaction under quasi-isothermal conditions at low temperatures (70 and 85°C), and under a non-isothermal condition applying a sufficiently small heating rate (0.2°C min^{-1}). The arrow in Figure 2.20 indicates the point where the reaction rate is still totally chemically controlled. The thin dashed line simulates a chemically controlled experiment using the same model, but with *DF* fixed to unity. From this indicated point, a drop in the reaction rate is observed, which is again perfectly described. Under non-isothermal conditions at 0.2°C min^{-1}, a shoulder of more or less constant reaction rate, which

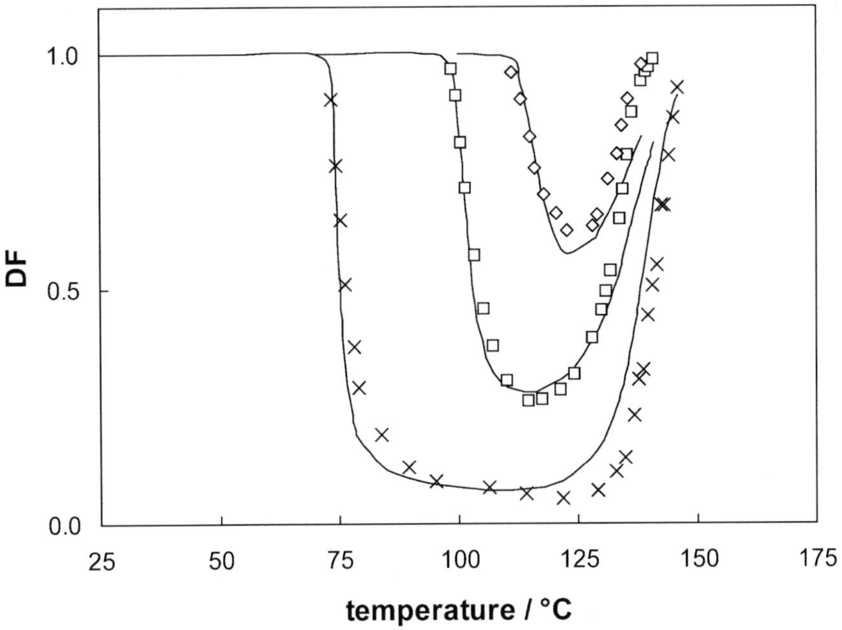

Figure 2.19. Diffusion factor for the non-isothermal cure of an epoxy-anhydride system at 0.04°C min^{-1} (×), 0.2°C min^{-1} (□), and 0.4°C min^{-1} (◇); Data points from MTDSC (*DF**) and optimised model (——).

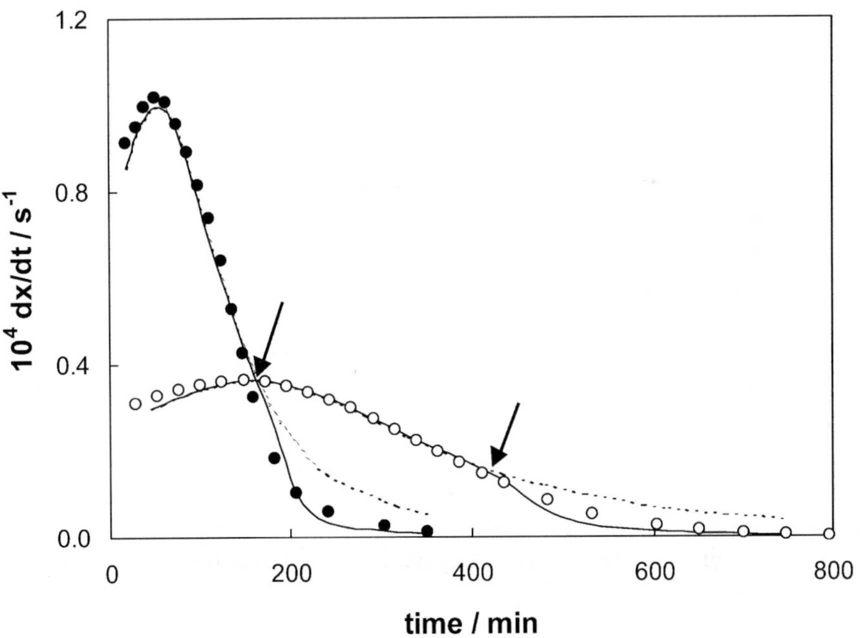

Figure 2.20. Reaction rate for the quasi-isothermal cure of an epoxy–anhydride system. Measured rate at 70°C (•) and 85°C (o), optimised rate (——) and simulated experiment for a chemically controlled rate (- - - -).

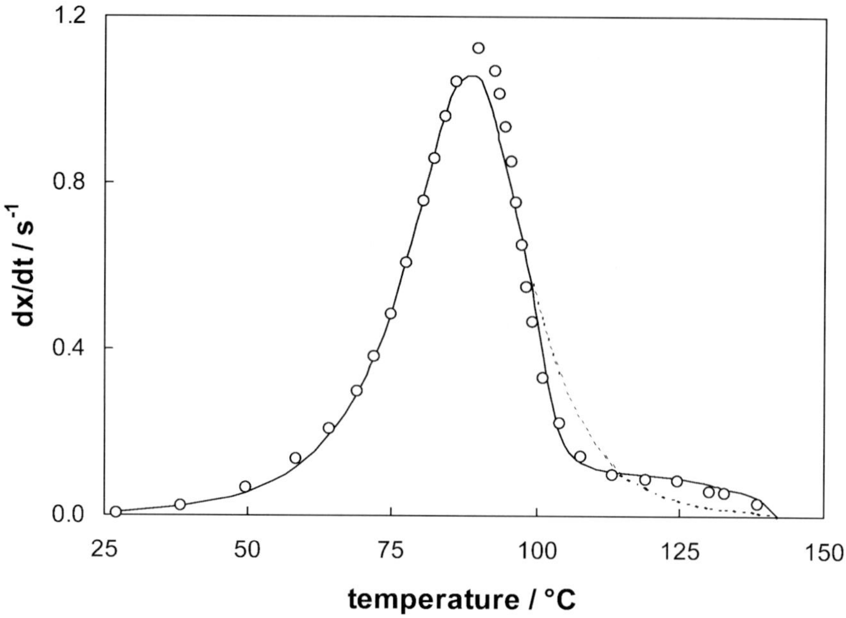

Figure 2.21. Reaction rate for the non-isothermal cure of an epoxy–anhydride system at 0.2°C min^{-1}. Measured rate (o), optimised rate including diffusion control (——) and simulated chemically controlled rate (- - - -).

spans a temperature interval of 40°C, is noticed (Figure 2.21). Simultaneously, a low value of the mobility factor is found (Figure 2.19). This shows that under mobility-controlled conditions the proposed model permits the estimation of the level of both reaction rate and diffusion (mobility) factor and thus provides a description of the overall cure kinetics.

More important is that the model clearly demonstrates that the most critical factor determining whether a reaction is diffusion-controlled or not is the difference between the reaction temperature and the glass transition temperature. When T_g rises up to the curing temperature, chain segments become less mobile, which results in a mobility-restricted reduced reaction rate. Prior to vitrification (when $T_g \ll T_{cure}$) the apparent rate constant equals the kinetic rate constant, k_{kin} ($k_D \gg k_{kin}$). The diffusion factor then equals unity. On the contrary, diffusion becomes a limiting step when $k_{kin} \gg k_D$, which is the case when T_g nears the reaction temperature (in the vitrification zone). The reaction becomes diffusion-controlled and *DF* drops to 0. *DF* remains unity as long as $T - T_g$ exceeds 25°C. When T_g becomes 15–20°C higher than T, *DF* drops toward 0. The latter condition occurs during isothermal curing over extended periods of time. For the anhydride–epoxy system, the isothermal cure temperature chosen has little influence

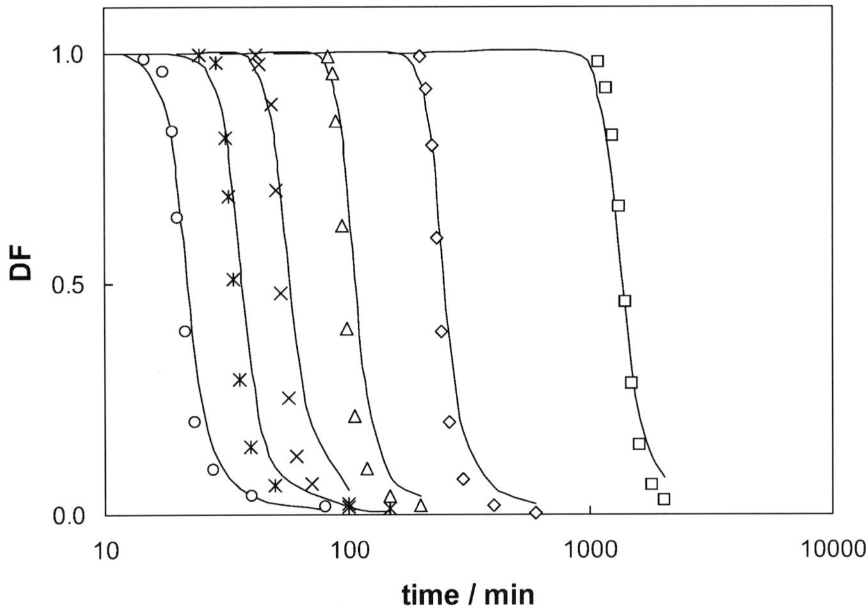

Figure 2.22. Diffusion factor for the quasi-isothermal cure of an epoxy($f = 4$)-amine($f = 4$) at 100°C (o), 90°C (∗), 80°C (×), 70°C (Δ), 52°C (◇), and 25°C (□); Data points from MTDSC(DF^*) and optimised model (——).

on the value of *DF* as a function of $T - T_g$. Hence, T_g almost equals T when *DF* or DF^* is 0.5.

6.2.2 Epoxy–Amine System

An analogous approach has been applied to the epoxy–amine system. The three sets of parameters were derived: one set for the chemical rate equation [Eq. (14)], one set for the diffusion rate constant according to Eq. (26), and one set for the $T_g - x$ relation [Eq. (27)]. As seen in Figure 2.22, the experimental and the calculated *DF* profiles agree very well for the quasi-isothermal cure at reaction temperatures ranging from 25 to 100°C.

Compared to the epoxy–anhydride system, where the deceleration of reaction rate caused by mobility restrictions was only noticed at the final stages of cure (Figures 2.20 and 2.21), a tremendous decrease in the rate of reaction is already observed early in the epoxy–amine cure process (Figures 2.23 and 2.24). For the isothermal cure at 90 and 100°C (Figure 2.23), the deviation from chemical kinetics modelling (dashed line) already becomes pronounced at the maximum of reaction rate. Looking at the non-isothermal cure at 0.2°C min^{-1} (Figure 2.24), chemical kinetics modelling predicts a higher reaction rate even at the peak temperature. The remarkable

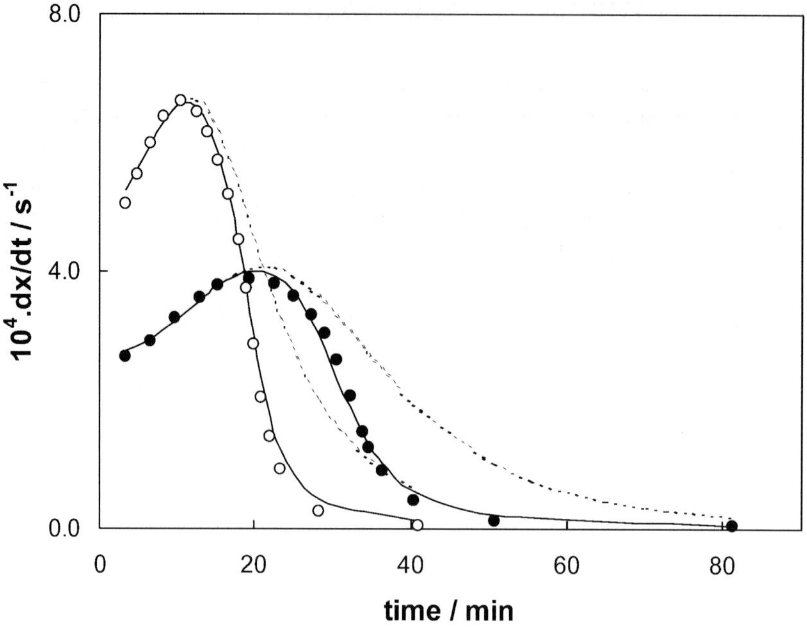

Figure 2.23. Reaction rate for the quasi-isothermal cure of an epoxy($f = 4$)-amine($f = 4$) system; measured rate at 100°C (o) and 90°C (•), optimised rate (——) and the simulated chemically controlled rate (- - - -).

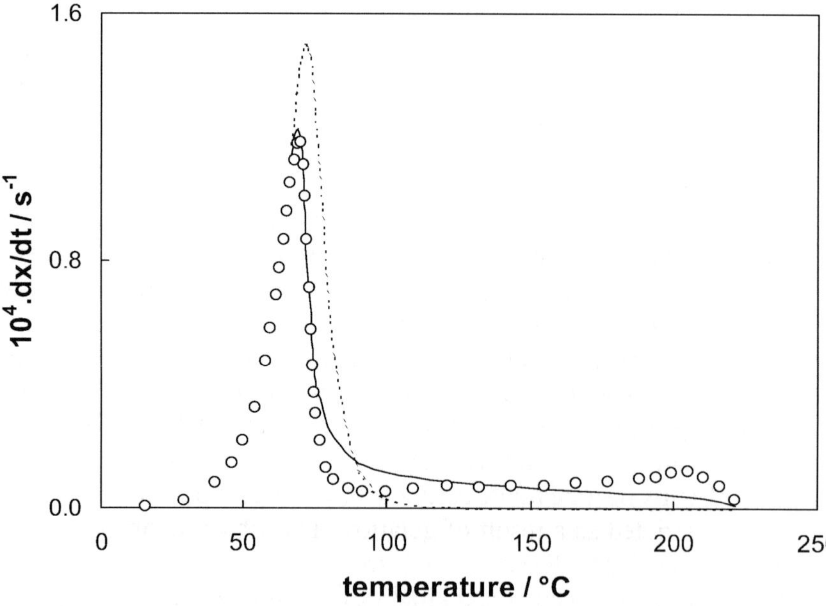

Figure 2.24. Reaction rate for the non-isothermal cure of an epoxy($f = 4$)–amine($f = 4$) system; measured rate at 0.2°C min^{-1} (o), optimised rate including diffusion control (——) and simulated chemically controlled rate (- - - -).

differences between the observed reaction rate and the calculated chemical reaction rate demonstrate the importance of the effect of mobility restrictions in these cases.

The epoxy–amine system is obviously more difficult to treat than the epoxy–anhydride system. The experimental conditions are more stringent (temperature interval of more than 200°C) and the empirical kinetic rate equation is probably not accurate enough. However, the proposed approach allows the model to be refined without too much difficulty.

6.3 REMARKS CONCERNING THE PROPOSED MODEL AND LITERATURE MODELS

The proposed approach has a few remarkable advantages.

(i) MTDSC is the only technique needed to model cure kinetics over the entire range of cure, as it quantitatively assesses conversion and reaction rate as well as the mobility factor. The latter allows predicting the decrease of reaction rate in the vitrification zone.

(ii) The optimisation procedure provides a unique set of kinetic, diffusion and '$T_g - x$' model parameters to describe all experimental profiles in a broad range of isothermal and non-isothermal conditions.

(iii) Experiments occurring in chemically controlled conditions only (for high cure temperatures or low conversion) can be simulated correctly using the general cure rate law. This proves that effects of chemical and diffusion control are well separated, even outside the range of conditions for which the model was optimised.

(iv) The mobility factor, derived from heat capacity, was proposed as a direct measurement (generated by MTDSC) of the diffusion factor for the epoxy resins studied. In other words, the mobility required for diffusion of reactive groups toward each other corresponds to the chain segment mobility displayed in the (normalised) heat capacity (see also section 6.4). In this way, the modelled MTDSC results confirm the overall diffusion control mechanism in curing systems obeying a step-growth (or a chain-growth without termination) polymerisation mechanism (see also section 2.4).

(v) The mobility needed for diffusion of reactive groups toward each other is not restricted as a result of gelation. This observation is characteristic of diffusion, which is not influenced by the existence of large-scale molecular structures until these constitute effective topological constraints. For the epoxy–anhydride system, gelation occurs at 25% conversion (determined by dynamic rheometry), corresponding to a T_g of −20°C.

(vi) More elaborated models concerning reaction kinetics, diffusion limitations and $T_g - x$ relationships can be introduced without much effort and models proposed in literature can be evaluated [39,47,117–119].

For the cure study of radical reactions, such as the unsaturated polyester resin–styrene copolymerisation, a different and more elaborated approach incorporating a molecular weight dependent diffusion coefficient, should be employed to take the Trommsdorff, or gel, effect into account.

(vii) An important advantage of the approach based on MTDSC measurements is the fact that the $T_g - x$ relationship does not need to be known in advance. The experimental determination of the $T_g - x$ relationship demands a lot of effort since it comprises $T_g - x$ data derived from a series of residual cure experiments.

6.4 REMARKS CONCERNING THE FREQUENCY DEPENDENCE OF HEAT CAPACITY DURING CURE [120]

The decrease in heat capacity due to vitrification, attributed to the loss of the co-operative mobility involved in the glass transition, occurs when the characteristic time-scale of the co-operative movements becomes longer than the modulation period (or the characteristic time-scale of the experiment) [121]. If the modulation frequency is increased, the corresponding T_g of the (curing) material will be higher for the same conversion (about 3–7°C per decade of frequency). Thus, at a higher frequency the reaction-induced vitrification will be observed at a lower degree of conversion. Note that the reaction rate itself depends on the average temperature and remains largely unaffected by the modulation frequency.

The observed coincidence of *DF* and *DF** (at 1/60 Hz) for the amine and the anhydride-cured epoxy systems indicates that the characteristic times for the (co-operative) chain segment mobility involved in the glass transition (region) measured at a frequency of 1/60 Hz, are comparable to those of the (reaction) rate-determining mobility upon transition to diffusion-controlled reaction conditions. Since only a limited frequency domain is available for MTDSC (about 0.01–0.05 Hz) and since a quantitative correspondence was found for 1/60 Hz, no attempt was made to fine-tune the frequency for each of the epoxy–hardener systems. However, for other systems with a different reaction mechanism and rate-controlling mobilities the frequency of correspondence might be totally different. Examples include unsaturated polyesters, for which the free radical polymerisation involves molecules with significantly different mobilities (e.g. monomers and growing polymer chains) [81] (see section 5.3.2), and the low-temperature synthesis of IPGs for which the reaction is unaffected by the main (glass) transition [41] (see section 5.3.4). Therefore, a study of the effect of frequency on the vitrification of resin systems related to their chemical structure and reaction mechanism might offer new insights concerning reaction kinetics, vitrification and diffusion control effects.

As mentioned above, one problem in studying the frequency dependence of the heat capacity (during cure) is the limited frequency interval available. *Light (heating) (temperature) modulated DSC (LMDSC)* [122–124] offers the advantage that the frequency range can be extended up to 1 Hz. In this technique, the temperature modulation is directly applied to the sample and the reference pans by irradiation with light beams of modulated intensity (keeping the furnace at a constant temperature). Similarly, frequencies up to 0.5 Hz were attained by passing an alternating gas flow over the sample and reference pans [125]. However, each frequency requires a new cure experiment with a fresh sample. A different approach, the *complex saw-tooth modulation method* [126], allows for multiple frequencies to be measured in a single experiment, thus eliminating effects of sample reproducibility. Alternative multi-frequency modulation methods for measuring the heat capacity with higher precision have been developed [127,128].

In an experimental exploration [120], all approaches show the expected decrease in vitrification time with increasing frequency (LMDSC results shown in Figure 2.25). However, even with the extended frequency range (ca. 2 decades for LMDSC) the simultaneously measured heat flow is not accurate enough to correlate a specific frequency with the reaction kinetics of the different epoxy thermosetting systems. Indeed, there is a considerable experimental error (see scatter in Figure 2.25) and the variation of the vitrification time and the conversion at vitrification associated with 2 decades in frequency is only about 15 min and 6%, respectively. The latter is below the accuracy of the (partial) reaction enthalpy determination (LMDSC).

A further extension of the frequency domain would be desirable in order to distinguish different kinds of (mobility-controlled) reaction mechanisms. Indeed, for example for the free radical polymerisation, the ratio of the (chemical) rate constants for termination and propagation is in the order of 10^7–10^{10}. Thus, in order to investigate how the co-operative mobility is evolving at the instants that either the termination reaction or the propagation reaction becomes diffusion-controlled, a frequency range of more than 7 orders of magnitude should be studied. Even for less demanding systems, this would require further developments in the measuring devices. However, two factors complicate a further extension of the frequency interval. If the heat flow of the (chemical) transformations needs to be simultaneously and quantitatively measured, the use of lower frequencies (below 0.01 Hz)

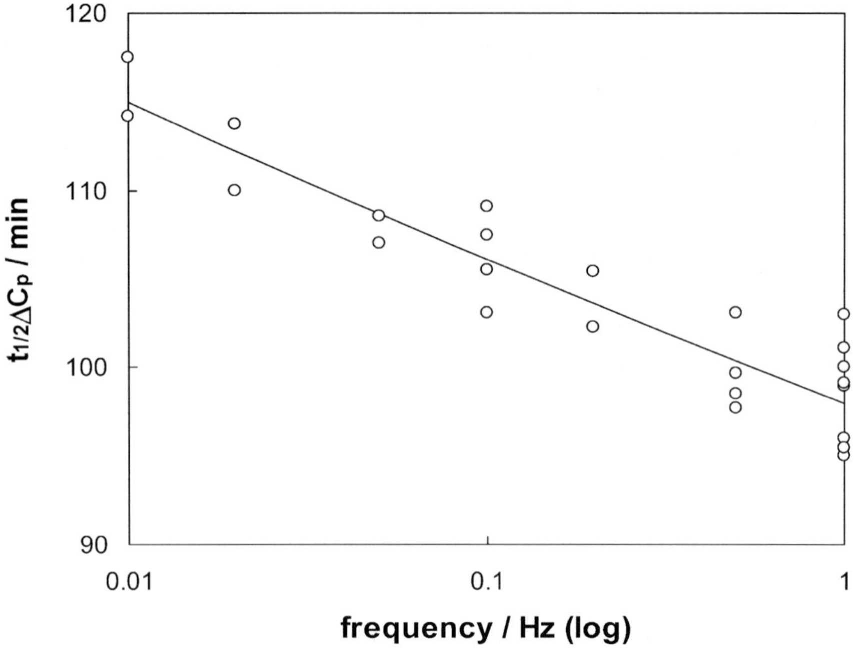

Figure 2.25. Vitrification times $t_{1/2}\Delta C_p$ as a function of the modulation frequency (from 0.01 to 1 Hz, logarithmic) for the quasi-isothermal cure of an epoxy–amine system at 80°C. Results from LMDSC [120].

is often in conflict with the desired invariance of the sample during at least one modulation. The use of higher frequencies is limited by thermal diffusivity effects, which become important for a film thickness of 100 μm (or more) at frequencies above 0.1 Hz. In this case, the amount of sample necessary for quantitative measurements restricts a further extension of the frequency window.

7 Glass Transition–Conversion Relationship

7.1 VALIDATION OF THE $T_g - x$ MODEL

The T_g of a series of partially reacted stoichiometric mixtures of the epoxy–anhydride system were determined independently. The standard procedure is to cool the sample at a relatively high rate to avoid further reaction after partial cure and to determine T_g and the residual reaction enthalpy in a subsequent heating experiment. Figure 2.26 displays the experimental T_g values as a function of cure conversion, x, determined after a partial cure at different cure temperatures. Within the experimental scatter, all the values

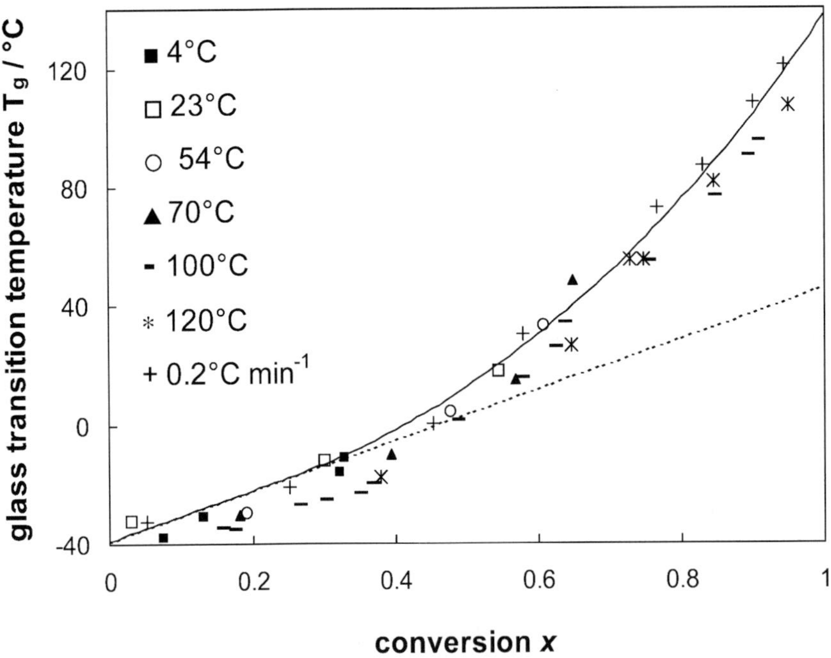

Figure 2.26. Glass transition temperature as a function of cure conversion for an epoxy–anhydride system. Experimental points were measured in a heating subsequent to a partial cure. $T_g - x$ and $T_{gu} - x$ curves [see Eq. (27)] are simulations by means of the optimised model for diffusion-controlled cure.

collapse to a single curve, indicating that T_g is a function of x only and not of the cure temperature. Note the sharper increase in T_g at high x, which is attributed to the increasing crosslink density.

As seen in Figure 2.26, the $T_g - x$ data obtained from the residual cure (MT)DSC experiments are well described by the optimised $T_g - x$ relationship [Eq. (27)] of the diffusion-controlled cure model (T_g is the solid line and T_{gu} is the dashed line). The departure of the experimental data from the continuous dashed line is due to the effect of increasing crosslinking beyond the gel point. The conversion at gelation correlates well with the value of 25% measured with dynamic rheometry (using the criterion $G' = G''$).

The approach to estimate the $T_g - x$ relationship directly from the diffusion-controlled cure is especially beneficial for thermosetting systems with a high functionality of the monomers. For these systems, it is impossible to obtain the $T_g - x$ data during the entire range of cure because of degradation reactions taking place at high temperatures. Besides, even when relatively high heating rates are applied, the high ultimate crosslink density and the high reactivity cause the cure reaction over an extended part to proceed under diffusion-controlled conditions. As a consequence, the residual

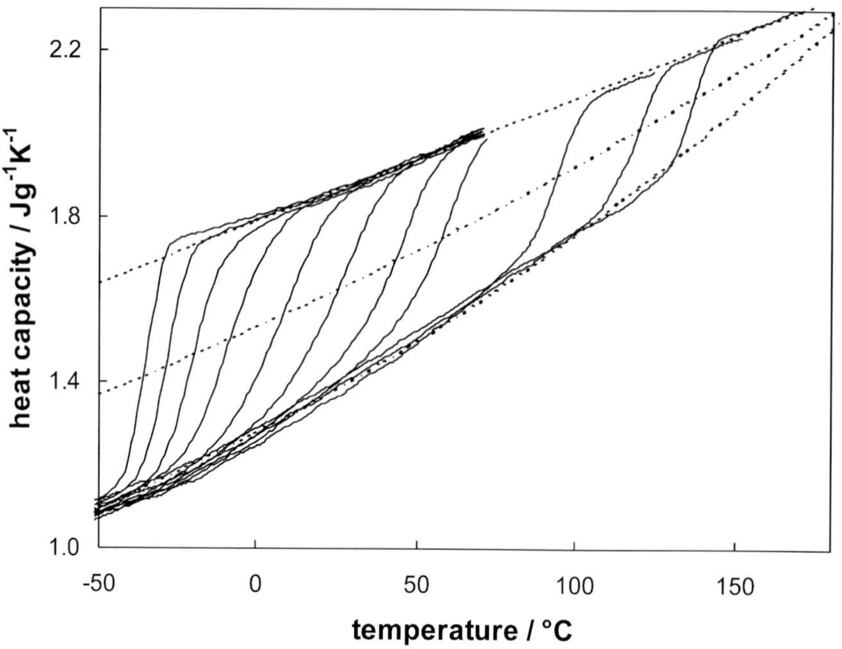

Figure 2.27. Progress of the glass transition in the heat capacity signal during cure of an epoxy–anhydride system. Reference lines are indicated (see text).

cure reaction after partial isothermal cure also proceeds under mobility restrictions and no previous transition from glass to rubber is observed (Figure 2.10, section 5.5).

7.2 ΔC_p AND ΔT_g AT T_g AS A FUNCTION OF CONVERSION

The power of MTDSC to evaluate ΔC_p and the width of the glass transition region (ΔT_g) during cure is illustrated with an experiment of the epoxy–anhydride system, consisting of several cooling and subsequent heating runs after partial cure segments. Figure 2.27 shows the heat capacity signal as a function of temperature (solid line) together with reference lines (dashed lines) representing the heat capacity of the glassy state, C_{pg}, of the liquid or rubbery state, C_{pl} and of a state in between where 50% of the material is in the glassy state. This last reference line is located at half the change between C_{pg} and C_{pl} and is used to calculate T_g. The difference between C_{pl} and C_{pg} gives the value of ΔC_p as a function of T_g, which can further be correlated to x by means of the $T_g - x$ model.

The $\Delta C_p - x$ data in Figure 2.28 demonstrate that ΔC_p remains almost constant up to 40% conversion, which is beyond the gel conversion, and then

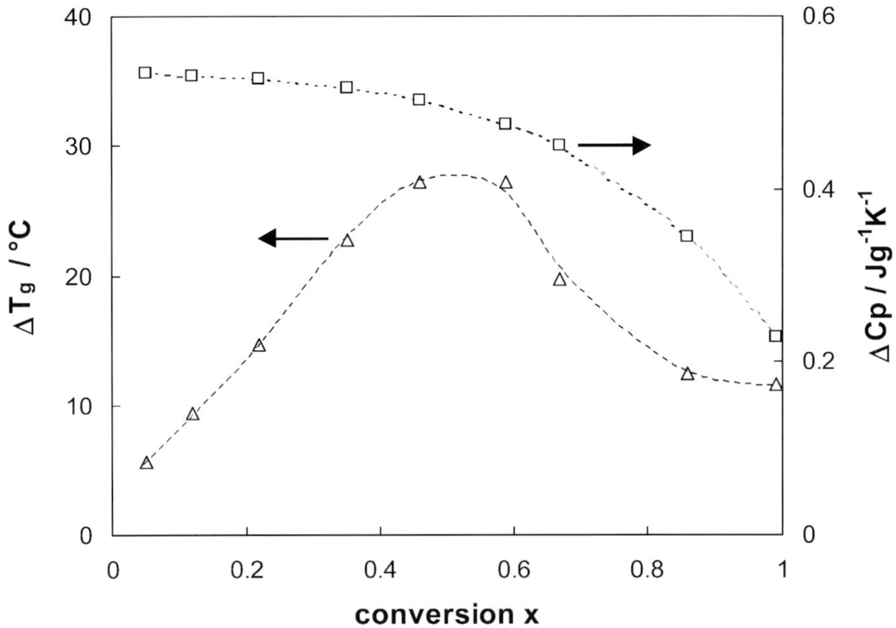

Figure 2.28. Heat capacity change at the glass transition (ΔC_p) and its width (ΔT_g) as a function of cure conversion for an epoxy–anhydride system.

steadily decreases as a result of further crosslinking. The results indicate that the $\Delta C_p - x$ relationship should not be considered as linear over the entire conversion range. A similar decrease of ΔC_p with x was already observed for another anhydride cured epoxy network [129].

ΔT_g, a measure of the width of the glass transition of the epoxy–anhydride system, is also displayed in Figure 2.28. The broadening of ΔT_g at the beginning of cure is attributed to the broadening of the molecular weight distribution, but because of further crosslinking ΔT_g decreases again. So, in the initial stages of cure for the epoxy–anhydride system, ΔC_p is almost constant whereas ΔT_g increases, but as the crosslinking reactions continue, ΔC_p and ΔT_g decrease. These trends cannot be generalised and depend on the reaction chemistry (mechanism) of the curing system.

8 TTT and CHT Cure Diagrams [12,13,20–29,130]

8.1 MTDSC CALCULATION PROCEDURE

The TTT diagram plots the cure temperature versus the time to reach different important events during 'isothermal' cure, e.g. gelation, vitrification, iso-T_g, iso-conversion, etc. (Figure 2.29).

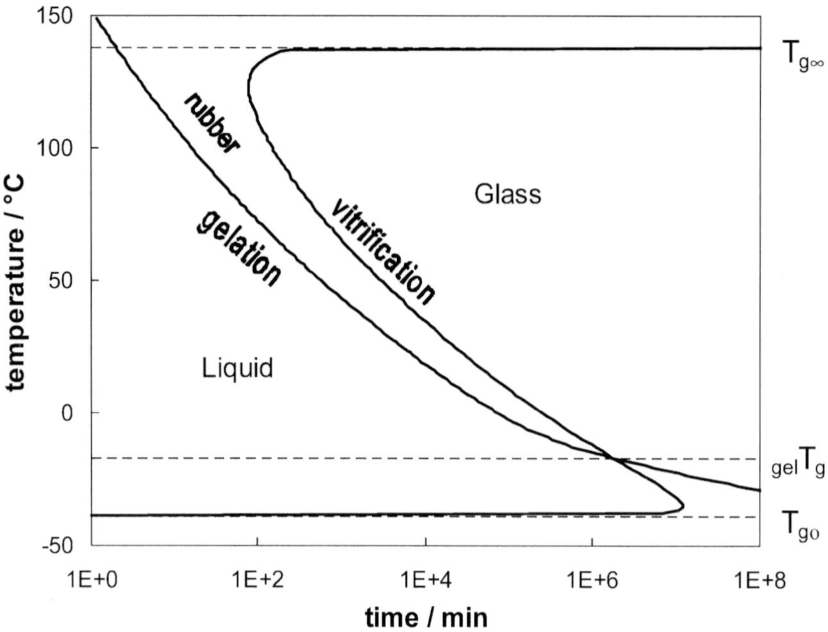

Figure 2.29. Schematic TTT cure diagram for a thermosetting system. The gelation and vitrification lines divide the liquid, rubbery and glassy state. T_{g0}, and $T_{g\infty}$ are the glass transition temperatures of the uncured and fully-cured resin, respectively. ${}_{\mathrm{gel}}T_{\mathrm{g}}$ is the temperature at which gelation coincides with vitrification.

It is divided into three major parts representing distinct rheological states encountered on isothermal cure. Before gelation, the material shows a viscous behaviour. In between gelation and vitrification, a sol/gel rubbery state exists. After vitrification, the material is in the glassy state. The gelation and vitrification curves intersect at ${}_{\mathrm{gel}}T_{\mathrm{g}}$, a critical temperature of simultaneous gelation and vitrification in between T_{g0} and $T_{g\infty}$. At a temperature above ${}_{\mathrm{gel}}T_{\mathrm{g}}$, gelation occurs before vitrification, while below ${}_{\mathrm{gel}}T_{\mathrm{g}}$ vitrification occurs first. The vitrification curve is generally S-shaped, passing through a maximum and a minimum in vitrification time when the temperature is raised. The gelation line is to be considered as an iso-conversion line [11] and also as an iso-T_g line if a unique $T_g - x$ relationship exists.

Similar to the TTT cure diagram, the CHT diagram [24,25,28] displays the temperature–time combinations required to reach the same events during non-isothermal cure at a series of constant heating rates. Both diagrams are very useful toward processing technology, since in the vicinity of gelation the material is loosing flow characteristics, and after vitrification the cure reactions are greatly reduced in rate.

The latter effect is frequently requested in practical conditions to lower the exothermic effect released during the cure of large parts. Besides, TTT and CHT cure diagrams permit time/temperature cure paths to be designed in order to improve processing and final material properties.

The experimental determination of the evolution of the mobility factor by MTDSC during quasi-isothermal cure at different reaction temperatures, or during non-isothermal cure at several heating rates permits the construction of lines of equal mobility restrictions on these TTT or CHT diagrams, including the vitrification line. Nevertheless, a direct experimental determination of the TTT and the CHT diagrams necessitates multiple independent experiments, using different techniques, and is very time-consuming. An alternative approach is to compute the TTT or the CHT cure diagram according to a modelling procedure essentially based on two relations: the conversion–time–temperature relation ($x - t - T$) or cure rate law (including diffusion control) and the glass transition temperature–conversion relation ($T_g - x$). It should be noted that no direct method is available to detect gelation by MTDSC, so information obtained using other techniques is needed (see section 3.1). Once all relationships are known, the times to gelation or vitrification are obtained via numerical integration of the rate equation until at a particular reaction time the specific condition is fulfilled ($x = x_{gel}$ or $T_g = T$, respectively). Together with the two key relations, additional models can be incorporated (e.g. for drawing iso-viscosity lines).

Based on the optimised parameters for the cure rate law of the bifunctional epoxy–anhydride and the tetrafunctional epoxy–diamine (see previous section), the TTT and the CHT cure diagrams for both systems can be calculated. Figures 2.30 and 2.31 represent the TTT diagrams for both systems.

The CHT diagram for the epoxy–amine system is given in Figure 2.32. The experimental points (symbols) in Figures 2.30–2.32 are data obtained with MTDSC and dynamic rheometry. The thick lines are the gelation lines, the vitrification contour (similar to the line of $DF^*_{0.5}$) and the isodiffusion contours $DF^*_{0.9}$ and $DF^*_{0.1}$. The thin lines display the T_g evolution as a function of time for selected isothermal (TTT) or non-isothermal cure paths (CHT).

For the TTT diagrams in Figures 2.30 and 2.31, and for the experimental results mentioned in Table 2.1, the following observations can be made.

1) Vitrification during isothermal cure, associated with $DF^*_{0.5}$, is attained at longer reaction times as the isothermal cure temperature is lowered.
2) The extent of conversion at $DF^*_{0.1}$, $DF^*_{0.5}$ or $DF^*_{0.9}$ decreases as the isothermal reaction temperature is lowered, which can be explained in terms of the one-to-one relationship between T_g and x. Close to complete conversion, only small changes in conversion result in large changes in T_g due to the influence of crosslink density. The conversion attained at

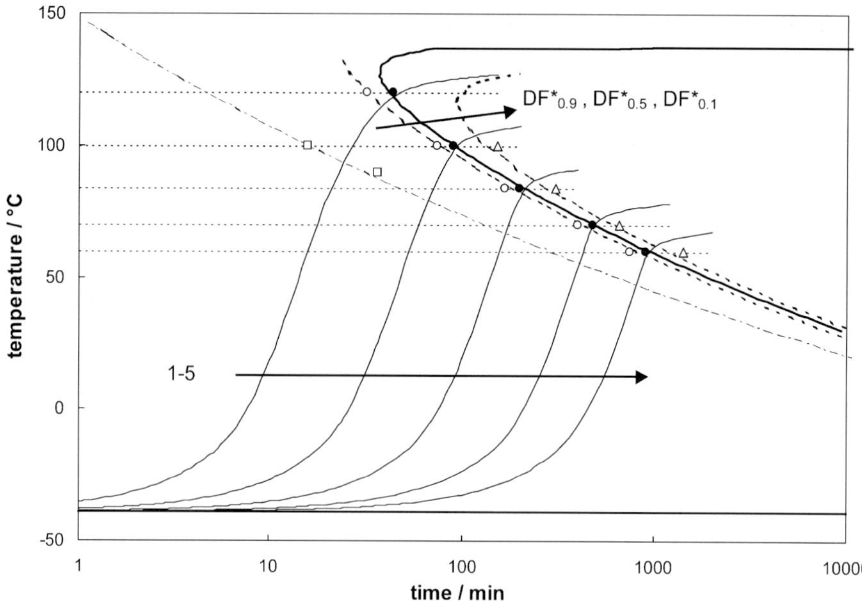

Figure 2.30. TTT cure diagram for an epoxy–anhydride thermosetting system. MTDSC results: (o) $DF^{*}_{0.9}$; (•) $DF^{*}_{0.5}$; (Δ) $DF^{*}_{0.1}$. Data obtained with dynamic rheometry: (□); Calculated profiles: vitrification curve (——); gelation line (– . –); isodiffusion curves (– –); T_g evolution (——) and their corresponding temperature profiles (- - - -) for the isothermal cure at 120°C (1), 100°C (2), 85°C (3), 70°C (4) and 60°C (5).

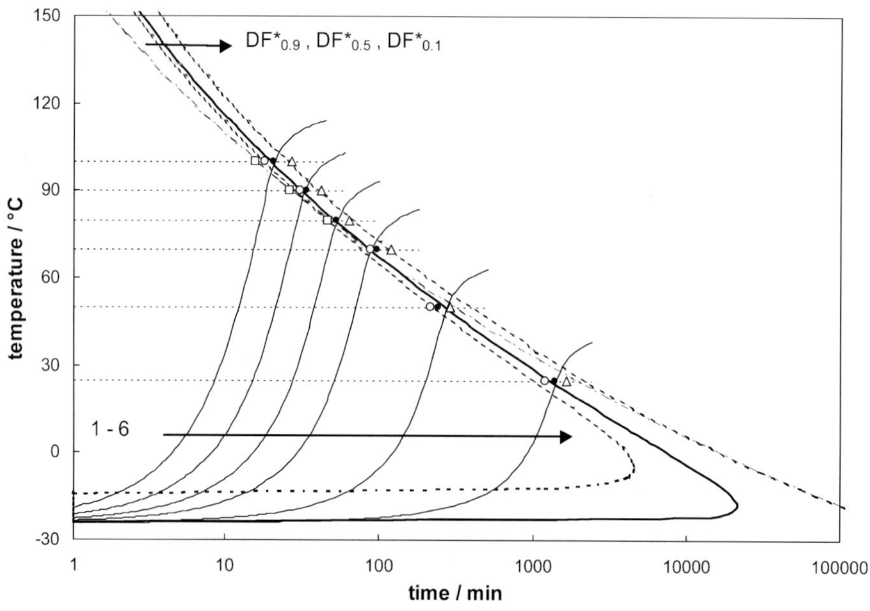

Figure 2.31. TTT cure diagram for an epoxy($f = 4$)-amine($f = 4$) thermosetting system. MTDSC results: (o) $DF^{*}_{0.9}$; (•) $DF^{*}_{0.5}$; (Δ) $DF^{*}_{0.1}$. Data obtained with dynamic rheometry: (□); Calculated profiles: vitrification curve (——); gelation line (– . –); isodiffusion curves (– –); T_g evolution (——) and their corresponding temperature profiles (- - - -) for the isothermal cure at 100°C (1), 90°C (2), 80°C (3), 70°C (4), 50°C (5) and 25°C (6).

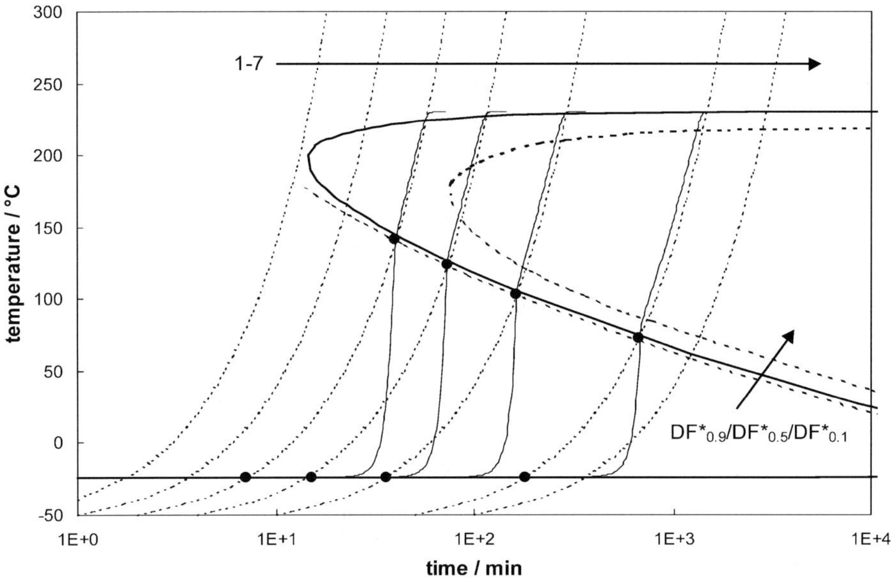

Figure 2.32. CHT cure diagram for an epoxy($f = 4$)-amine($f = 4$) thermosetting system. MTDSC results: (•) $DF^*_{0.5}$. Calculated profiles: vitrification curve (——); isodiffusion curves (– –); T_g evolution (——) and their corresponding temperature profiles (- - - -) for the non-isothermal cure at 20°C min^{-1} (1), 10°C min^{-1} (2), 5°C min^{-1} (3), 2.5°C min^{-1} (4), 1°C min^{-1} (5), 0.2°C min^{-1} (6) and 0.1°C min^{-1} (7).

vitrification for the epoxy–amine is approximately 20% lower than for the epoxy–anhydride.

It is worth noting that the experiments given in the TTT diagrams for the two epoxy systems cover different characteristic regions of the diagram. For the epoxy–anhydride, the experimental region studied was relatively closer to $T_{g\infty}$, well above the ${}_{gel}T_g$. Thus, gelation occurs first. Vitrification occurs at an advanced conversion (closer to full cure). For the epoxy–amine, the cure temperatures are close to the ${}_{gel}T_g$. Gelation and vitrification occur close to each other, with gelation occurring after vitrification for temperatures of 50°C and lower.

3) The reaction rate at $DF^*_{0.5}$ is greatly reduced compared to the reaction rate at $DF^*_{0.9}$, although a relatively small change in conversion is observed. This is an indication for the effect of vitrification on cure kinetics.

For the CHT diagram (Figure 2.32), for heating scans starting from below T_{g0} to above $T_{g\infty}$, devitrification initially occurs when T_{cure} first passes through T_{g0} of the reaction mixture. Vitrification corresponds to T_g becoming equal to the increasing T_{cure}. After vitrification, the reaction proceeds in the glassy state with T_g rising parallel to T_{cure}. Devitrification occurs when the rate of T_g-rise slows down (compared to the constant heating rate) due

Table 2.1. MTDSC and dynamic rheometry results for the epoxy systems studied. The symbols are defined in the text.

MTDSC: quasi-isothermal cure of the epoxy–anhydrid system

	$DF^* = 0.9$			$DF^* = 0.5$ (vitrification)				$DF^* = 0.1$		
T_{iso} (°C)	t (min)	x	dx/dt (10^{-5} s^{-1})	t (min)	x	dx/dt (10^{-5} s^{-1})	r	t (min)	x	dx/dt (10^{-5} s^{-1})
120	32	0.87	5.6	44	0.89	2.2	1:16			
100	73	0.85	4.5	89	0.88	2.0	1:8	141	0.91	0.67
84	165	0.78	2.8	196	0.81	1.4	1:5	303	0.85	0.24
70	400	0.75	1.0	475	0.79	0.47	1:6	655	0.82	0.13
60	742			897				1413		

MTDSC: quasi-isothermal cure of the tetrafunctional epoxy–diamine system

	$DF^* = 0.9$			$DF^* = 0.5$ (vitrification)				$DF^* = 0.1$		
T_{iso} (°C)	t (min)	X	dx/dt (10^{-5} s^{-1})	t (min)	x	dx/dt (10^{-5} s^{-1})	r	t (min)	x	dx/dt (10^{-5} s^{-1})
100	18	0.62	42	21	0.67	23	1:2.4	27	0.71	3.5
90	30	0.61	27	34	0.65	16	1:2.0	42	0.69	3.4
80	48	0.53	16	53	0.56	10	1:1.8	64	0.60	2.3
70	88	0.50	10	97	0.55	6.6	1:1.4	119	0.59	1.6
50	216	0.45	3.8	239	0.50	2.8	1:1.2	286	0.55	0.95
25	1183			1370				1650		

MTDSC: non-isothermal cure of the tetrafunctional epoxy–diamine system

Heating	$DF^* = 0.9$			$DF^* = 0.5$ (vitrification)		
rate(°C min^{-1})	T(°C)	x	dx/dt 10^{-5} s^{-1})	T(°C)	x	dx/dt 10^{-5} s^{-1})
2.5	119.7	0.72	79	123.9	0.78	33
1	100.7	0.66	47	103.5	0.73	22
0.2	71.4	0.51	11	72.9	0.56	8.0

Dynamic rheometry: isothermal cure of epoxy systems

Epoxy–anhydride			**Epoxy–amine**		
T_{iso} (°C)	$t_{G'=G''}$ (min)	x	T_{iso} (°C)	$t_{G'=G''}$ (min)	x
100	15.8	0.24	100	15.7	0.56
90	36.3	0.28	90	25.9	0.54
			80	46.6	0.54

to the depletion of reactive groups. Thus, this devitrification is not related to thermal degradation.

There are two critical heating rates worth mentioning: the lower critical heating rate and the upper critical heating rate. When material is heated more rapidly than the upper critical heating rate, cure proceeds entirely in the kinetically controlled regime because the material does not enter the glassy state. Curing below T_g, on the contrary, is accomplished by using the lower critical heating rate. Maximum density and minimum internal stresses are then achieved because no change of state occurs and because the rate of physical ageing is maximum just below T_g. Note that in order to retain this maximum density it is important to end the heating below $T_{g\infty}$, so that T_g remains above T_{cure} at the end of the cure.

The following additional observations can be made.

1) As the heating rate is decreased under non-isothermal conditions, vitrification takes place at lower reaction temperatures and lower conversions. In contrast, devitrification occurs at higher reaction temperatures (see section 5.4.3).
2) The amount of vitrified material is raised from 10% to 50% over less than 2°C in the case of the epoxy–amine cured at 0.2°C min^{-1}; a more than 4°C temperature rise is necessary for the epoxy–anhydride cured at the same heating rate.
3) For the amine system under non-isothermal cure at 0.2°C min^{-1}, $T_{g\infty}$ of 245°C causes devitrification to occur at a temperature more than 150°C above vitrification. The importance of this extended mobility-restricted cure on the final material's properties should be emphasised. For this tetrafunctional epoxy-diamine system, an increase in T_g of ca. 170°C, corresponding with a residual cure of ca. 44% and a reaction enthalpy of more than 230 J g^{-1}, is caused by diffusion-controlled reactions and drastically influences the final network structure (crosslink density).
4) Heating rates of about 20°C min^{-1} are not high enough to prevent diffusion-controlled curing of the epoxy–amine system, whereas a heating rate above 0.4°C min^{-1} already fulfils this requirement for the epoxy–anhydride system. The whole cure process proceeds in the glassy state when a heating rate below 210^{-6} °C min^{-1} is applied for the epoxy–anhydride system, compared to 210^{-4} °C min^{-1} for the epoxy–amine system.

To evaluate and to compare the extension of the region of vitrification or restricted mobility, the mobility factor is plotted as a function of the reduced time, equal to the ratio $t/t_{DF^{*}0.5}$, for the quasi-isothermal cure at several temperatures for the anhydride and amine–epoxy system (Figures 2.33a and b). As seen from Figure 2.33, all the experimental profiles

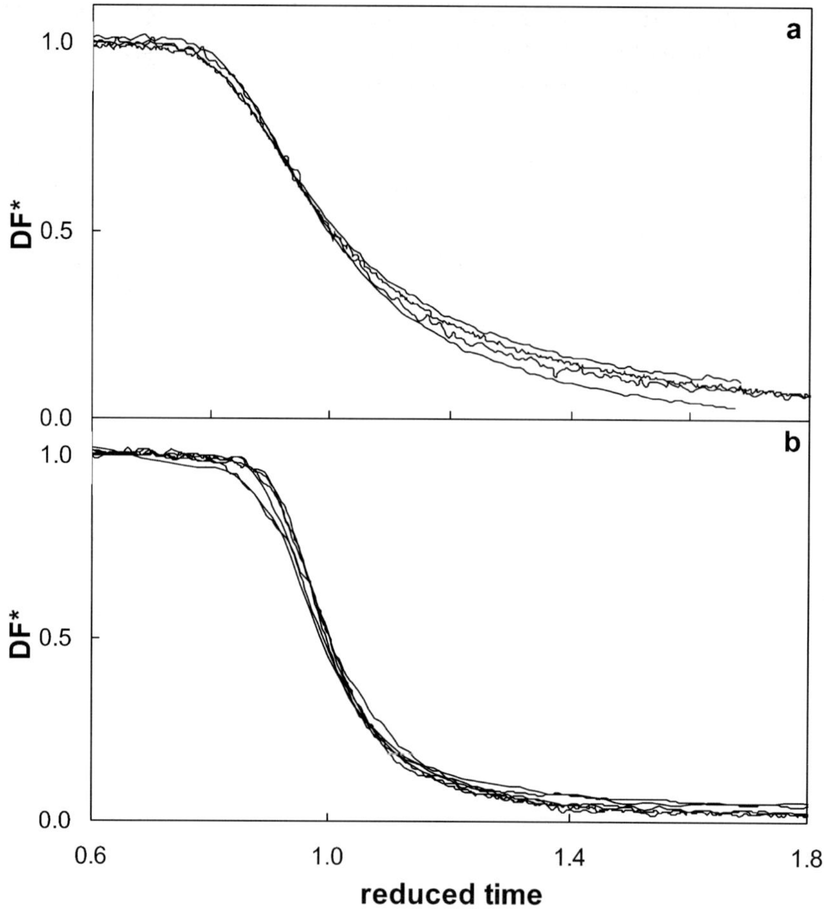

Figure 2.33. DF^* as a function of the reduced time ($t/t_{DF^*0.5}$): (a) for the isothermal cure at 100, 85, 70 and 60°C for an epoxy–anhydride system; (b) for the isothermal cure at 100, 90, 80, 70, 50 and 25°C for an epoxy($f = 4$)–amine($f = 4$) system.

collapse to one single curve (master curve). For a particular thermosetting system, the extension of this region varies little over the temperature range studied. Comparing the two systems, a more expanded vitrification region is observed for the epoxy–anhydride system than for the epoxy–amine system, especially toward the end of the vitrification process. This steeper vitrification behaviour for the epoxy–amine system is probably due to vitrification at a relatively low reaction conversion (because the epoxy–amine is cured closer to ${}_{\mathrm{gel}}T_{\mathrm{g}}$, well below $T_{\mathrm{g}\infty}$) and to more rapid cure kinetics (which is reflected in a vitrification time that is a fivefold lower at 70°C compared to the epoxy–anhydride system).

8.2 INFLUENCE OF CHEMICAL STRUCTURE ON THE (DE)VITRIFICATION BEHAVIOUR OF THE THERMOSETTING SYSTEMS

In the course of the previous sections, several striking differences between the two epoxy systems were noted concerning the conversion at vitrification, the diffusion-controlled region in non-isothermal experiments, the critical heating rates, etc. These differences can be related to the chemical structure of the monomers, which influence the reactivity and the growing network structure.

Vitrification (e.g. at 70°C) takes place at a much lower conversion for the epoxy–amine system. This can be attributed to a different evolution of T_g with x. For the amine–epoxy system, both resin and hardener are tetrafunctional, whereas a bifunctional epoxy is used with the anhydride. Therefore, $T_{g\infty}$ for the amine–epoxy is higher due to a higher final crosslink density, and the increasing glass transition temperature while curing this system will reach the isothermal reaction temperature at a lower conversion than for the anhydride system. The higher conversion at vitrification, $x_{DF^{*}0.5}$, for the anhydride system causes the onset of mobility limitations to be seen near the end of the reaction exotherm, whereas for the amine system the vitrification process starts near the maximum reaction rate. This influences the shape of the vitrification transition, as discussed in the previous paragraph (Figure 2.33).

To evaluate, in more detail, the effect of the chemical structure of the reactants upon isothermal curing, the rate of conversion at vitrification $(\mathrm{d}x/\mathrm{d}t)_{DF^{*}0.5}$ can be compared to the average rate before vitrification, $\langle \mathrm{d}x/\mathrm{d}t\rangle$, which equals $x_{DF^{*}0.5}/t_{DF^{*}0.5}$. It is necessary to work with ratios or relative rates r (Table 2.1) because the amine–epoxy system is much more reactive than the anhydride–epoxy system. For the latter system, the ratio r of $(\mathrm{d}x/\mathrm{d}t)_{DF^{*}0.5}$ to $\langle \mathrm{d}x/\mathrm{d}t\rangle$ is lower than 1:5 over the temperature range considered, which is much smaller than the lowest ratio of 1:2.4 for the epoxy–amine system. The ratio r also decreases with increasing cure temperature.

The variations of this ratio correlate to the differences in final isothermal cure state. Since the rate of conversion at vitrification is non-zero, conversion and T_g further increase in the (partially) glassy state with a rate dependent on the relative rate at vitrification. A relatively lower $(\mathrm{d}x/\mathrm{d}t)_{DF^{*}0.5}$ or ratio r results in a smaller increase in conversion and T_g after vitrification. For example, T_g at the end of the isothermal cure at 70°C for the epoxy–anhydride system amounts to 85°C, whereas a value of 103°C is determined for the amine system under similar isothermal cure conditions.

These results seem to indicate that the epoxy–amine reaction is less hindered by the occurrence of vitrification than the epoxy–anhydride reaction. This can be corroborated from a more chemical point of view. For a step-growth polymerisation reaction to take place, the two reacting entities have to move toward each other by translational diffusion and segmental diffusion and rotation. Besides other parameters, the mobility needed for reaction is determined by the distances separating the functional groups. At vitrification, the concentration of reactive units in the epoxy–amine is approximately 4 times higher compared to the epoxy–anhydride, due to a higher initial concentration and a lower conversion at vitrification. Postulating an equal distribution of non-reacted units, the mean distance to the nearest 'reaction partner' at vitrification, termed the reaction distance, is 8Å for the epoxy–amine and 13Å for the epoxy–anhydride. In the glass transition region, a mobility of about 30–50 chain segments is available, depending on the specific chemical structure, allowing for reaction while the material is vitrifying. The reaction can continue in the glassy state, reaching a T_g higher than T_{cure}, until the increasing crosslink density restricts the mobility of shorter chain segments (or until reaction distances become bigger than mobility allows for). Because of the smaller reaction distances, the epoxy–amine requires less mobility to react, and, since crosslink density is coupled to T_g, a relatively higher T_g can be reached before the mobility is so strongly restricted that the reaction is halted.

Because of the structures of the tetrafunctional epoxy and amine, small cycles are formed [131]. These short distance intramolecular cyclisation reactions probably continue while the material is vitrifying. This will decrease the number of chain ends, without decreasing the absolute number of mechanically effective crosslinks already formed. Thus, it will increase the stiffness of the macromolecules (on a molecular scale) and contribute to a further increase of conversion and T_g. These points again correlate well with the higher increase in conversion and T_g of the amine-cured epoxy in the glassy state.

9 Conclusions and Future Developments

MTDSC is a powerful thermal analysis technique to characterise important events along the reaction path of reacting polymer systems. An empirical modelling of both heat flow and heat capacity MTDSC signals in quasi-isothermal and/or non-isothermal reaction conditions enables the quantification of the influence of vitrification and devitrification on the reaction kinetics. In this way, the cure kinetics can be determined more accurately than with conventional DSC, even up to high overall reaction conversion.

The information available in the heat capacity evolution is a key factor for the correct interpretation of the heat flow signal. The results indicate that the heat flow phase angle contains interesting information regarding the rheological state of the reacting material and especially about the occurrence of relaxation phenomena.

A mobility factor based on heat capacity, DF^*, was proposed in our work. The points for which DF^* equals 0.5 can be used to quantify the times and temperatures of vitrification and devitrification (for the organic systems studied). Moreover, the DF^* curve gives information on the degree of vitrification while the reaction occurs in mobility-restricted conditions. If an isothermal cure experiment is performed close to the glass transition of the fully-cured resin, partial vitrification occurs and the fully glassy state will never be reached at that temperature.

For the epoxy resins studied, the mobility factor based on heat capacity coincides very well with the diffusion factor, calculated from the non-reversing heat flow via chemical kinetics modelling, and describing the effects of diffusion control on the rate of conversion of the cure reaction. Although the two resins behave quite differently, this coincidence between the mobility factor and diffusion factor is valid for both systems. Therefore, the mobility factor can be used for a quantitative description of their altered rate of conversion in the (partially) vitrified state: for the decrease in rate during vitrification, the increase in rate during devitrification and the diffusion-controlled rate in the (partially) vitrified region in between both processes.

For a free radical polymerisation system, an unsaturated polyester resin, an auto-acceleration was observed close to the onset of vitrification. To model the curing kinetics for these systems, including the mobility-controlled regions, a specific diffusion control model will need to be incorporated in a mechanistic reaction model. The heat capacity and the mobility factor can still give information about how vitrification is occurring, and how it is related to the auto-acceleration effect.

The combined information of heat capacity, heat flow and heat flow phase also provides an excellent tool for more detailed mechanistic studies of reacting polymer systems. The change in heat capacity due to chemical reactions, measured as a function of the conversion and/or the composition of the initial reaction mixture, gives valuable constraints in determining the rate constants of important reactive species involved in the mechanism. The effects of the type of reaction mechanism, e.g. step growth versus radical chain growth, addition versus condensation or organic versus inorganic, can be investigated in a systematic way.

MTDSC allows the real-time monitoring of phase separation induced by cure in epoxy–amine reactive mixtures modified with an amorphous high-T_g engineering thermoplastic, such as polyethersulphone. Reaction-induced phase separation was observed by a vitrification of a segregating PES-rich phase and of the reacting epoxy–amine rich matrix. If the cure is performed sufficiently below the full cure glass transition of the unmodified epoxy–amine, both the vitrification of the dispersed PES-rich phase and vitrification of the epoxy-rich matrix can be observed. The relaxation peaks in the heat flow phase signal prove to be very valuable to monitor these effects, especially in a thermosetting system. Non-isothermal (post-cure) MTDSC experiments contain information on the in situ formed phases and on the effect of temperature on phase separation, giving support to the construction of a phase diagram.

It can be concluded that MTDSC, in combination with other established techniques for studying multi-phase materials, is an excellent new analytical tool for the real-time monitoring of morphology development in complicated reacting systems. The excellent control of temperature during cure enables reproducible and meaningful results on kinetics of phase separation and cure.

The MTDSC method will further be explored and extended to topics, such as:

(i) The effect of additives (thermoplastic modifiers, fibres in composites) on the kinetics of imposed cure schedules via the simultaneous information of the non-reversing heat flow.
(ii) The balance between the kinetics of cure, reaction-induced phase separation and crystallisation of reacting polymer systems in the presence of crystallisable thermoplastic modifiers.
(iii) The formation of liquid crystalline thermosets is another one. In these cases, the influence of the dispersed phase on the reaction kinetics of the matrix can be investigated with an improved sensitivity. This is of extreme importance when the additive (or reinforcing fibre) itself is able to react with the matrix components and interphase regions are developing.
(iv) A very interesting extension is the study of interpenetrating networks. In this case, not only the final properties of the network structure can be evaluated using heat capacity and derivative signals, but also the influence of the in situ production.

References

[1] M. Reading, A. Luget and R. Wilson, Thermochim. Acta, 238 (1994) 295.
[2] B. Wunderlich, Y. Jin and A. Boller, Thermochim. Acta, 238 (1994) 277.
[3] M. Reading, Trends Polym. Sci., 1 (1993) 248.

[4] P.S. Gill, S.R. Sauerbrunn and M. Reading, J. Thermal Anal., 40 (1993) 931.
[5] M. Reading, D. Elliot and V.L. Hill, J. Thermal Anal., 40 (1993) 949.
[6] C. Schick and G.W.H. Hohne (Eds.), Special issue on temperature modulated calorimetry, Thermochimica Acta, 304/305 (1997).
[7] C. Schick and G.W.H. Hohne (Eds.), Special issue on temperature modulated calorimetry, Thermochimica Acta, 330(1–2) (1999).
[8] J.D. Menczel and L. Judovits (Eds.), Special issue on temperature-modulated differential scanning calorimetry, J. Thermal Anal., 54 (1998).
[9] A.A. Lacey, C. Nikolopoulos and M. Reading, J. Thermal. Anal., 50 (1997) 279.
[10] J.E.K. Schawe, Thermochim. Acta, 261 (1995) 183.
[11] P.J. Flory, Principles of Polymer Chemistry, Cornell University Press, Ithaca, New York (1953).
[12] J.K. Gillham, In Encyclopedia of Polymer Science and Engineering, Vol. 4, Herman F. Mark, Norbert Bikales, Charles G. Overberger, George Menges, and Jacqueline I. Kroschwitz Eds., Wiley-Interscience, New York (1986), p. 519.
[13] G. Wisanrakkit and J.K. Gillham, J. Coating Technol., 62 (1990) 35.
[14] S. Montserrat, J. Appl. Polym. Sci., 44 (1992) 545.
[15] S. Montserrat, J. Polym. Sci. Polym. Phys., 32 (1994) 509.
[16] G. Wisanrakkit and J.K. Gillham, J. Appl. Polym. Sci., 42 (1991) 2465.
[17] X. Wang and J.K. Gillham, J. Appl. Polym. Sci., 47 (1993) 447.
[18] M.B.M. Mangion, M. Wang and G.P. Johari, J. Polym. Sci. Polym. Phys., 30 (1992) 445.
[19] M.B.M. Mangion and G.P. Johari, J. Polym. Sci. Polym. Phys., 29 (1991) 437.
[20] J.B. Enns and J.K. Gillham, J. Appl. Polym. Sci., 28 (1983) 2567.
[21] J.K. Gillham, Polym. Eng. Sci., 26 (1986) 1429.
[22] S.L. Simon and J.K. Gillham, J. Appl. Polym. Sci., 53 (1994) 709.
[23] M.T. Aronhime and J.K. Gillham, J. Coating Technol., 56 (1984) 35.
[24] G. Wisanrakkit and J.K. Gillham, J. Appl. Polym. Sci., 42 (1991) 2453.
[25] J.K. Gillham and J.B. Enns, Trends Polym. Sci., 2 (1994) 406.
[26] M.T. Aronhime and J.K. Gillham, Adv. Polym. Sci., 78 (1986) 83.
[27] X. Wang and J.K. Gillham, J. Appl. Polym. Sci., 47 (1993) 425.
[28] G. Wisanrakkit, J.K. Gillham and J.B. Enns, J. Appl. Polym. Sci., 41 (1990) 1895.
[29] K.P. Pang and J.K. Gillham, J. Appl. Polym. Sci., 39 (1990) 909.
[30] B. A. Rozenberg, Adv. Polym. Sci., 72 (1985) 113.
[31] L. Matejka, J. Lövy, S. Pokorny, K. Bonchal and K. Dusek, J. Polym. Sci. Polym. Chem. Ed., 21 (1983) 2873.
[32] K.C. Cole, J.J. Hechler and D. Noël, Macromolecules, 24 (1991) 3098.
[33] N. Bouillon, J.P. Pascault and L. Tighzert, J. Appl. Polym. Sci., 38 (1989) 2103.
[34] B. Steinmann, J. Appl. Polym. Sci., 39 (1990) 2005.
[35] B. Van Mele and E. Verdonck, J. Adhes., 57 (1996) 245.
[36] U. Khanna and M. Chanda, J. Appl. Polym. Sci., 50 (1993) 1635.
[37] V. Trappe, B. Burchard and B. Steinmann, Macromolecules, 24 (1991) 4738.
[38] M.R. Kamal, Polym. Eng. Sci., 14 (1974) 231.
[39] K. Dusek and I. Havlisek, Prog. Org. Coatings, 22 (1993) 145.
[40] S. Lunak, J. Vladyka and K. Dusek, Polymer, 19 (1978) 931.
[41] H. Rahier, B. Van Mele and J. Wastiels, J. Mater. Sci., 31 (1996) 80.
[42] G. Van Assche, A. Van Hemelrijck, H. Rahier and B. Van Mele, Thermochim. Acta, 268 (1995) 121.
[43] J.P. Pascault and R.J.J. Williams, J. Polym. Sci. Polym. Phys., 28 (1990) 85.
[44] A. Hale, C.W. Macosko and H.E. Bair, Macromolecules, 24 (1991) 2610.

[45] H. Stutz, K.-H. Illers and J. Mertes, J. Polym. Sci. Polym. Phys., 28 (1990) 1483.
[46] T.G. Fox and S. Loshaek, J. Polym. Sci., 25 (1955) 371.
[47] G. Wisanrakkit and J.K. Gillham, J. Appl. Polym. Sci., 41 (1990) 2885.
[48] T.G. Fox and P.J. Flory, J. Appl. Phys., 21 (1950) 581; J. Polym. Sci., 14 (1954) 315.
[49] C.Y.M. Tung and P.J. Dynes, J. Appl. Polym. Sci., 27 (1982) 569.
[50] H.H. Winter, Polym. Eng. Sci., 27 (1987) 1698.
[51] G.P. Johari, In Chemistry and Technology of Epoxy Resins (Chapter 6), B. Ellis, Ed., Blackie Academic & Professional, London (1994).
[52] A.Ya Malkin and S.G. Kulichikhin, Adv. Polym. Sci., 1001 (1992) 217.
[53] M. Reading and P.J. Haines, In Thermal Methods of Analysis; Principles, Applications and Problems (Chapter 4), P.J. Haines, Ed., Blackie Academic & Professional, London (1995).
[54] J.K. Gillham, In Structural Adhesives: Developments in Resins and Primers, A.J. Kinloch, Ed. (1986) p. 1.
[55] R.B. Prime, In Thermal Characterisation of Polymeric Materials (Chapter 5), E.A. Turi, Ed., Academic Press, New York (1981) p. 435.
[56] K.A. Nass and J.C. Seferis, Polym. Eng. Sci., 29 (1989) 315.
[57] S.D. Senturia and N.F. Sheppard, Jr., Adv. Polym. Sci., 80 (1986) 1.
[58] R.J. Roe, In Encyclopedia of Polymer Science and Engineering, Vol. 7, Herman F. Mark, Norbert Bikales, Charles G. Overberger, George Menges and Jacqueline I. Kroschwitz, Eds., Wiley-Interscience, New York (1986), p. 531.
[59] B. Ellis, In Chemistry and Technology of Epoxy Resins (Chapter 3), B. Ellis, Ed., Blackie Academic & Professional, Glasgow (1994), p. 72.
[60] A.C. Grillet, J. Galy, J.P. Pascault and I. Bardin, Polymer, 30 (1989) 2094.
[61] E. Mertzel and J.L. Koenig, Adv. Polym. Sci., 75 (1986) 73.
[62] J.M. Barton, Adv. Polym. Sci., 72 (1985) 111.
[63] R.A. Fava, Polymer, 9 (1968) 137.
[64] S. Vyazovkin and N. Sbirrazzuoli, Macromolecules, 29 (1996) 1867.
[65] W. Yu and E.D. von Meerwall, Macromolecules, 23 (1990) 882.
[66] Y. Deng and G.C. Martin, Macromolecules, 27 (1994) 5141.
[67] Y. Deng and G.C. Martin, Macromolecules, 27 (1994) 5147.
[68] G. Van Assche, A. Van Hemelrijck, H. Rahier and B. Van Mele, Thermochim. Acta, 304/305 (1997) 317.
[69] G. Van Assche, A. Van Hemelrijck and B. Van Mele, J. Thermal Anal., 49 (1997) 443.
[70] J.H. Flynn, J. Thermal Anal., 34 (1988) 367.
[71] J. Borchardt and F. Daniels, J. Am. Chem. Soc., 79 (1957) 41.
[72] R.B. Prime, Polym. Eng. Sci., 13 (1973) 365.
[73] T.W. Char, G.D. Shyu and A.I. Isayev, Rubber Chem. Technol., 66 (1994) 849.
[74] P.J. Haines and F.W. Wilburn, Thermal Methods of Analysis; Principles, Applications and Problems, 1st Ed. (Chapter 3, section 3.10.3), Blackie Academic & Professional, Glasgow (1995).
[75] I.H. Updegraff, In Encyclopedia of Polymer Science and Engineering, Vol. 1, Herman F. Mark, Norbert Bikales, Charles G. Overberger, George Menges, and Jacqueline I. Kroschwitz, Eds., Wiley-Interscience, New York (1985) p. 752.
[76] H. Rahier, B Van Mele, M. Biesemans, J. Wastiels and X. Wu, J. Mater. Sci., 31 (1996) 71.
[77] A. Toda, T. Arita and M. Hikosaka, J. Therm. Anal. Calorim., 60 (2000) 821.
[78] J.E.K. Schawe, Thermochim. Acta, 361 (2000) 97.

[79] G.A. O'Neil and J.M. Torkelson, Trends Polym. Sci., 5 (1997) 349.
[80] S. Swier, G. Van Assche, A. Van Hemelrijck, H. Rahier, E. Verdonck and B. Van Mele, J. Thermal Anal., 54 (1998) 585.
[81] G. Van Assche, E. Verdonck and B. Van Mele, J. Therm. Anal. Calorim., 39 (2000) 305–318.
[82] G. Van Assche, E. Verdonck and B. Van Mele, Polymer, 42(7) (2001) 2959.
[83] A. Van Hemelrijck, PhD Thesis, Vrije Universiteit Brussel, Brussel (1996).
[84] G. Van Assche, A. Van Hemelrijck, H. Rahier and B. Van Mele, Thermochim. Acta, 286 (1996) 209.
[85] S. Montserrat and I. Cima, Thermochim. Acta, 330 (1999) 189.
[86] K. Horie, I. Mita and H. Kambe, J. Polym. Sci.: Part A-1, 6 (1968) 2663; J. Polym. Sci.: Part A-1, 8 (1970) 2839.
[87] C.H. Bamford, Radical polymerisation, In Encyclopedia of Polymer Science and Engineering, Vol. 13, Herman F. Mark, Norbert Bikales, Charles G. Overberger, George Menges, and Jacqueline I. Kroschwitz, Eds., Wiley-Interscience, New York (1988).
[88] R.G.W. Norrish and R.R Smith, Nature, 150 (1942) 336.
[89] K. Horie, I. Mita and H. Kambe, J. Polym. Sci.: Part A-1, 6 (1968) 2663; J. Polym. Sci.: Part A-1, 8 (1970) 2839.
[90] R.M. Noyes, J. Am. Chem. Soc., 77 (1955) 2042.
[91] A. Van Hemelrijck, R. Scherrenberg, J.J. Nusselder, M. Scheepers and B. Van Mele, Personal communication.
[92] K. Wark, In Thermodynamics, R.H. Klas and J.W. Maisel, Eds., McGraw-Hill Book Company, New York, 1983, p. 799.
[93] H. Rahier, W. Simons, B. Van Mele and M. Biesemans, J. Mater. Sci., 32 (1997) 2237.
[94] M. Cassettari, G. Salvetti, E. Tombari, S. Veronesi and G.P. Johari, J. Non-Cryst. Solids, 172–174 (1994) 554.
[95] M. Cassettari, G. Salvetti, E. Tombari, S. Veronesi and G.P. Johari, J. Polym. Sci. Part B: Polym. Phys., 31 (1993) 199.
[96] M. Cassettari, F. Papucci, G. Salvetti, E. Tombari, S. Veronesi and G.P. Johari, Rev. Sci. Instrum., 64 (1993) 1076.
[97] J. Mijovic, A. Fishbain and J. Wijaya, Macromolecules, 25 (1992) 979.
[98] B. Fitz, S. Andjelic and J. Mijovic, Macromolecules, 30 (1997) 5227.
[99] S. Andjelic, B. Fitz and J. Mijovic, Macromolecules, 30 (1997) 5239.
[100] E.S. Domalski and E.D. Hearing, J. Phys. Chem. Ref. Data, 22 (1993) 805.
[101] R.J.J. Williams, B.A. Rozenberg and J.-P. Pascault, Adv. Polym. Sci., 128 (1997) 95.
[102] K. Dusek and J.P. Pascault, in The Wiley Polymer Networks Group Review, Vol. 1, Chemical and Physical Networks: Formation & Control of Properties, K. te Nijenhuis and W.J. Mijs, Eds, John Wiley & Sons, Chichester (1998) p. 277.
[103] T. Araki, Q. Tran-Cong and M. Shibayama, Structure and Properties of Multiphase Polymeric Materials, Marcel Dekker: New York (1998) p. 155.
[104] S. Swier and B. Van Mele, Thermochim. Acta, 330 (1999) 175.
[105] I. Mita and K. Horie, J. Macromol. Sci. Rev. Macromol. Chem. Phys., C27 (1987) 91.
[106] K. Horie and I. Mita, Adv. Polym. Sci., 88 (1989) 77.
[107] E. Rabinowitch, Trans. Faraday Soc., 33 (1937) 1225.
[108] M. Gordon and W. Simpson, Polymer, 2 (1961) 383.
[109] J.D. Ferry, Viscoelastic Properties of Polymers, 3rd Ed. (Chapter 11), John Wiley & Sons, New York (1961).
[110] I.M. Ward, Mechanical Properties of Solid Polymers, 2nd Ed. (Chapter 7, section 7.4), John Wiley & Sons, New York (1983).

[111] J.M.G. Cowie, Polymers: Chemistry and Physics of Modern Materials, 2nd Ed. (Chapter 12, section 12.11), Blackie Academic & Professional, New York (1991).
[112] M.L. Williams, R.F. Landel and J.D. Ferry, J. Am. Chem. Soc., 77 (1955) 3701.
[113] E.A. DiMarzio, J. Res. Natl. Bur. Stand. A, 68A (1964) 611.
[114] G. Huybrechts and G. Van Assche, Comput. Chem., 22 (1998) 413.
[115] Subroutine VAO5A from Harwell Subroutine Library.
[116] G. Huybrechts, Y. Hubin and B. Van Mele, Int. J. Chem. Kinet., 21 (1989) 575.
[117] D.H. Kim and S.C. Kim, Polym. Bull., 18 (1987) 533.
[118] H. Stutz, J. Mertes and K. Neubecker, J. Polym. Sci. Polym. Chem. Ed., 31 (1993) 1879.
[119] H. Stutz and J. Mertes, J. Polym. Sci. Polym. Chem. Ed., 31 (1993) 2031.
[120] G. Van Assche, B. Van Mele and Y. Saruyama, Thermochim. Acta, 377 (2001) 125.
[121] R. Scherrenberg, V. Mathot and P. Steeman, J. Therm. Anal. Calorim., 54(2) (1998) 477.
[122] M. Nishikawa and Y. Saruyama, Thermochim. Acta, 267 (1995) 75.
[123] Y. Saruyama, Thermochim. Acta, 283 (1996) 157.
[124] Y. Saruyama, J. Therm. Anal. Calorim., 54(2) (1998) 687.
[125] P.G. Royall, D.Q.M. Craig, M. Reading and T.J. Lever, J. Thermal Anal. Calorim., 60 (2000) 795.
[126] B. Wunderlich, R. Androsh, M. Merzlyakov, M. Pyda and Y.K. Kwon, Thermochim. Acta, 348 (2000) 181.
[127] P. Kamasa, C. Schick and B. Wunderlich, Thermochim Acta, 392 (2002) 195.
[128] M. Merzlyakov and C. Schick, Thermochim. Acta, 380 (2001) 5.
[129] S. Monserrat, Polymer, 36 (1995) 435.
[130] A. Van Hemelrijck and B. Van Mele, J. Thermal Anal., 49 (1997) 437.
[131] L. Matejka and K. Dusek, Macromolecules, 22 (1989) 2902.

Chapter 3

APPLICATIONS OF MODULATED TEMPERATURE DIFFERENTIAL SCANNING CALORIMETRY TO POLYMER BLENDS AND RELATED SYSTEMS

Douglas J. Hourston and Mo Song
IPTME, Loughborough University, Loughborough LE11 3TU, UK

1 Introduction

The characterisation of multi-component polymer materials [1] has been pursued vigorously in recent years. Many types of such materials (including polymer blends, block copolymers, structured latexes and interpenetrating polymer networks) are now commercially available [2,3] and their ever better characterisation remains important. It is necessary to obtain morphological parameters such as the thickness and weight fraction of interfaces/interphases[1] and to understand the relationships between morphology and mechanical properties of such multi-component polymeric materials [2–8]. A common feature across the spectrum of multi-component polymeric materials is the presence of interfaces [2,5,7,8]. The properties of the interface are invariably central to the properties of the composite and the ability to understand and optimise the interface is recognised as a key feature in the development of improved polymeric materials. Most polymer pairs are immiscible [5,6]. Thus, the majority of blends are two-phase and their morphology depends on the type of molecular interaction, the rheology of the components and the processing history. Models used to describe

[1] The term interface implies a two-dimensional structure. It is clear in nearly all practical cases in polymer science that the regions between phases are three-dimensional in nature. These regions are also often likely not to be isotropic, but of a compositionally graded nature which means they do not meet the strict definition of a phase. In this chapter, the terms interface and interphase will be used essentially interchangeably.

M. Reading and D. Hourston (eds.), Theory and Practice of Modulated Temperature Differential Scanning Calorimetry, 161–216.

multi-component materials show that certain properties can be correlated with the interphase volume fraction [7,8]. Many techniques have been used to characterise the morphology of multi-phase polymeric materials. Porod's analysis [9] of small-angle X-ray (SAXS) and neutron scattering (SANS) data has been used to estimate interfacial thickness and domain size [10,11]. Dynamic mechanical thermal analysis (DMTA) data have been modelled [12] by assuming interfacial profiles. A technique that can yield both interfacial thickness and composition gradient across the interface is transmission electron microscopy (TEM) [10]. Results that are in good agreement with SAXS and DMTA [12] have been obtained for highly ordered systems, such as ABA-type block copolymers.

In the characterisation of the morphology of multi-component polymeric materials, the glass transition temperature, the composition distribution in the phases, phase size and shape and the thickness and volume (or weight) fraction of the interface are clearly important. DMTA and differential scanning calorimetry (DSC) are suitable for the measurements of the glass transition temperature. It is conventional, simple and rapid to use DSC to study polymer blends. However, because the sensitivity and resolution of DSC are usually not good enough, overlapping thermal events, including T_gs from pure phases and any interface resulting from partial miscibility, cannot usually be separated [13,14].

A basic limitation exists on the use of glass transition determinations in ascertaining the extent of polymer–polymer miscibility in blends composed of components which have similar (<15°C difference) T_gs. In these cases, resolution by the DSC technique [5] is not possible. Also, for small concentrations (less than 10%), the transition signal is difficult to resolve [5,15]. Structural relaxation at the T_g [15] can also distort the shape of the transition. Although DSC has been used extensively to characterise IPNs [16–18], it fails when IPNs show complex phase structure. Most researchers have turned to DMTA to observe the transitions in IPNs, for example, because it is more sensitive [1,5,6]. An interesting morphological parameter, the degree of segregation in IPNs can be obtained from DMTA data using Lipatov's method [19]. The DMTA characterisation method developed by Annighofer and Gronskin [12] is only suitable for the study of the morphology of block copolymers with a high degree of orientation. In fact, it is difficult, quantitatively, to obtain either the weight (or volume) fraction of each phase or information on composition distribution in multi-phase polymeric materials from DMTA data. It is always necessary to make some assumptions regarding the nature of the interface.

Microscopies and scattering techniques [10,12] are used to study the micro-domain size, shape and interface content. TEM has been used in many instances in order to determine the miscibility, or phase segregation, of IPNs. Detailed information about polymer blend morphology can be gained from this technique. This includes information about the continuous phase and the size and shape of the domains and their distribution. At very high magnifications, domains in the order of 1 nm can, in theory, be investigated [20]. The preparation of the samples can sometimes be difficult, since ultra-thin sections have to be cut. Specimen preparation [21] and the interpretation [22], and possible artefacts caused by electron beam [20] and sectioning damage [22] have been described.

It is not difficult to study the micro-domain size and the interfacial thickness of block copolymers using SAXS. The volume fraction of interface in such a multi-phase system [10,12] has been obtained using this technique. However, it is not easy to determine the fraction of interphase in partially miscible, or essentially immiscible, polymer blends. Regarding the application of SAXS to measurements of the morphological parameters of multi-phase polymeric materials, Ruland [23] has fully analysed the experimental difficulties. He indicated that the determination of the width of domain boundaries by the SAXS method can contain substantial errors if the boundary region is not represented by a smooth homogeneous density transition, but by a statistical structure of a certain coarseness. It has been shown that these errors, in general, lead to an under-estimation of the values of the boundary widths in the case of block copolymers. Samples with a highly preferred orientation of the interface planes can be used to minimise the errors and to obtain information on the coarseness of the domain boundaries [23].

Although the existence of a diffuse interfacial region in multi-phase systems has been detected by solid-state NMR spectroscopy [24–27] and by dynamic relaxation measurements [12,28], to date only SAXS and SANS are capable of providing interfacial thickness values. Scattering techniques, especially SANS, are rather specialised and are not widely available. DMTA can be used to study interfaces [12] by assuming interfacial profiles. TEM results [10] from highly ordered systems are in good agreement with SAXS and DMTA [12] data. SANS has been used by McGarey [29] to study IPNs, but IPNs are far from the ideal system for study by this technique.

To help summarise the above discussion, Table 3.1 gives a comparison of the applicabilities of the DSC, DMTA, SAXS, SANS, microscopies and solid-state NMR techniques to the study of multi-component polymeric materials. It can be seen that if one wants to obtain detailed morphological information, several characterisation techniques must be used. It is also obvious that even when the above characterisation techniques are available, one cannot obtain all the morphological parameters such as the weight fraction of each phase and the concentration distribution in multi-phase polymeric materials.

Table 3.1. Comparison of the abilities of various characterisation methods for multi-component polymeric materials

	DSC	SAXS	SANS	DMTA	LM	SEM	TEM	NMR
Resolution (nm)	20	2	1	15	1000	20	1	1
Specimen preparation	Easy	Easy	Difficult	Easy	Easy	Easy	Difficult	Easy
T_g	Quant	No	No	Quant	No	No	No	No
Multi-phase information	Yes	Yes	Yes	Yes	Yes	Yes	Yes	Yes
Interfacial information	Yes*	Yes	Yes	Yes*	No	No	Yes*	Yes
Interfacial thickness	No	Quant	Quant	No	No	No	Qual	No
Weight fraction	Qual	No	No	No	No	No	No	Yes*
Domain size	No	Yes	Yes	No	Yes	Yes	Yes	No

LM: light microscopy; Qual = qualitative; Quant = quantitative.
*Not always possible (1,5,10,26).

It is very desirable to establish widely applicable, and readily available, methods for the characterisation of multi-phase polymeric materials that can overcome the disadvantages of the above techniques.

The possibilities arising from the advent of MTDSC will now be discussed. Complex thermal histories affect the ease with which it is possible to make determinations of the increment of heat capacity, ΔC_p, at T_g because of structure relaxation. If a thermal analysis apparatus that can separate the structure relaxation part from the total heat flow signal can be developed, ΔC_p could be determined accurately. It is well known that ΔC_p is related to the weight fraction of each component in a heterogeneous system such as a polymer blend. In multi-phase polymeric materials, each phase has its own characteristic glass transition temperature and ΔC_p. Thus, important information may be obtained from ΔC_p and glass transition measurements, allowing such materials to be analysed quantitatively.

For pure, fully annealed polymers, the glass transition is approximately symmetrical [5]. For partially miscible systems in which there are interfaces, the transition will be asymmetric and become broadened [5]. This asymmetry and broadening may provide a wealth of information of both practical and theoretical value that has not yet been fully extracted.

Because modulated temperature DSC (MTDSC) can separate overlapping thermal events and separate the total heat flow into two parts: the reversing (proportional to heating rate) and the non-reversing (dependent on temperature) components, it allows the study of the asymmetry and broadening of the glass transition. Important information can be obtained from the differential of heat capacity, dC_p/dT, signal over the glass transition region. Using this signal, multi-component polymeric materials may be analysed quantitatively. In this chapter, we will discuss the dC_p/dT signal and its use in the quantitative characterisation of such materials. The MTDSC technique leads to an improvement in the detection of the glass transition, readily provides a measure of ΔC_p and indicates the extent of polymer–polymer miscibility. Based on this new signal, symmetric and asymmetric interdiffusion, interface development in bilayer and structured latex films, and the morphology of IPN materials will be discussed.

2 Heat Capacity and its Differential with Temperature Signal Over the Glass Transition Region

In chapter 1, a full theoretical treatment of the behaviour of the MTDSC signals over the glass transition region [30] has been presented.

The following equations arise from this treatment discussed in Chapter 1.

$$C_p' = A + BT + \Delta C_p / \left(1 + \omega^2 \tau_g^2 \exp\left(-2\Delta h^* / (RT_g^2)(T - T_g)\right)\right) \quad (1)$$

$$C_p'' = \Delta C_p \omega \tau_g \exp\left(-\Delta h^* / RT_g^2 (T - T_g)\right) / \left(1 + \omega^2 \tau_g^2 \exp\left(-2\Delta h^* / (RT_g^2)(T - T_g)\right)\right) \quad (2)$$

Figures 3.1 and 3.2 compare the dC_p/dT versus temperature data from experiments with theoretical data (using the above equations) and a Gaussian function for, respectively, polystyrene and a (50/50 by weight) miscible blend of poly(methyl methacrylate) and poly(styrene-*co*-acrylonitrile) [30]. Clearly, the experimental data can be described by both the theory, and also by a Gaussian function, G, of the glass transition temperature, the width

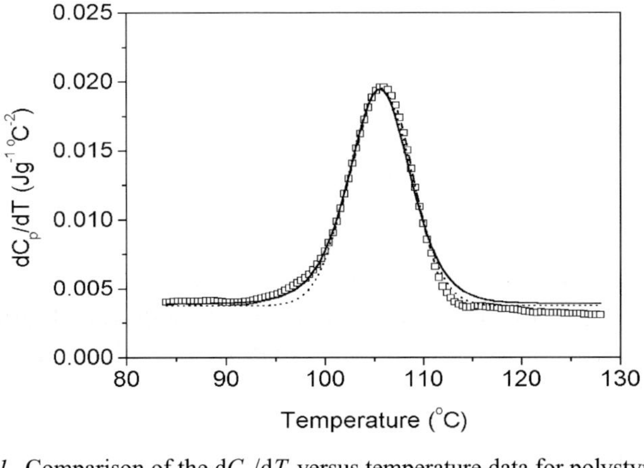

Figure 3.1. Comparison of the dC_p/dT versus temperature data for polystyrene from experiment (square points), theory (solid line) and from the Gaussian function (dots).

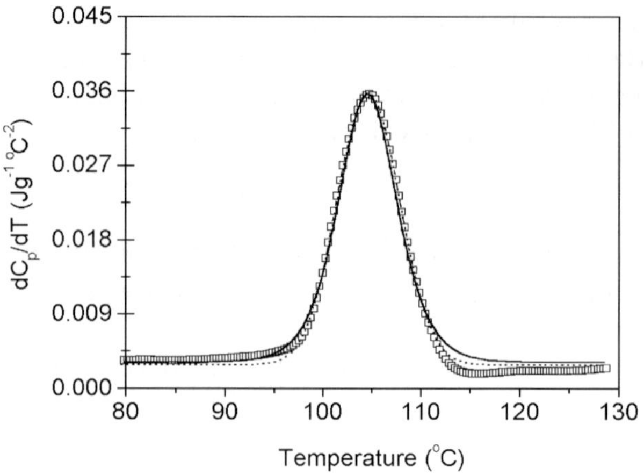

Figure 3.2. Comparison of the dC_p/dT versus temperature data for a PMMA/SAN(50:50) blend from experiment (square points), theory (solid line) and from the Gaussian function (dots).

of the transition at half height, ω_d, the increment of heat capacity and the temperature.

$$G = f(T, T_g, \omega_d, \Delta C_p) = \Delta C_p/\left[\omega_d(\pi/2)^{1/2}\right]\exp\left[-2(T - T_g)^2/\omega_d^2\right] \quad (3)$$

In this chapter, the Gaussian function description of the change of dC_p/dT versus temperature at the glass transition will be used in the analysis of various polymer blend systems. The Gaussian function approach to modelling the glass transition is chosen over theory [30] because in Eq. (1), the τ_g and Δh^* terms are generally unavailable for polymers.

3 Measurements of the Glass Transition Temperature and Increment of Heat Capacity

As mentioned above, the commonly occurring complex thermal histories experienced by polymeric artefacts during manufacture affect the ease with which it is possible to make determinations of the glass transition temperature accurately by conventional DSC. Thermograms with different shapes in the glass transition region often make the conventional extrapolations ambiguous. It is also often the case that the measurement of ΔC_p in the glass transition region is highly subjective, not to mention time-consuming, using conventional DSC.

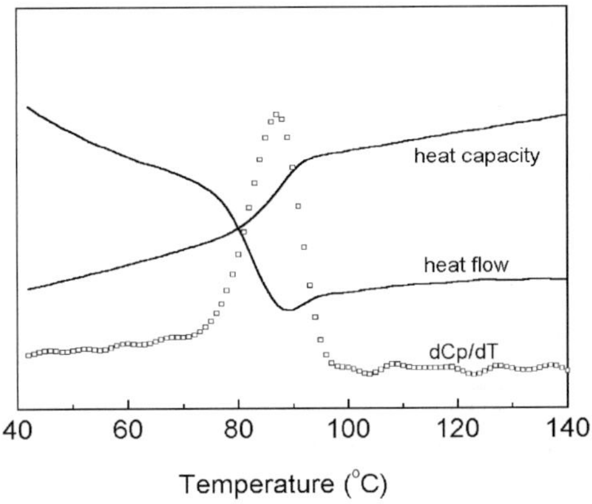

Figure 3.3. Heat flow, heat capacity and dC_p/dT versus temperature data for polystyrene.

Figure 3.3 shows the changes of total heat flow, heat capacity and dC_p/dT with temperature for a PS sample [31]. Because of the effect of thermal history, the relaxation event appears in the total heat flow signal. It can be seen that the peak position of the dC_p/dT versus temperature signal corresponds to the point of inflection of the heat capacity curve between the glassy and liquid states. If the peak position, as is often done for a melting point, is used to determine the T_g, it will be very easy and reproducible to use in subsequent analyses.

Figure 3.4 gives another example of MTDSC output. In this case, data for an interpenetrating polymer network are reported. Obviously, it is very difficult to obtain the T_g values with any accuracy from the total heat flow signal, which is very complex. However, it is very easy, using the dC_p/dT

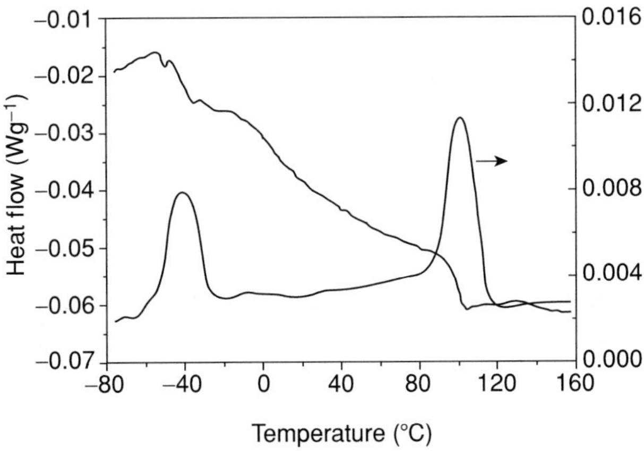

Figure 3.4. Heat flow and dC_p/dT versus temperature data for a polyurethane/polystyrene IPN.

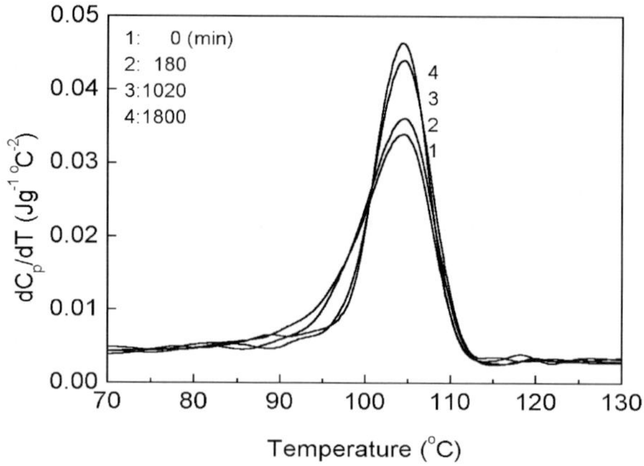

Figure 3.5. dC_p/dT versus temperature data for different annealing times at 80°C for a SAN/PMMA blend (50/50).

signal, to obtain both of these T_gs accurately and simply. Figure 3.5 again shows the change of dC_p/dT with temperature for a PMMA/SAN (50/50, wt/wt) compatible blend [31], but for different annealing times at 80°C. The peak position is almost constant with time. However, the onset point shifts to higher temperature with increasing annealing time. Figure 3.6 gives the result of a heat/cool experiment for polystyrene [31]. The T_g is 85°C on cooling and 86°C on re-heating showing the measurement to be robust. Figure 3.7 gives another example for polystyrene, this time annealed at different temperatures for 1 hour. The T_g was 86 ± 1°C for the different annealing temperatures. Figure 3.8 shows the changes of T_g with annealing time for a polyvinyl acetate sample. With increasing time, the value of T_g

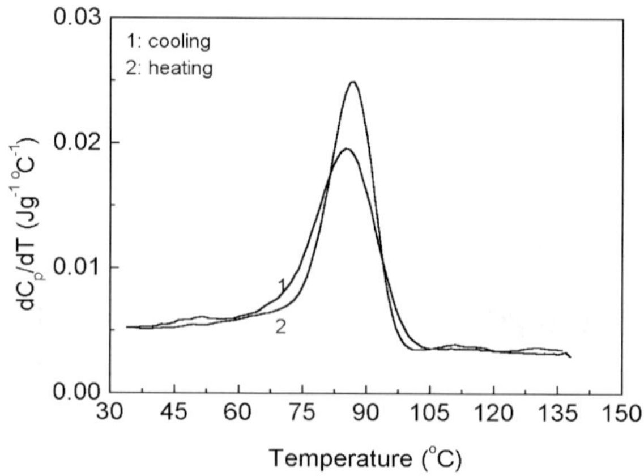

Figure 3.6. dC_p/dT versus temperature data for polystyrene in a cyclic experiment.

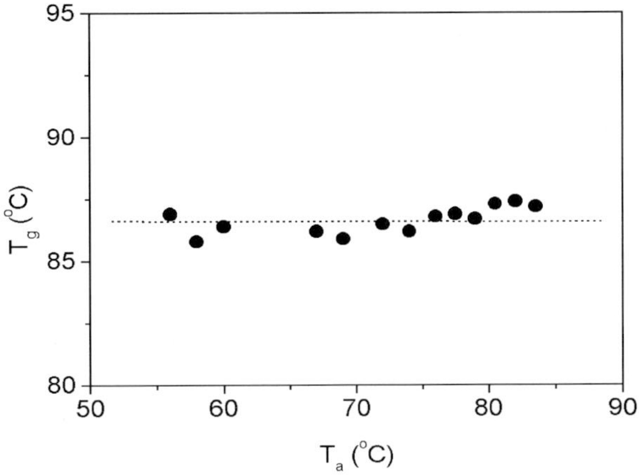

Figure 3.7. T_g versus annealing temperature for polystyrene. Annealing time was 1 h.

increased. However, even for long times, the difference was only 1.6°C. These changes are relatively small. These last few figures illustrate that the dC_p/dT signal is a sensitive, and, therefore, a valuable one with which to probe the glass transition.

For small concentrations of a given component in a polymer blend (less than 10 wt%), the resulting weak transition is typically very difficult to resolve using conventional DSC or DMTA [5,15]. Using MTDSC, T_g determinations were performed [32] on a physical blend containing four components: pure PS plus PPO-30 (a PS/polyphenylene oxide (PPO) blend at a composition ratio of 70/30) plus PPO-70 (a PS/PPO blend at a composition ratio of 30/70) plus pure PPO. The amount of each component was 44.0:7.1:13.4:34.5, by weight. Figure 3.9 shows both the heat capacity and

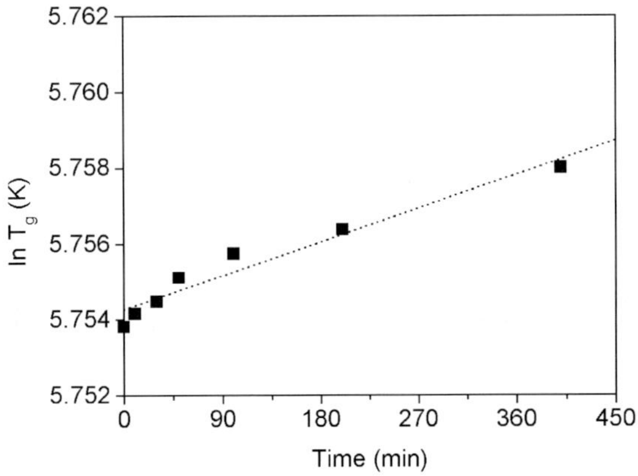

Figure 3.8. ln T_g versus annealing time at 30°C for polyvinyl acetate.

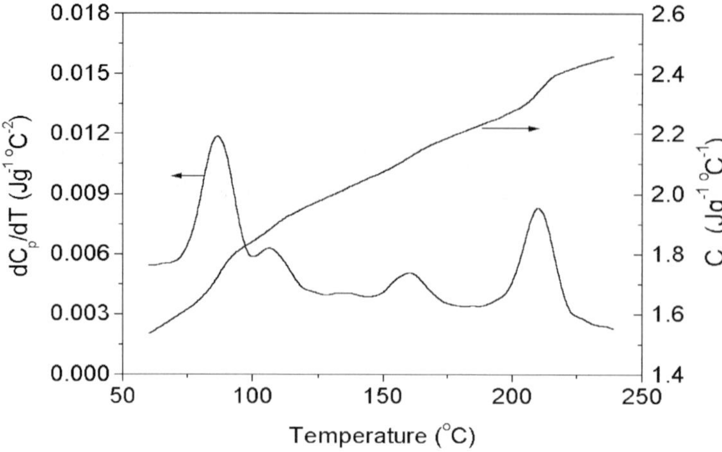

Figure 3.9. dC_p/dT versus temperature data for a PS + PPO-30 + PPO-70 + PPO physical blend.

the dC_p/dT with temperature signals. From the heat capacity signal, not all the transitions are clear. However, four transitions are clearly evident in the dC_p/dT signal, despite the fact that the PPO-30 is only present at 7.1% by weight.

In summary, the dC_p/dT signal is a very useful tool to determine T_g values. The benefits of using dC_p/dT to measure T_g are as follows.

(i) The position and shape of the glass transition are much less affected by thermal history and experimental conditions than is the case with conventional DSC.

(ii) Glass transitions can be represented as Gaussian curves.

(iii) Events such as the loss of small amounts of residual solvent, which can occur when studying blends, affect the reversing signal very little (see Chapter 1), but can have significant effects on the heat flow signal in conventional DSC.

(iv) Resolution is improved in MTDSC because both the step at T_g in the reversing signal is sharper than that in conventional DSC and low underlying heating rates can be used while still retaining a high signal-to-noise ratio in the reversing heat capacity measurement.

The value of apparent heat capacity, C_p^a, (not calibrated) may be written as follows [31].

$$C_p^a = A + BT + f(T) \tag{4}$$

A and B are constants and $f(T)$ is a function of temperature. Outside the glass transition region, $f(T) = 0$. The following relation holds for the $\mathrm{d}C_p^a/\mathrm{d}T$ value.

$$\mathrm{d}C_p^a/\mathrm{d}T = B + \mathrm{d}f(T)/\mathrm{d}T \tag{5}$$

To obtain the required ΔC_p values, it is only necessary to integrate the signal over the region of interest, which in this case is the glass transition.

$$\Delta C_p = \int_{C_p(i)^a}^{C_p(e)^a} (\mathrm{d}C_p^a/\mathrm{d}T)\mathrm{d}T \tag{6}$$

$C_p(i)^a$ and $C_p(e)^a$ are the initial and final values of the apparent heat capacity in the glass transition region. It is assumed that the integration constant is independent of temperature. The above equation to calculate ΔC_p only needs a one-point calibration for heat capacity selected in the transition region. The reason for this is that if it is assumed that the calibration constant of heat capacity is K_1 at the onset point of the glass transition and is K_2 at the final point, ΔC_p is given as follows.

$$\Delta C_p = K_2 C_p(e)^a - K_1 C_p(i)^a \tag{7}$$

The value of the one-point calibration constant, K, is given approximately by Eq. (8)

$$K = (K_1 + K_2)/2 \tag{8}$$

Consider that

$$K = K_1 + \delta = K_2 - \delta \tag{9}$$

δ is a small increment. Then, Eq. (10) can be rewritten as follows.

$$\Delta C_p = K\left[\Delta C_p^a + \delta/K(C_p(e)^a + C_p(i)^a)\right] \tag{10}$$

Table 3.2 lists how the calibration constants change with temperature. According to the experimental results, it was found that $\delta/K \sim 10^{-3}$. Thus,

$$\Delta C_p = K\Delta C_p^a \tag{11}$$

The difference between the results from Eq. (11) and those from Eq. (7) is small. The error resulting from using Eq. (11) is about 3%.

Table 3.2. Change of heat capacity calibration constant with temperature

Temperature (°C)	Calibration constant
36.85	1.1947
56.85	1.1846
76.85	1.1764
96.85	1.1654
116.85	1.1573
136.85	1.1522
156.85	1.1507
166.85	1.1459

There is considerable interest in the values of ΔC_p at the T_g and various generalisations [33,34] have been suggested either for ΔC_p or for the product $\Delta C_p T_g$. ΔC_p measurement is complex and time-consuming by conventional DSC [13,35]. Heat capacity values at T_g from conventional DSC studies have been obtained [36] by extrapolation of the linear equations used to describe the glass and liquid states. Based on the new MTDSC method, the determination becomes very simple and rapid. Later, we will discuss how this makes it a convenient way to analyse multi-phase polymeric materials.

Figures 3.10 and 3.11 show the changes of ΔC_p for PS [31] and for a 50/50 SAN/PMMA blend at different annealing temperatures and for different annealing times, respectively. For the PS sample, the annealing time was 60 min. The results show that the values for PS are almost constant

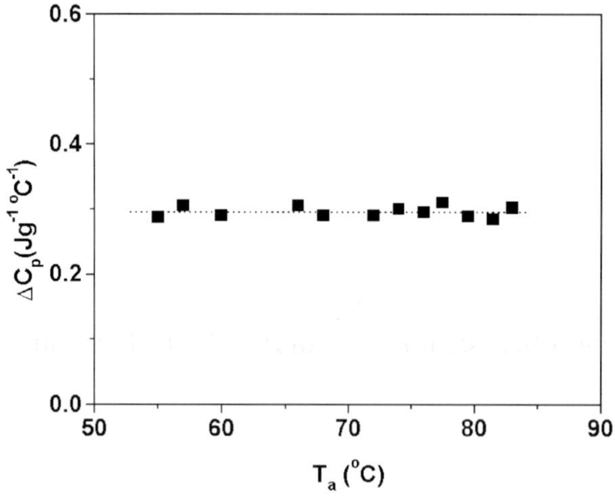

Figure 3.10. ΔC_p versus annealing temperature for polystyrene. Annealing time was 1 h.

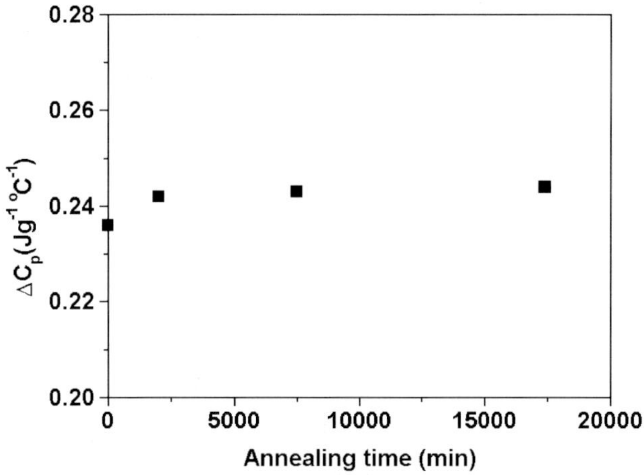

Figure 3.11. ΔC_p versus annealing time at 80°C for an SAN/PMMA blend (50/50 wt/wt).

for these different thermal histories. The average value of ΔC_p is 0.293 Jg^{-1} $°C^{-1}$. Comparison with values in the literature [37] indicates that the average difference is about 3%.

4 Multi-Component Polymer Materials

4.1 IMPROVEMENT IN THE MEASUREMENT OF POLYMER–POLYMER MISCIBILITY

Polymer–polymer miscibility is usually characterised [1,5,6] by investigating the optical appearance, morphology, glass transition temperature or the crystalline melting behaviour of the blend [38,39]. A blend of two amorphous polymers with different refractive indices will be judged to be miscible if it is optically clear. Measurement of the glass transition temperature, or temperatures, of a polymer blend is the most convenient and popular way of investigating polymer–polymer miscibility.

T_g is commonly measured by the DSC technique, but the use of T_g determination for studying polymer–polymer miscibility has its limitations. The glass transition region for a given polymer can cover at least a 15°C range [5,15] and often significantly more. Thus, if the difference of the glass transition temperatures between the two polymers in a blend is less than about 15°C, it has been almost impossible to detect the extent of mixing by DSC [5,15].

It is known [6] that poly(styrene-*co*-acrylonitrile), SAN, is miscible with PMMA when the acrylonitrile content is between 10 and 30 wt%. To check

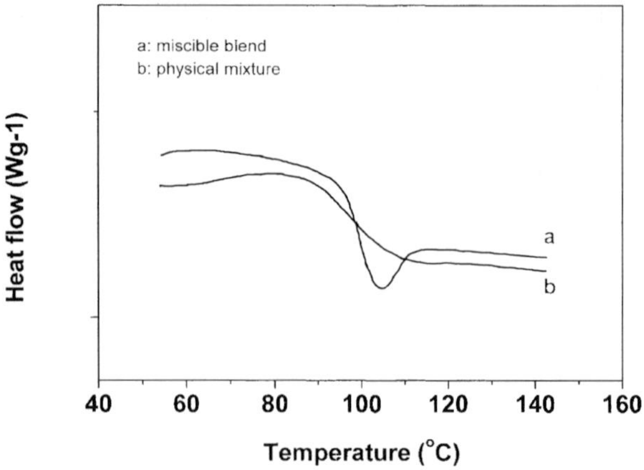

Figure 3.12. Heat flow versus temperature data for (a) the miscible blend and (b) the physical mixture (PMMA/SAN, 50/50 (wt/wt)).

the usefulness of the dC_p/dT signal in studying polymer–polymer miscibility in situations with similar T_gs, miscible and physical blends of SAN and PMMA were designed.

Figures 3.12 and 3.13 show the heat flow and the heat capacity data for the blend and for a physical mixture of PMMA and SAN [39]. From these data, it was not possible to draw any conclusions about miscibility because only one glass transition was observed for both the miscible blend and for the physical mixture. The T_g difference between the two constituent polymers is only about 10°C. However, it is clear from the dC_p/dT versus

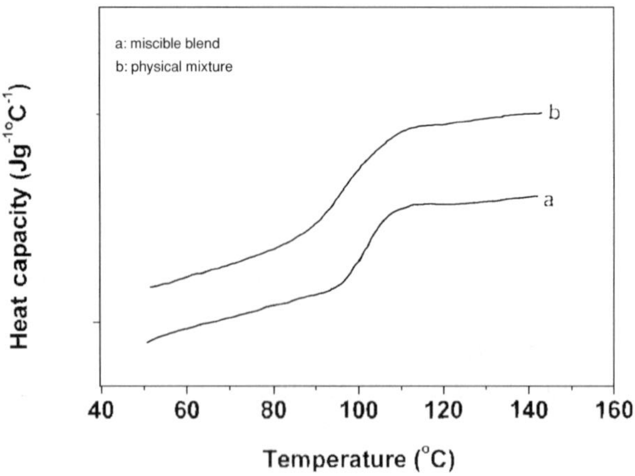

Figure 3.13. Heat capacity versus temperature data for the same (a) miscible blend and (b) physical mixture. The data are shifted vertically for clarity.

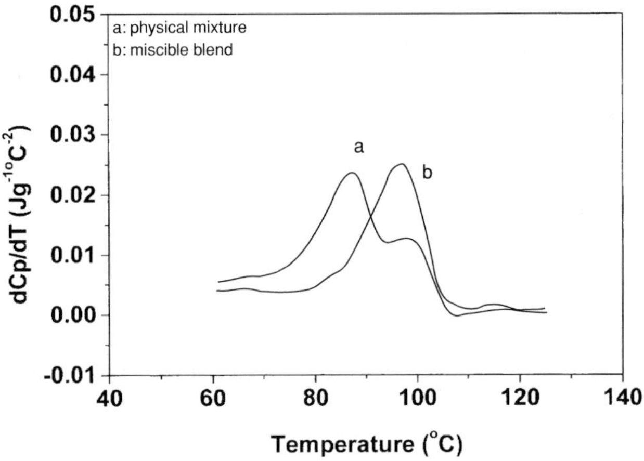

Figure 3.14. Differential of heat capacity versus temperature data for the physical mixture (PMMA/SAN, 50/50 (wt/wt)) and for the blend.

temperature data, shown in Figure 3.14 for both the miscible blend and the physical mixture, that there are differences. The physical mixture shows two clearly resolved transitions which appear to be the result of a simple linear addition of the $\mathrm{d}C_\mathrm{p}/\mathrm{d}T$ signals of the constituent polymers. The miscible blend shows the expected single glass transition.

Figure 3.15 shows the glass transition temperatures plotted versus composition for these PMMA/SAN blends. This shows a positive deviation from linearity often observed for miscible blends and ascribed to specific interactions between segments [6,38].

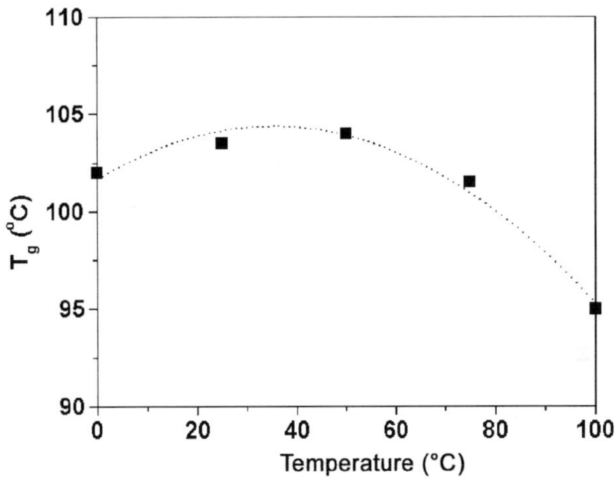

Figure 3.15. Glass transition temperature versus composition for PMMA/SAN (18 wt% AN) miscible blends.

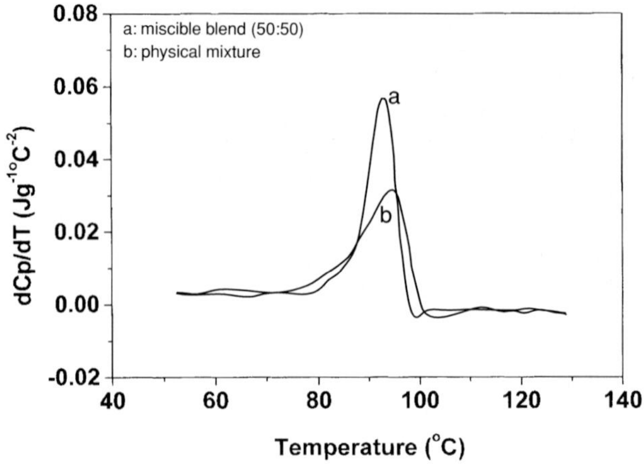

Figure 3.16. Comparison of (a) the miscible blend and (b) the physical mixture (PMMA/SAN, 50/50).

Figures 3.16 and 3.17 show results for a miscible blend and its equivalent physical mixture based on PMMA and a SAN with a 25 wt% AN content. The T_g difference for PMMA and this SAN is approximately 5°C. A single peak in the dC_p/dT signal is very clear for the, by definition, phase separated physical mixture, indicating that it is very difficult to detect miscibility in blends if the difference of T_gs is around this value. However, it is the case that the physical blends show broader transitions than do the miscible ones.

For most polymer pairs to be miscible, an exothermic interaction is required. Nandi *et al.* [40] studied the miscibility of poly(methyl acrylate) (PMA) and poly(vinyl acetate) (PVAc) in several solvents by the inverse

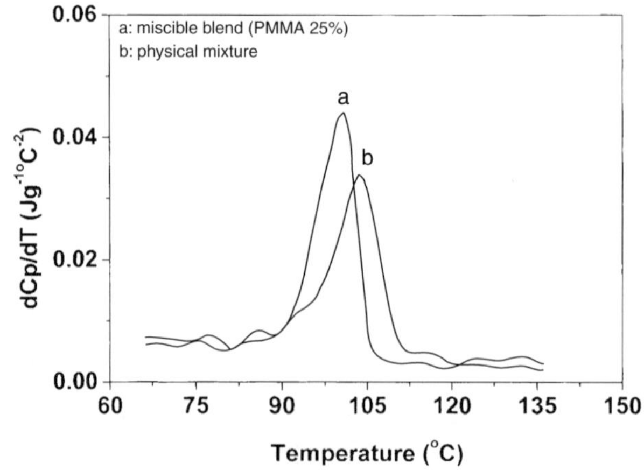

Figure 3.17. Comparison of (a) the miscible blend with (b) the physical mixture (PMMA/SAN, 25/75).

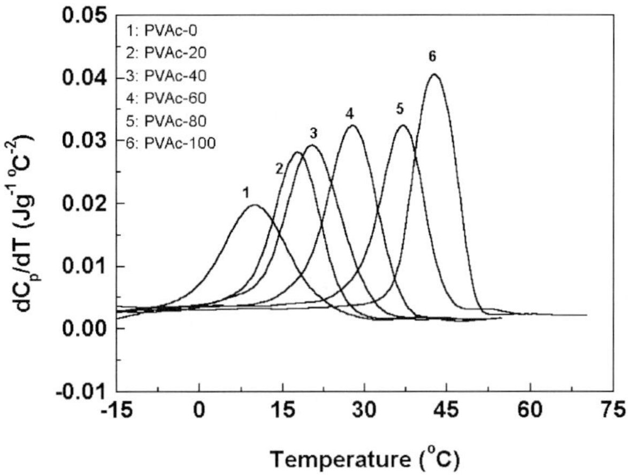

Figure 3.18. dC_p/dT versus temperature data for different PMA/PVAc blend compositions.

gas chromatography method. They concluded that the PMA/PVAc blend is miscible, and that no specific interactions are operative.

Figure 3.18 shows the dC_p/dT signal versus temperature for different PMA/PVAc blend compositions. The dC_p/dT signal showed a high degree of symmetry, which implies that the miscibility level is high. Compare this with the behaviour of PVC/poly(ethyl methacrylate) (PEMA) blends.

Perrin and Prud'homme [41] studied, by means of conventional DSC, the miscibility of PVC blended with PEMA. They showed this system to be miscible. The T_g difference was about 12°C. Using their experimental conditions [41], the miscibility of this blend was studied again by means of MTDSC. Figures 3.19 and 3.20 show, respectively, the changes of heat

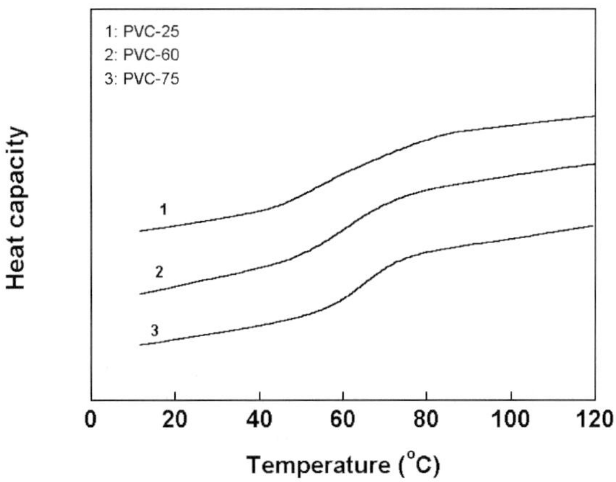

Figure 3.19. Heat capacity versus temperature for PVC/PEMA blends.

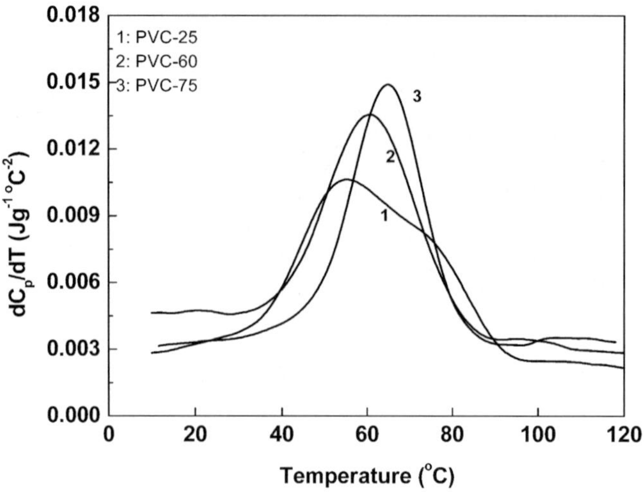

Figure 3.20. dC_p/dT versus temperature for PVC/PEMA blends.

capacity and dC_p/dT versus temperature for the PVC/PEMA blends with 25/75, 60/40 and 75/25 (by weight) compositions. The heat capacity signals show that this blend system may be miscible. However, the dC_p/dT signal for the 25/75 PVC/PEMA blend showed that this blend was not fully miscible. The dC_p/dT signals show that the levels of miscibility of the 60/40 and 75/25 PVC/PEMA blends were higher than that of the 25/75 PVC/PEMA blend. This further emphasises that polymer–polymer miscibility can be checked sensitively using the dC_p/dT signal.

Figures 3.21 and 3.22 show the changes of T_gs and ΔC_p versus composition for some PMA/PVAc blends. The following relations hold for T_g and ΔC_p.

$$T_g = w_1 T_{g1} + w_1 T_{g2} \tag{12}$$

$$\Delta C_p = w_1 \Delta C_{p1} + w_1 \Delta C_{p2} \tag{13}$$

The ΔC_p term is a significant parameter because it appears in the Ehrenfest equation [42]. Perhaps, in polymer blends, the intermolecular contribution to ΔC_p plays a more important role than in many common homopolymers and copolymers.

To date, many supposedly miscible polymer pairs [5,6,13,14,42] have been reported in the literature. However, in some cases [13,14], the breadth of the glass transition region, ΔT_g, taken as the difference between the onset and completion temperatures, is quite broad. For some blend systems, ΔT_g values approach 100°C [13,14]. The transition region may also be

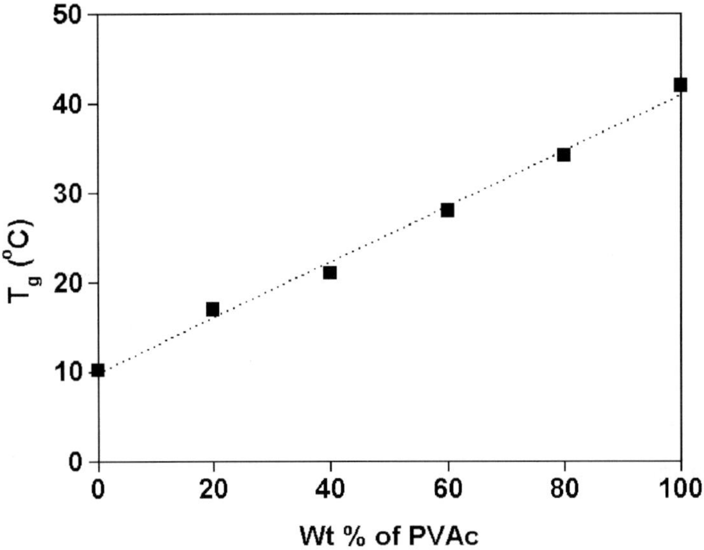

Figure 3.21. Glass transition temperature versus composition for PMA/PVAc blends.

asymmetrical. Because conventional DSC is not sensitive enough and lacks good resolution, overlapping T_gs and interfaces resulting from partial miscibility, cannot be separated. It is possible that some incorrect conclusions have been reached [13,14] for polymer blends that have quite large ΔT_gs [13,14]. To study this problem, the poly(epichlorohydrin) (PECH)/PMMA blend system was chosen for further investigation using MTDSC.

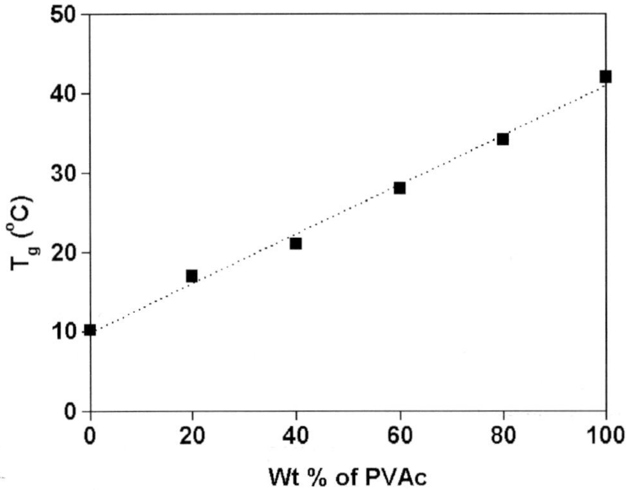

Figure 3.22. Plot of ΔC_p versus composition for PMA/PVAc blends.

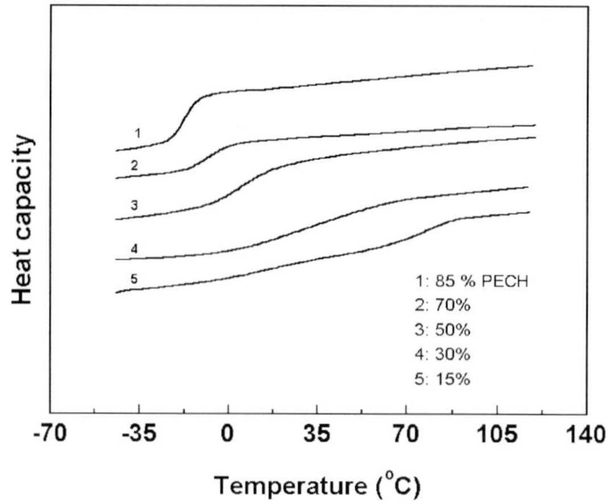

Figure 3.23. Heat capacity (arbitrary scale) versus temperature data for PECH/PMMA blends.

Figure 3.23 shows the change of heat capacity with temperature for five different compositions. These thermograms appear to offer essentially the same interpretation as the results presented by Higgins and co-workers [43] and by Fernandes *et al.* [13,14]. A single and broad T_g transition is seen indicating that the blend is miscible. However, SANS results reported by Higgins and co-workers [43] showed the blend system to possess two phases, indicating that it is essentially immiscible. The dC_p/dT versus temperature data for PECH/PMMA blends at 100/0, 85/15, 70/30, 50/50, 30/70, 15/85 and 0/100 (wt/wt) compositions were checked. The results are shown in Figures 3.24(a)–(g). The dC_p/dT signals give detailed and clear information about miscibility. For pure PECH and PMMA, the transitions are highly symmetrical. For the 85/15 PECH/PMMA blend, the transition peak shows the same behaviour as PECH, or PMMA, in that it is highly symmetrical. This implies that at this composition the polymers are miscible. For the 70/30 PECH/PMMA blend, there is a weak transition between 40 and 100°C. For the 30/70 PECH/PMMA blend, there is obviously phase separation. The dC_p/dT signal shows two transitions. Because the two components have very similar refractive indices [43], it is very difficult to check the phase separation behaviour using optical methods. For the 50/50 blend, the transition peak is markedly asymmetrical, and exhibits a shoulder. At the 15/85 composition, the dC_p/dT signal shows two separated transition peaks clearly confirming immiscibility. Table 3.3 shows the T_g and the ΔT_g values, which were defined as shown in Figure 3.24(g). The correlation lengths shown in Table 3.3 were obtained from the literature [43]. The value for the 15/85 PECH/PMMA blend was omitted because the dC_p/dT signal from this system showed two clear transitions indicating that this correlation length had

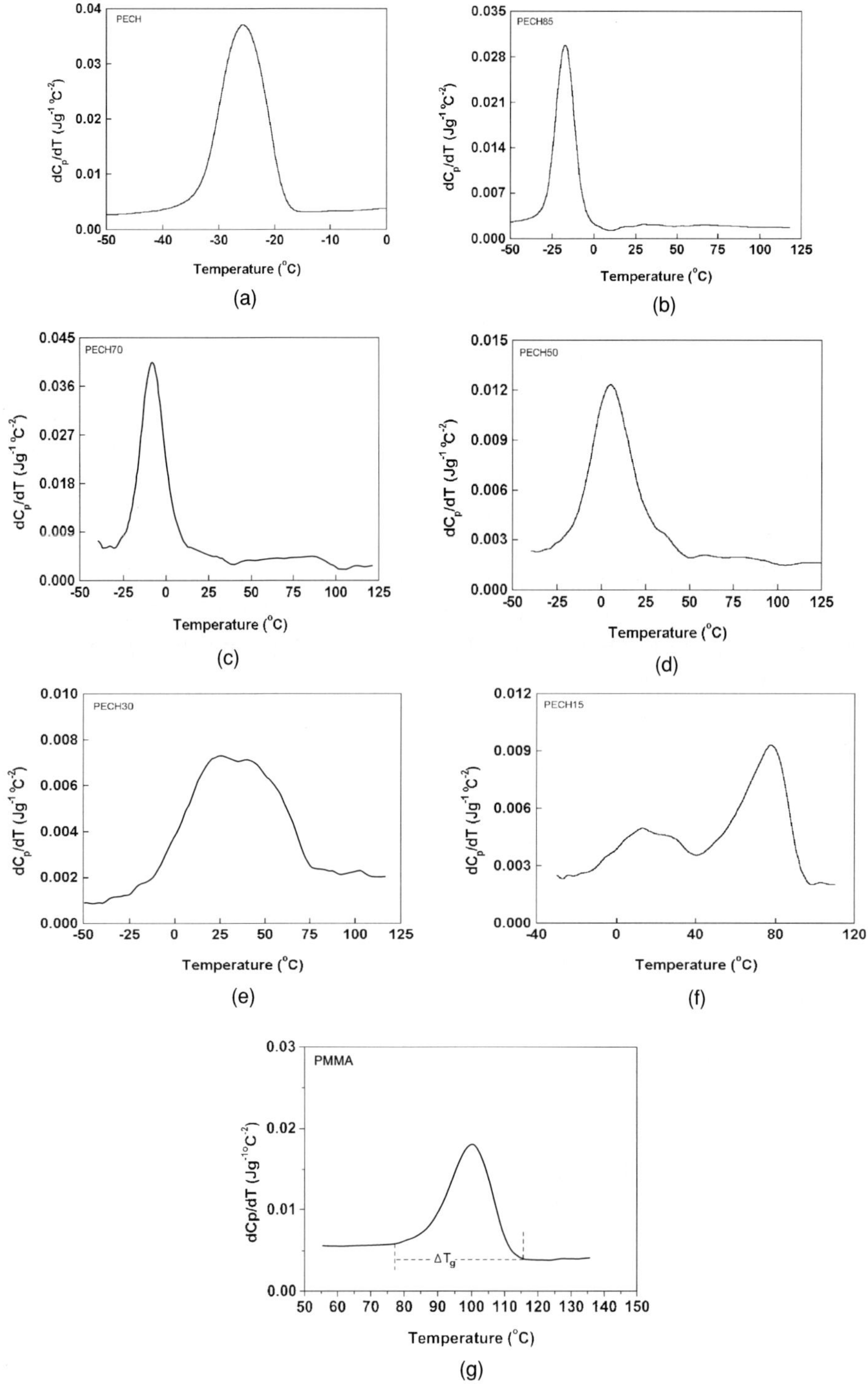

Figure 3.24. (a–f) dC_p/dT versus temperature data for PECH and the PECH/PMMA blends; (g) definition of ΔT_g.

Table 3.3. Glass transition and ΔT_g for the PECH/PMMA blends

PECH/PMMA	T_g (°C)	ΔT_g (°C)	Correlation length (nm) (Ref. [43])
100/0	−26	20	–
85/15	−17	26	–
70/30	–	65	14
50/50	–	80	37
30/70	–	100	47
15/85	Two phase transition signals		
0/100	102	40	–

no physical meaning. It can be seen that the ΔT_g values of the PECH/PMMA blends are quite large and increase with increasing correlation length.

For the PMMA homopolymer, the onset temperature was about 80°C. For PECH, the completion temperature was about −18°C. Obviously, the large ΔT_g values are not due to the fact that the completion of the lower transition and the onset of the higher transition cannot be resolved [43]. The conclusion is that these blend systems exhibit interfaces. The PECH/PMMA blends are, therefore, partially miscible. It is this partial miscibility that causes the large ΔT_g values. It is concluded that most of the PECH forms a mixed phase with PMMA for the 50/50 and 70/30 PECH/PMMA blends. However, for the 30/70 PECH/PMMA blend, there are predominantly PECH-rich and PMMA-rich phases.

For fully miscible systems, the deviation, δT_g, defined as $\delta T_g = \Delta T_g - (w_1 \Delta T_{g1} - w_2 \Delta T_{g2})$, is, by definition, very small. Table 3.4 shows δT_g

Table 3.4. δT_g values for PECH/PMMA, PS/PPO and PMA/PVAc blends

PECH/PMMA	δT_g (°C)
100/0	0
85/15	5
70/30	39
50/50	50
30/70	66
0/100	0
PS/PPO	
100/0	0
75/25	1
50/50	−1
25/75	−1
0/100	0
PMA/PVAc	
100/0	0
75/25	0.5
50/50	1
25/75	−0.5
0/100	0

values for PECH/PMMA, PS/PPO and for PVAc/PMA blends over a range of compositions. Clearly, the immiscible system shows the largest δT_g value.

SANS is able to distinguish between micro-phase separation and concentration fluctuations [43]. However, SANS results showed curves for four blend compositions (PECH/PMMA: 70/30, 50/50, 30/70 and 15/85) which were very similar (Ref. [43]). There was no obvious trend in scattered intensity with composition. These data were fitted by a two function scattering law assuming that the sample was phase separated, but that within the domains, a single-phase scattering law prevailed. Higgins' results [43] showed that it is more probable that the very large concentration fluctuations which gives rise to the Debye–Bueche neutron scattering are also responsible for the extraordinarily broad ΔT_g in this blend. Because the curves for the four blend compositions [43] were very similar, it is difficult to obtain more detailed information about morphology and the concentration distribution in domains from these SANS results. Checking the dC_p/dT signal versus temperature for the four blend compositions, it was found that the four dC_p/dT signals versus temperature were very different, indicating that this approach could prove useful in obtaining a fuller understanding of phase morphology.

For different domains, the concentration distribution will be different. These different domains will show different glass transition behaviour. The system may be divided into many sub-systems, 1, 2, 3, ..., n each with a corresponding T_g: T_{g1}, T_{g2}, T_{g3}, ..., T_{gn}. When the difference in concentration between domains is small, the glass transition may be considered to arise from a continuous distribution of such sub-systems.

From the above discussion, it is concluded that the dC_p/dT signal from MTDSC can give very useful information about polymer–polymer miscibility more directly than can the scattered intensity signal from SANS experiments.

4.2 INTERFACE DEVELOPMENT BETWEEN COMPATIBLE POLYMER FILMS

The interface between two polymers, whether compatible or incompatible, is a region of finite thickness within which the composition varies continuously from one bulk phase to the other [44]. This interfacial region is formed by interdiffusion of the two continuous phases, driven by the chemical potential gradient. In an incompatible system, the equilibrium interfacial thickness is attained when the entropy effect equals the enthalpy effect [45–48], giving a thickness of typically 1–20 nm, depending on the degree of compatibility [45–49]. The formation of a diffuse interface is important in adhesion [45–48,50,51], phase separation and the consequent morphology in polymer blends [52–54], welding and crack healing [55,56], and co-extrusion [57]. In these applications, the final properties are determined by the thickness of the interface and the concentration profile of the two polymers across that interface. Interdiffusion at polymer–polymer interfaces is a strong function of temperature, mutual compatibility, molecular weight, molecular weight distribution, chain orientation and the molecular structure of the polymers concerned [58–62]. For example, Brochard-Wyart and de Gennes [62,63] showed that under asymmetrical conditions polymers reptate in a set of moving tubes. Brochard-Wyart and co-workers [64,65] showed that the initial asymmetry in the kinetics induced by the chain end segregation is healed after a characteristic Rouse time. Jabbari and Peppas [66] showed experimentally that for polymer pairs with dissimilar physical properties the concentration profile is highly asymmetric.

To describe the effect of the above parameters on interdiffusion, de Gennes [67] used the chemical potential gradient as the driving force for interdiffusion. Assuming that the fluxes of the two components were equal, but opposite, Brochard-Wyart *et al.* [68] derived the slow-mode theory for interdiffusion at polymer interfaces.

$$D = \Lambda_A \Lambda_B / (\Lambda_A + \Lambda_B)[1/(N_A \phi_A) + 1/(N_B \phi_B) + 2\chi] \tag{14}$$

D is the interdiffusion coefficient, Λ_A and Λ_B are the segment mobilities of polymers A and B, respectively, N_A and N_B are the number of repeat units in each polymer, ϕ_A and ϕ_B are the molar fractions of each polymer and χ is the Flory–Huggins interaction parameter. The slow-mode theory predicts that interdiffusion is dominated by the slow-diffusing polymer. Later, de Gennes [69] showed that the mobility was directly related to the diffusion coefficient of each polymer. The limitation of this theory is that it assumes that the fluxes of the two polymers are equal and opposite, which means that the interface remains symmetrical as interdiffusion proceeds.

On the other hand, Kramer and co-workers [70,71] showed that, for polymer pairs with different molecular weights, the interface moves towards the polymer with the lower molecular weight as interdiffusion proceeds. Kramer *et al.* [72] and Sillescu [73] described interdiffusion in systems with a moving interface by unequal fluxes of polymers A and B, which were balanced by a net flux of vacancies across the interface. By assuming that the chemical potential of these vacancies was zero in the melt state, but the flux of vacancies was finite, they derived the following equation for the interdiffusion coefficient.

$$D = \phi_A\phi_B/(\phi_B/\phi_A\Lambda_A + \phi_A/\phi\Lambda_B)[1/(N_A\phi_A) + 1/(N_B\phi_B) + 2\chi] \quad (15)$$

In the fast-mode theory, the overall mobility is linearly related to the mobility of each component, indicating that the interdiffusion coefficient is dominated by the faster-moving component.

Akcasu *et al.* [74] attempted to identify the fast and slow modes with the two modes observed in dynamic scattering experiments from ternary polymer solutions. They defined the vacancies as the third component in a mixture of A and B polymers and concluded that the slow mode was obtained when vacancies were gradually removed, resulting in an incompressible binary mixture of A and B. The fast mode was obtained in the opposite limit of high vacancy concentration or a matrix with very high mobility. Since the polymer mobility and the vacancy concentration are small below, and high above, T_g, this suggested that the slow and fast-mode theories described interdiffusion below and above T_g, respectively.

In fact, most of the interdiffusion data in the literature [69–72,75,76] that were collected above T_g, are consistent with the fast-mode theory of interdiffusion. Kramer *et al.* [72] used Rutherford back-scattering spectroscopy to follow the movement of a gold marker at the interface between PS and deuterated PS (d-PS) with different molecular weights. They observed movement of the interface towards the fast-diffusing component. Reiter *et al.* [77] used X-ray reflection spectrometry also to follow the movement of a gold marker placed at the interface between PS and d-PS. They were able to detect a delay in the onset of interface movement, which depended on molecular weight, and there was a strong indication of a correlation between this induction time and the reptation time of the chain. Wu *et al.* [78] investigated the structure and kinetics of the diffuse interface between PMMA and poly(vinylidene fluoride) in the melt. They too detected interface movement using a gold marker. The structure and kinetics confirmed the predictions of the reptation theory [55]. The interfacial thickness was seen to grow with $t^{1/2}$, where t is the diffusion time.

Interdiffusion between two compatible polymers has also been studied by means of X-ray reflection spectrometry [77], TEM [78], Rutherford backscattering spectrometry [72] and forward recoil spectrometry [79]. We will now report on the use MTDSC to study symmetrical and asymmetrical interdiffusion between two compatible polymers. The main aim is to provide a relatively accessible method to investigate symmetrical and asymmetrical interdiffusion. Conclusions on whether symmetrical or asymmetrical interdiffusion occurs between two compatible polymers have been based on

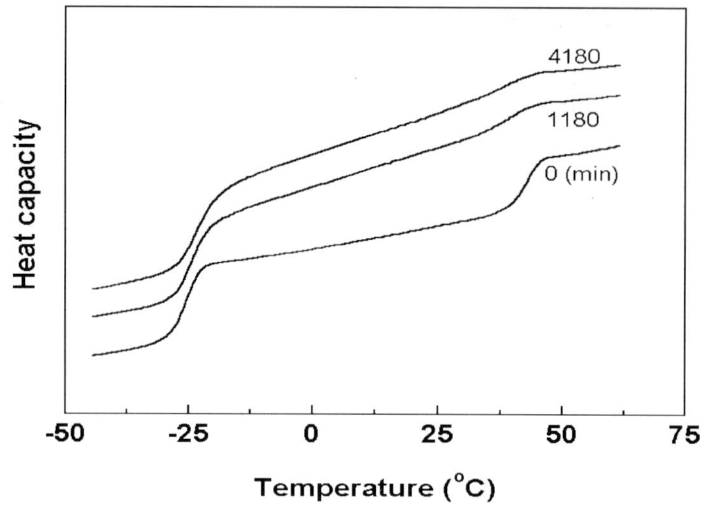

Figure 3.25. Heat capacity (arbitrary units) versus temperature at different diffusion times for the PECH–PVAc combination.

the diffusion coefficients of the two polymers and on measurements of the diffusion profile [78,79].

4.2.1 *Asymmetrical Interdiffusion: Polyepichlorohydrin/Poly(vinyl acetate)*

Figure 3.25 shows the changes of heat capacity with temperature for the polyepichlorohydrin (PECH)/poly(vinyl acetate) (PVAc) combination at different diffusion times. In the glass transition region, the heat capacity traces are different for the different diffusion times.However, it is difficult to draw out more detailed information from these traces. The dC_p/dT curves, however, clearly showed that an interface is formed by thermal diffusion. (see Figure 3.26). This is shown by the increase in the dC_p/dT signal between the two glass transitions. With increasing diffusion time, the concentration of the interface will change and its thickness will increase.

When a system exhibits an interface, the following equations hold.

$$\Delta C_p = \Delta C_{p^1} + \Delta C_{p^2} + \Delta C_{p^i} \tag{16a}$$

$$\Delta C_{p^1} = \omega_1 \Delta C_{p^{10}} \tag{16b}$$

$$\Delta C_{p^2} = \omega_1 \Delta C_{p^{20}} \tag{16c}$$

ω_1 and ω_2 are the weight fractions of components 1 and 2, respectively, in the mixed phases. ΔC_{p^i} is the increment of heat capacity of the diffuse interface in its glass transition region, and δ_1 and δ_2 in the following equations are

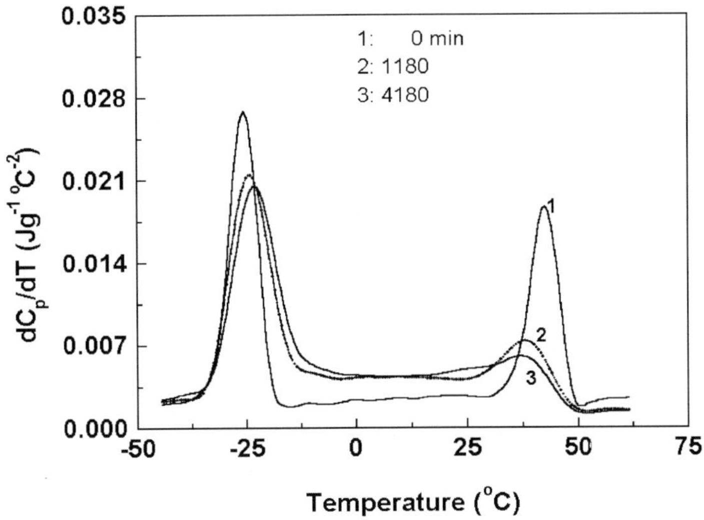

Figure 3.26. dC_p/dT versus temperature at different diffusion times for the PECH–PVAc combination.

the weight fractions in the diffuse interface for polymer 1 and polymer 2, respectively, which can be obtained from these equations.

$$\delta_1 = \omega_{10} - \Delta C_{p^1}/\Delta C_{p^{10}} \tag{17a}$$

$$\delta_2 = \omega_{20} - \Delta C_{p^2}/\Delta C_{p^{20}} \tag{17b}$$

ω_{i0} and ΔC_{pi0} are the weight fraction and the increment of heat capacity of the polymers before mixing.

Using Eqs. (16) and (17), the weight fraction of interface can be calculated.

Figure 3.27 shows the change of weight fraction of the interface with time and Figure 3.28 shows the changes of weight fraction, ω_A and ω_B, of the PECH and PVAc components in the interface with time. Clearly, the change of ω_A and of ω_B with time are different. This indicates that the diffusion rate for PVAc is faster than that for PECH. The interdiffusion for this polymer pair is, thus, asymmetrical.

Now, consider the average value, ρ, of the density of PECH and PVAc in the diffuse interface. Assuming ρ approximates to the linear sum of ρ_A and ρ_B,

$$\rho = (\rho_A \omega_A W_{PECH} + \rho_B \omega_B W)/(\omega_A W_{PECH} + \omega_B W_{PVAC}) \tag{18}$$

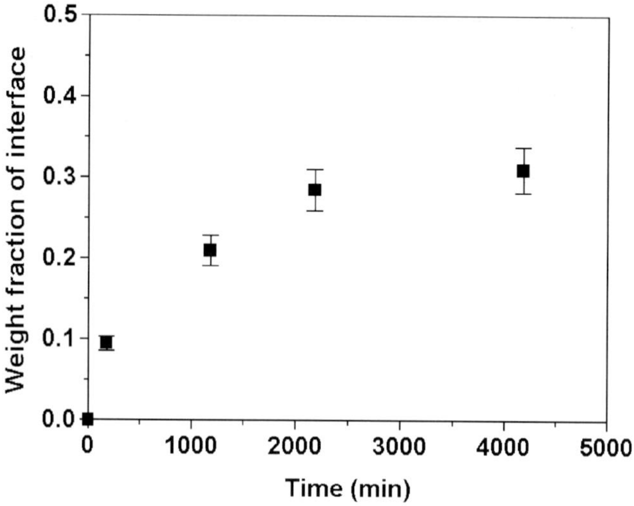

Figure 3.27. Weight fraction of the interface versus diffusion time for the PECH–PVAc combination.

W_{PECH} and W_{PVAC} are the weights of PECH and PVAc, respectively, in the pure phases before mixing. The volume of the interface, V, is given as follows.

$$V = W/\rho \tag{19}$$

W is the mass of the polymers in the interface.

$$W = \phi(W_{\mathrm{PECH}} + W_{\mathrm{PVAC}}) \tag{20}$$

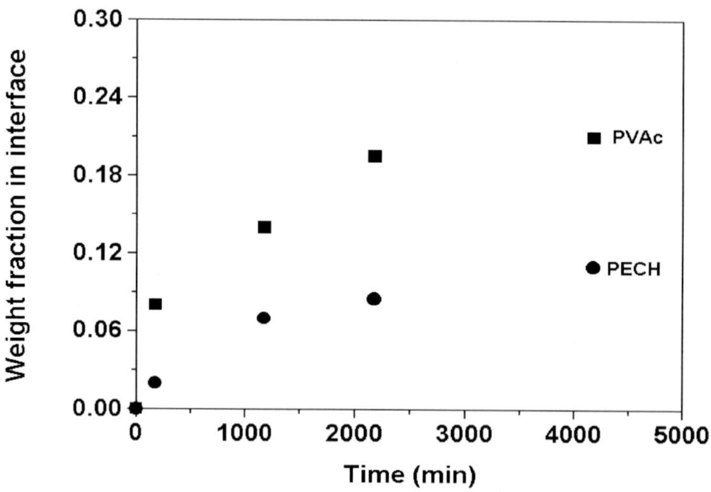

Figure 3.28. Weight fraction of the PECH and PVAc in the interface versus diffusion time.

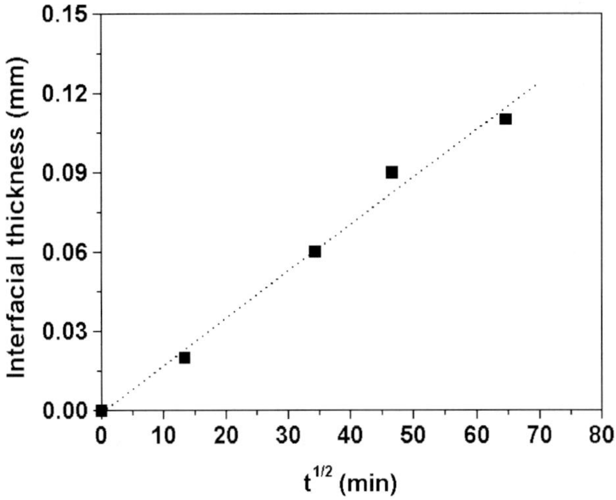

Figure 3.29. Thickness of the interface versus diffusion time for the PECH–PVAc combination.

ϕ is the weight fraction of interface. The average thickness of the interface, d, can be obtained as shown in Eq. (21).

$$d = \phi(W_{\mathrm{PECH}} + W_{\mathrm{PVAC}})/(S\rho) \tag{21}$$

S is the area of the sample, and, therefore, also of the diffuse interface, when considering two superimposed films. The change of thickness of the interface with diffusion time is shown in Figure 3.29. Here, the densities of PECH and PVAc at room temperature were used to calculate the average density, ρ. Obviously, the thickness of interface is a function of diffusion time, t. The interfacial thickness grows according to the following rule which is consistent with the reptation analysis [80] of Wool and Kim [55], Prager and Tirrell [81], Adolf and co-workers [82,83] and Wu *et al.* [78].

$$d \propto t^{1/2} \tag{22}$$

Here, we only give an estimate of the interdiffusion coefficient of the PECH/PVAc pair at 100°C. Based on Fick's diffusion theory [84], the mean-square interfacial thickness, r_{eff}, is given by Eq. (23).

$$r_{\mathrm{eff}} = (d)^2 = (2Dt)^{1/2} \tag{23}$$

From Figure 3.30, which shows the change of r_{eff} with time, it can be calculated that D is approximately 6.25×10^{-11} cm^2/s.

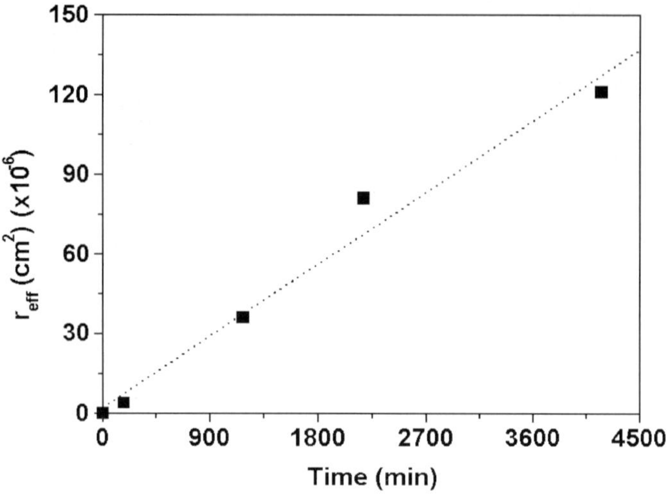

Figure 3.30. Mean-square interfacial thickness versus diffusion time for the PECH–PVAc combination.

4.2.2 *Symmetrical Interdiffusion: Poly(methyl acrylate)/ Poly(vinyl acetate)*

The PMA–PVAc blends are miscible, but show no specific interactions. The interdiffusion coefficient will be as follows.

$$D = D_A = D_B \tag{24}$$

Figure 3.31 shows dC_p/dT versus time at 100°C for the PMA/PVAc combination. The dC_p/dT signal shows clearly that an interface is formed by thermal diffusion. This is shown by the increase in the dC_p/dT signal between the two glass transitions. It can also be seen that the PMA, PVAc and interface signals overlap. A peak-resolution technique, with the condition that ΔC_p (observed) $= \Delta C_p$ (calculated), can be used to deal with this problem. Figure 3.32 shows the result for the sample annealed for 130 h.

Figure 3.33 shows how the weight fraction of the interface increases with time, whilst Figure 3.34 shows how ω_A and ω_B, the weight fractions of PMA and PVAc, respectively, in the interface change with time. The changes of ω_A and ω_B with time are similar, which indicates that interdiffusion in this particular polymer pair is symmetrical. The change of thickness of the interface with diffusion time is shown in Figure 3.35. Here, the room temperature densities of PMA and PVAc were used to calculate the average density, ρ. Thus, for both symmetrical and asymmetrical interfaces, the growth of interfacial thickness can be described by Eq. (22).

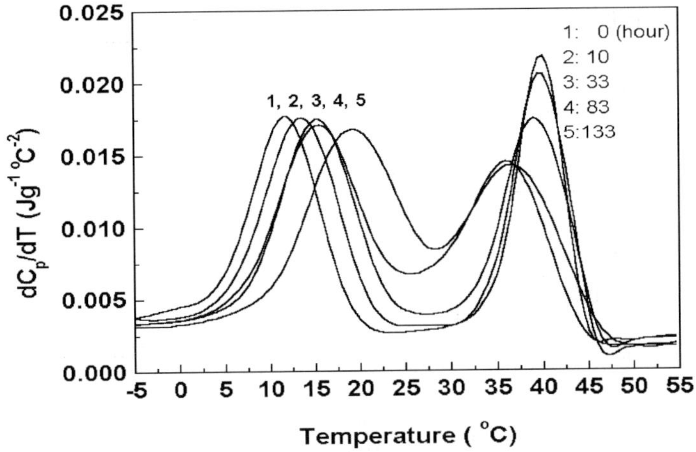

Figure 3.31. dC_p/dT versus temperature data at different diffusion times for the PMA–PVAc combination.

For symmetrical diffusion, the diffusion equation can be solved analytically [84] to give the following solution.

$$\Pi_A(x,t) = 1/2\{1 - \mathrm{erf}[x/(2(Dt)^{1/2})]\} \tag{25a}$$

$$\Pi_B(x,t) = 1/2\{1 + \mathrm{erf}[x/(2(Dt)^{1/2})]\} \tag{25b}$$

Figure 3.36 shows how r_{eff} changes with time. The calculated D value is approximately 4.1×10^{-11} cm^{2}s^{-1}.

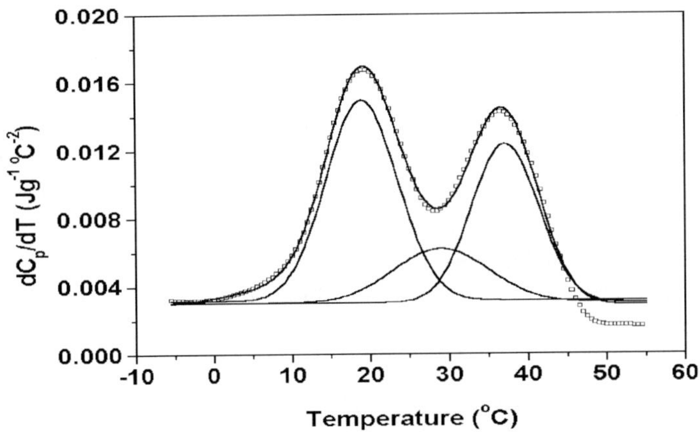

Figure 3.32. Comparison of the multi-peak resolution results with the experimental data (□) for the PMA–PVAc combination annealed at 100°C for 130 h.

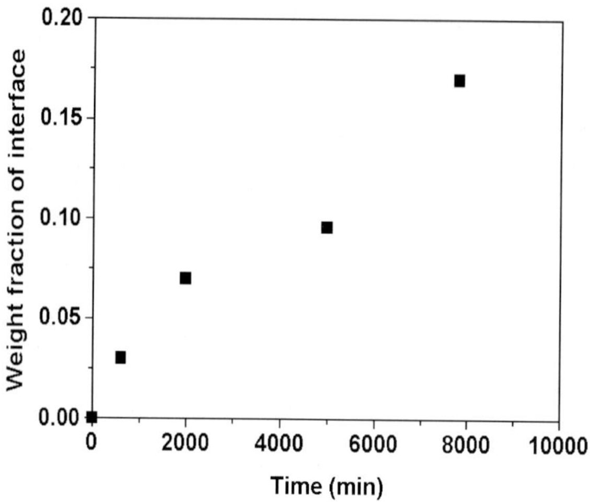

Figure 3.33. Weight fraction of the interface versus diffusion time for the PMA–PVAc combination.

From the above discussion, the symmetrical and asymmetrical interdiffusion between two compatible polymers can be followed based on measurements of the component weight fractions in the interface region.

$\omega_A = \omega_B$ symmetrical interdiffusion

$\omega_A \neq \omega_B$ asymmetrical interdiffusion

The difficulty in a full test of Eqs. (14) and (15) lies in the considerable amount of data required. Tracer diffusion coefficients [72], which are related

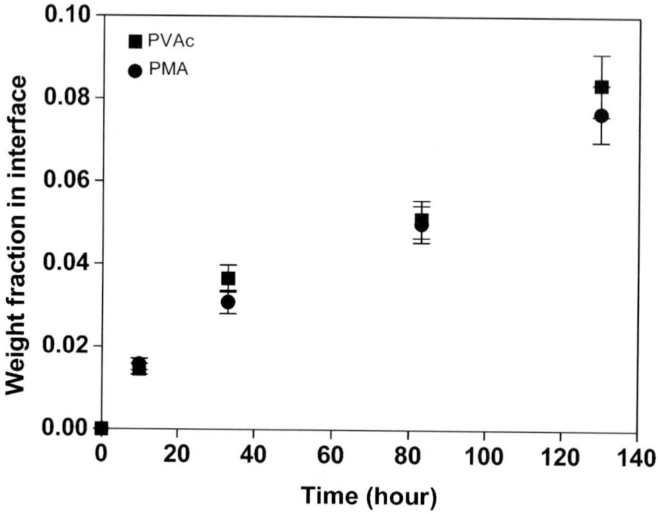

Figure 3.34. Weight fraction of the PMA and PVAc in the interface versus diffusion time.

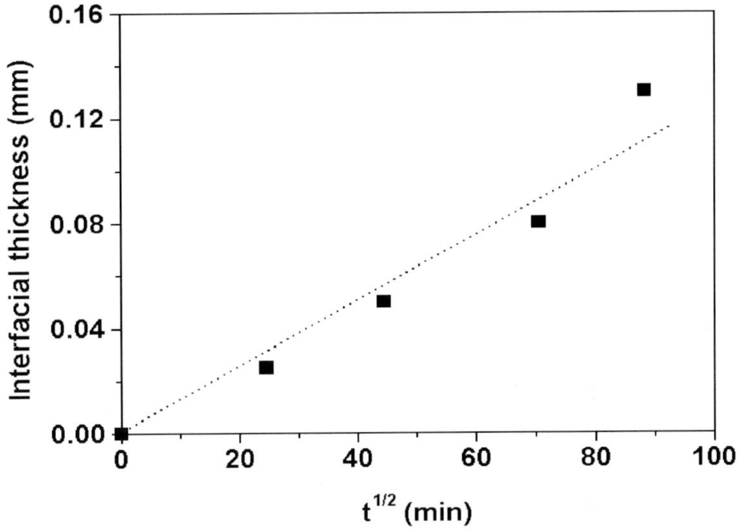

Figure 3.35. Thickness of the interface versus diffusion time for the PMA–PVAc combination.

to Λ_A and Λ_B as a function of composition, as well as the Flory interaction parameter, will, in general, be needed to predict *D*. These quantities are not easy to measure, so that experimental data are quite scarce.

Equation (14) always predicts a lower value of *D* than does Eq. (15). In a system where one of the tracer diffusion coefficients is very small, Eq. (14) predicts that *D*, will also be small, leading to the notion that interdiffusion

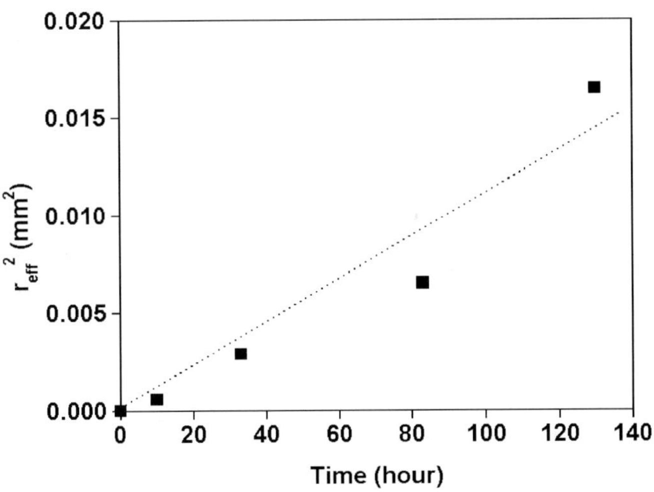

Figure 3.36. Mean-square interfacial thickness versus diffusion time for the PMA–PVAc combination.

is "controlled" by the less mobile species. Equation (15) makes the opposite prediction. Murschall *et al.* [85] have investigated the temperature dependence of D using light scattering and found that the parameters describing interdiffusion as a function of temperature are very close to those describing self-diffusion of the less-mobile species. They concluded that this fact implies that Eq. (14) accurately describes interdiffusion in polymer–polymer systems.

On the other hand, results from recent experiments where the displacement of markers across a polymer–polymer interface has been observed [75,81] have been interpreted to favour Eq. (15). This conclusion has been based largely on arguments concerning the compressibility of the system. Equation (14) implies an incompressible system, whereas Eq. (15) implies a compressible one.

A better approach is to measure the molecular weight dependence of D in entangled polymer mixtures as was done by Gilmore *et al.* [86]. These authors found that, at constant N_A, the dependence of D on N_B could be represented by Eq. (26).

$$D = \alpha + \beta/N_B \tag{26}$$

Assuming a reptation-type behaviour for D_A and D_B, this result is in good agreement with Eq. (15), where α and β will be functions of composition. Equation (14) is not consistent with Eq. (26).

Figure 3.37 shows the changes of the weight fraction of PVAc and PECH in the interface with $t^{1/2}$.

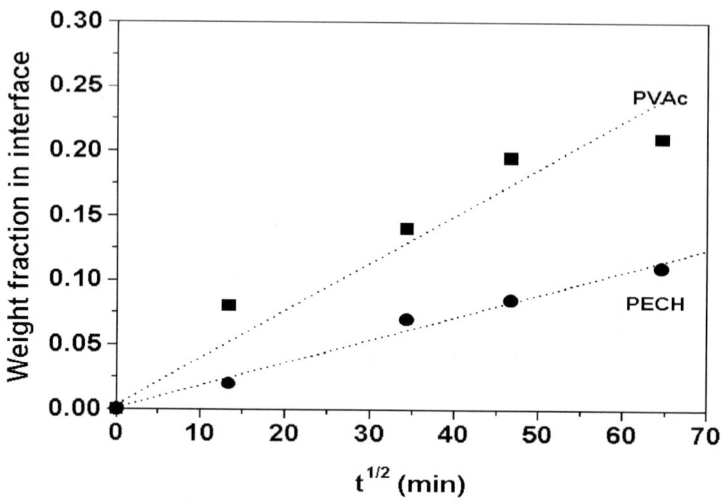

Figure 3.37. Weight fraction of PECH and PVAc in the interface versus diffusion time. (Dashed line is the best fit to the experimental data).

Because d_{PECH} (diffusion thickness) is proportional to ω_{PECH} (in the interface) and because D_{PECH} (or D_A) is proportional to d_{PECH} (diffusion thickness) and, in addition, because d_{PVAC} (diffusion thickness) is proportional to ω_{PVAC} (in interface) and because D_{PVAC} (or D_B) is proportional to d_{PVAC} (diffusion thickness) the reptation behaviour for D_A and D_B is confirmed experimentally. This is evidence in support of Eq. (15).

4.3 STRUCTURED LATEX FILMS

Over the past several years, concern for the environment has generated many instances where there is a need to turn from a polluting technology to one that is more benign. Since these changes are driven by factors outside the technology, this can have the result that the new system has poorer performance characteristics than the technology being replaced. Under these circumstances, it becomes important to understand the origins of good performance, so that adequate, or even improved, performance can be achieved with a new technology that is safer to the environment.

One current example of this situation is the impact on coatings technology of the stricter regulations on volatile organic compound emissions. Because of these restrictions, the use of waterborne latex-based coatings is expanding into areas such as automotive and industrial coatings, traditionally reserved for organic solvent-based systems. The industrial and automotive markets have resisted this change because the waterborne latex coatings are as yet unable to achieve the same high level of performance as the traditional solvent-based systems [87].

In solvent-based coatings, the polymer molecules are entangled and fully interpenetrating as they are applied to a surface. Solvent evaporation leaves a uniform film of low permeability. In latex coatings, the polymers are in the form of discrete (latex) particles that must coalesce during drying and subsequent ageing to form a protective film. Such films are more permeable, especially to moisture, than the corresponding solvent-based films [88] and they provide somewhat poorer protection of the underlying substrate. There are many reasons for the differences in properties between the two types of coating, but it is clear that the "quality of coalescence" of latex coatings has an important effect on the final film properties. This process of coalescence is one of the most important aspects of latex film formation. An understanding of the mechanism by which coalescence occurs is crucial for further advances in this area.

Film formation from polymer latexes is a complicated, multi-stage phenomenon and has been the subject of much theoretical and experimental attention. Many studies of the individual stages, utilising a variety of different techniques, have been published. The use of latex films to investigate molecular interdiffusion is important in terms of theory development in situations such as coatings coalescence, welding and crack healing. There are two basic methods of studying the diffusion of polymer molecules across the boundary between particles in a latex system: SANS [89] and fluorescence techniques [90]. The advantages of SANS lie in its high sensitivity and its ability to determine, easily, the diffusion coefficient and the chain interpenetration depth [89].

The other interesting method utilises fluorescence measurements. This approach has been mainly applied to latex film formation by Winnik and Wang [90]. In this technique, latex is prepared in two different batches. In one batch, the chains contain a "donor" group, while in the other, an "acceptor" group is attached. The interdiffusion of polymer chains between neighbouring latex particles is then studied by direct non-radiative energy transfer measurements.

AFM and TEM techniques can also give information about the change of particle size during coalescence. Goh *et al.* [91] and Hourston and coworkers [92] have studied the integration of a latex film using AFM. They calculated the surface diffusion coefficient based on the classical diffusion model and found it to be 1×10^{-13} cm^2 s^{-1}, which is three to four orders of magnitude larger than that obtained by SANS [89] (10^{-16} to 10^{-17} cm^2 s^{-1}). The difference was attributed to the extra driving force from the surface free energy, which causes faster diffusion near the surface than is the case in the bulk.

Molecular interdiffusion in a core (poly(butyl methacrylate)–shell (poly(butyl methacrylate-*co*-butyl acrylate) latex, which exhibits miscibility between the core and shell polymers, has been studied [93]. The volume fraction of mixing and the inter-particle penetration distance increased with annealing time [93]. In other core–shell latex films, phase separation can occur upon annealing, because of immiscibility of the core and shell phases.

As has already been made clear, interdiffusion is of great importance for the development of the physical properties of latex films [94]. In order to learn how to optimise the performance of a wide variety of coatings formulations, a deeper understanding of the coalescence process is needed. The essential feature that one needs to understand is the role of inter-particle polymer diffusion once the water has evaporated and the nascent film has formed. Although, as reported above, latex film coalescence processes have been studied [90–94], a much better understanding of these processes is needed. In this section, the process of core–shell latex film coalescence and the dynamics of surface structure development of latex films will be discussed in the light of recent MTDSC studies by the authors.

It has already been shown above that the dC_p/dT signal readily provides fruitful information about multi-phase polymer materials. Measurement of the ΔC_p values of the pure shell and core phases at their T_gs leads to

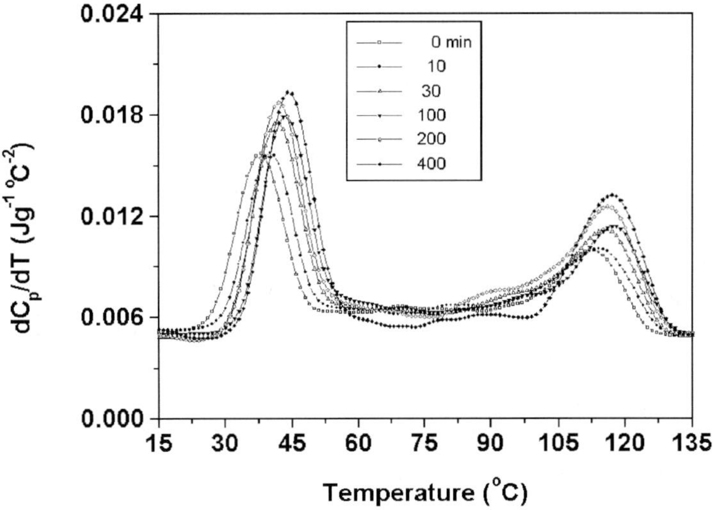

Figure 3.38. dC_p/dT versus temperature data for a PMMA/PVAc latex film annealed at 140°C for different times.

information about the interface between these regions. For core–shell latex particles, interfacial thickness and the weight fraction of that interface are two important property-influencing parameters. However, it is difficult to estimate these parameters for core–shell latexes from TEM and DMTA experiments. However, based on MTDSC measurements, these parameters can be obtained.

Figure 3.38 shows the dC_p/dT versus temperature signals for a PMMA/PVAc core–shell (50/50) latex film after different annealing times at 140°C and Figure 3.39 shows the same signal for the PMMA phase when annealed at 150°C for different times [95]. With increasing time, the dC_p/dT signal obviously changes. The magnitudes of the dC_p/dT signals for the pure PMMA and PVAc components increase, i.e. the ΔC_p values increase indicating that the weight fractions of the pure PMMA and PVAc components increase. The densities of PMMA and PVAc are about 1.19 and 1.192 g cm^{-3} [96], respectively. For an ideal PMMA/PVAc core–shell latex particle, the following relationship holds between the radius, R, of the core and the thickness, ΔR, of the shell.

$$3R^2\Delta R + 3R\Delta R^2 + \Delta R^3 = R^2 \quad (27)$$

For the films cast from the PMMA/PVAc core–shell latex, $R + \Delta R$ was found to be 100 nm. Then, R is 79 and ΔR is 21 nm.

Based on MTDSC measurements, the amount of interface in the unannealed PMMA/PVAc core–shell latex was about 44 wt%, a quite large value.

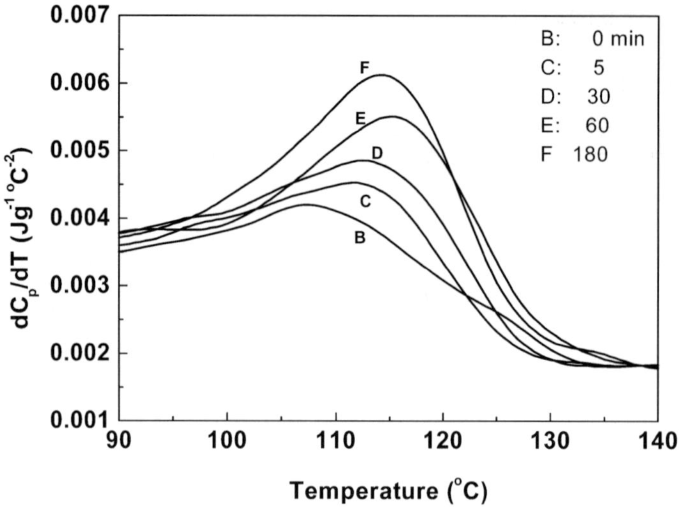

Figure 3.39. dC_p/dT versus temperature data for the PMMA phase in a PMMA/PVAc latex film annealed at 150°C for different times.

This is not surprising because the system is at least partially miscible [35]. When an interfacial phase exists, the shell phase will become thinner and the radius of the core phase will also decrease. For this latex, the interfacial region has a thickness of about 27 nm. This is taking zero annealing time as being a true reflection of the morphology in the original latex particle state. With increasing annealing time, the interfacial thickness decreases. Figure 3.40 shows the change of weight fraction of the interface with annealing time at 150°C. With increasing annealing time, the weight fraction of the

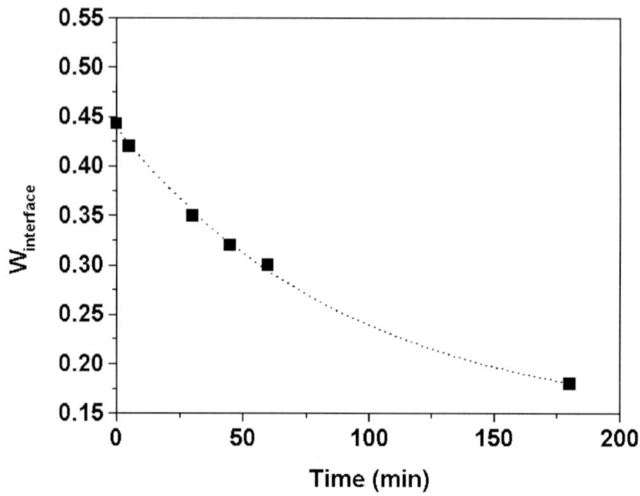

Figure 3.40. Weight fraction of interface versus annealing time at 150°*C* for the PMMA/PVAc core–shell latex.

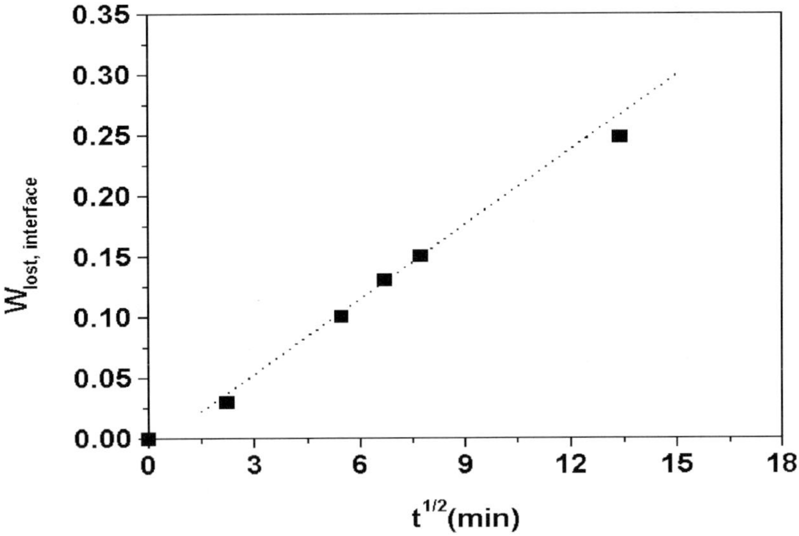

Figure 3.41. Lost weight fraction of the interface versus square root of time.

interface decreases. Figures 3.41 and 3.42 show the changes in the weight fraction, ω_{lost}, of interface for the total and individual parts, respectively [95]. The change with time can be described by Eq. (28).

It is, therefore, confirmed that the macromolecular diffusion during phase separation can be described by the reptation model, i.e. the mechanism of

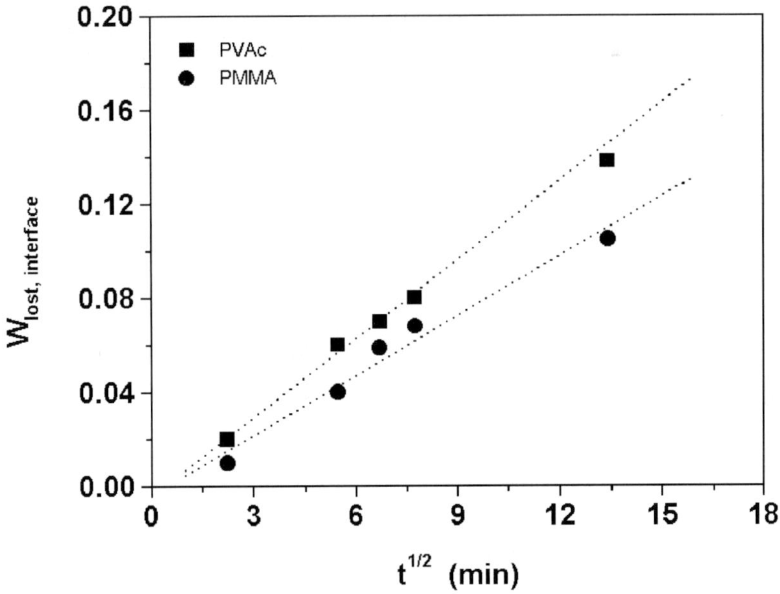

Figure 3.42. Lost weight fraction of the individual components in the interface versus square root of time.

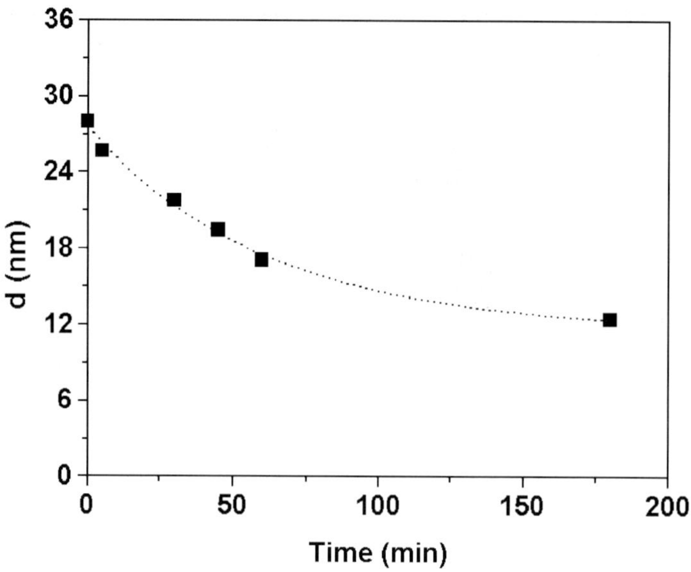

Figure 3.43. Interfacial thickness of the PMMA/PVAc core–shell latex versus time of annealing at 150°C.

phase separation is the same as that in the interdiffusion, discussed previously, of two compatible polymer films.

Figure 3.43 shows the change of interfacial thickness of the PMMA/PVAc core–shell latexes with time [95].

Macromolecular diffusion in the interface between the core and shell phases can be illustrated by a model composed of three parts: core, *A*, the interface between the core and the shell, *AB* and the shell phase, B as shown in Figure 3.44. It is assumed here that the core phase is totally covered by the shell phase. The PMMA/PVAc latex is phase-separated at high temperature [97]. During phase separation of the interfacial phase, polymer A in the core does not diffuse out and polymer B in the shell does not diffuse into the *AB* and core phases. The parameters $C(r, t)$ and $\omega(r, t)$ are the concentrations of polymer A and polymer B which diffuse into the core and shell phases, respectively.

According to Fick's second law,

$$\nabla(DC) = \partial C/\partial t \tag{28}$$

$$D_{\mathrm{A}}[\partial^2 C/\partial r^2 + 2/r\ \partial C/\partial r) = \partial C/\partial t \tag{29}$$

$$D_{\mathrm{B}}[\partial^2 \omega/\partial r^2 + 2/r\ \partial \omega/\partial r) = \partial \omega/\partial t \tag{30}$$

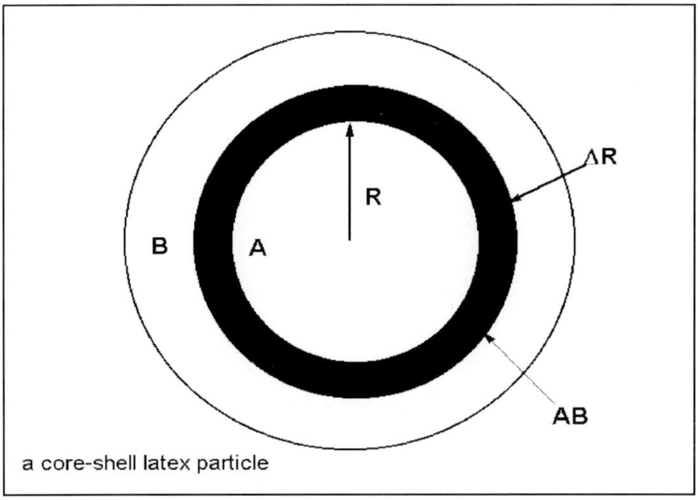

Figure 3.44. Model of a core–shell latex particle with an interphase.

The initial conditions are

$$C(r, 0) = 0 \tag{31}$$

$$\omega(r, 0) = 0 \tag{32}$$

D_A and D_B are the diffusion coefficients of polymers A and B, respectively.

Let $C = rY$, then

$$D_A \partial^2 Y/\partial r^2 = \partial Y/\partial t \tag{33}$$

$$Y(r, 0) = 0 \tag{34}$$

And let $\omega = r\mathrm{Z}$

$$D_B \partial^2 Z/\partial r^2 = \partial Z/\partial t \tag{35}$$

$$Z(r, 0) = 0 \tag{36}$$

Taking the Laplace transforms of Eqs. (33) and (35) yields Eqs. (37) and (38).

$$D_A \mathrm{d}^2 Y(r, p)/\mathrm{d}r^2 = pY(r, p) \tag{37}$$

$$D_B \mathrm{d}^2 Z(r, p)/\mathrm{d}r^2 = pZ(r, p) \tag{38}$$

Let

$$F(t) = \int_0^R 4\pi r^2 \; C(r,t)\mathrm{d}r \tag{39}$$

and

$$\Phi(t) = \int_R^{R+\Delta R} 4\pi r^2 \omega(r,t)\mathrm{d}r \tag{40}$$

$F(t)$ and $\Phi(t)$ are the weight fractions of polymer A and polymer B which have diffused into core and shell phases, respectively, at time t. Taking Laplace transforms [95],

$$F(t) = A_o\{R/(\pi D_A t)^{1/2}\exp[-R^2/(4D_A t)] - \mathrm{erf}(R/(4D_A t)^{1/2}) + 1\} \tag{41}$$

A_o is a constant. Equation (41) can be used to simulate the process of phase separation of the interfacial phase and to estimate the diffusion coefficients.

Figure 3.45 compares the calculated and experimental results. $D_A \approx 4.2 \times 10^{-14}$ cm^2 s^{-1}. This valuc is similar to that obtained [97] by the light scattering technique for the phase separation of PMMA/PVAc blends.

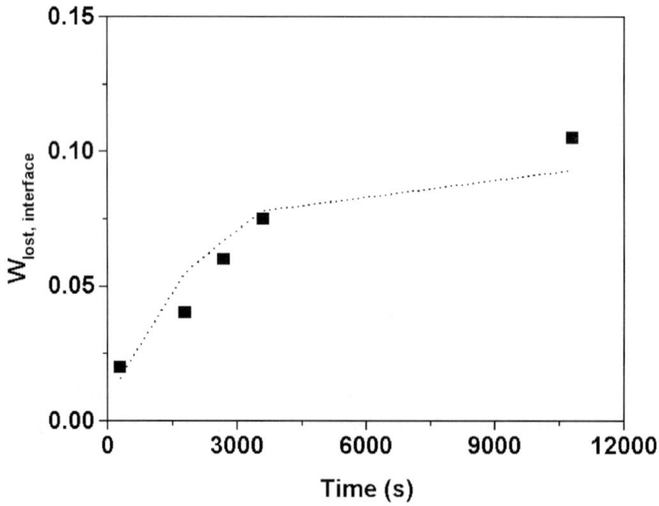

Figure 3.45. Weight fraction of PMMA which has diffused into the core phase versus annealing time at 150°C. [Dotted line is Eq. (41)].

4.4 MORPHOLOGY ANALYSIS OF INTERPENETRATING POLYMER NETWORKS

An interpenetrating polymer network (IPN) is defined as a combination of two crosslinked polymers, at least one of which has been synthesised [98] and/or crosslinked in the immediate presence of the other. From the topological point of view, IPNs are closely related to polymer blends and to block, graft and crosslinked copolymers. From the synthesis point of view, IPNs can be classified, broadly, into two general types: (a) sequential IPNs where a polymer network is formed which is then swollen by the monomer, plus a crosslinking agent and an activator, which is then polymerised in situ to form the second network; and (b) simultaneous IPNs (SIPN) where the components necessary to form both networks are mixed and polymerised, at the same time, by non-competing mechanisms. If one of the two polymers is linear (uncrosslinked), a semi-IPN results. A homo-IPN results if both the network polymers are identical in chemical composition [98].

Since the second polymer is still in monomeric form when it is mixed with the first polymer, there is still a considerable entropy of mixing and many monomer–polymer combinations are possible. Upon polymerisation, however, the entropy of mixing is greatly decreased and phase separation [98] usually occurs. The vast majority of IPNs are phase separated multiphase materials. The networks limit the extent of phase separation and give a degree of control of the phase size and extent of mixing of the two components.

Since the historic synthesis of an IPN by Millar [99] in 1960, many papers, including reviews, on IPNs, have been published, and around 20 different products are offered on the market [100]. Most of the papers describe the synthesis and morphological behaviour [98,101–109], status and developments [110,111], properties [112] and industrial applications [113–116] and self-organisation [117] of IPNs. In recent years, a significantly increasing number of commercial IPN products ranging from false teeth to ion-exchange resins, high impact plastics, thermoplastics, adhesives, vibration damping materials and high temperature alloys have been developed.

It is often important to know the morphology of IPNs and the factors influencing it, since phase size, shape and connectivity and the nature of the interphase boundary determine the physical and mechanical properties of such materials. Together, these parameters combine to describe the morphology of the IPN. IPN morphology can be particularly complicated and has been the subject of many studies [118,119]. Most show that during polymerisation, two competing processes take place simultaneously. Phase separation of the forming polymer chains proceeds by diffusion through an increasingly viscous medium to form the domains. The formation of crosslinks restricts this diffusion and, at gelation, the then present situation is frozen in. Consequently, phase separation in IPNs depends primarily on (i) the miscibility of the constituent polymers, (ii) the crosslink density in both polymer networks and any inter-network grafting, (iii) the reaction conditions (temperature, pressure) and (iv) the relative reaction rates of network formation. With highly incompatible polymers, the thermodynamic driving force for phase separation is so powerful that gross phase separation occurs before gelation [98].

Among the techniques that have been used to investigate IPN morphology are DSC [16,120], TEM [121], SEM [122], DMTA [19], SANS [29], SAXS [123] and dielectric measurements [124]. Inevitably there have been disagreements about the levels of miscibility in particular systems. The reader who wants further background should refer to Refs. [125–127]. To address this problem of the degree of mixing in IPNs, there is a continuing need for new techniques. Two approaches reported recently by Meyer co-workers [128] and Winnik *et al.* [129] involve solid-state NMR spin-diffusion [128] and direct non-radiative energy transfer [130] experiments, respectively. Can the MTDSC developments already introduced in this chapter play a role in revealing, in more detail, the morphologies of IPNs?

4.4.1 Characterisation of Glass Transition Behaviour in Interpenetrating Polymer Networks

The multi-phase nature of IPNs results in complicated glass transition behaviour [101]. Figure 3.46 shows that heat capacity changes with temperature for a series 60:40 polyurethane (PU)/ polystyrene (PS) IPNs (see Table 3.5 for the compositional details) [131,132]. It is, however, not possible to obtain much detailed information from these heat capacity signals.

Figures 3.47(a)–(e) show dC_p/dT versus temperature data for IPN2, IPN3, IPN4, IPN6 and IPN8. The dC_p/dT signal is much more sensitive to the transitions. Figure 3.48 gives a comparison of the dC_p/dT versus temperature plots of a 40% PS + 60% PU physical blend, a situation where no interphase can exist, and IPN9. It is obvious from these figures that the morphologies of these samples are quite complex. The transition region is very broad covering a span of about 180°C. For IPN4 and IPN8, there are broad transitions from 20 to 120°C [131,132].

The crosslink density in the PU component in this series of IPNs was varied by changing the diol/triol ratio. The crosslink level and the glass transition temperature, obtained via MTDSC, are listed in Table 3.6.

It can be seen that with increasing crosslink density in the PU network, the PU T_g shifted towards higher temperature. Figure 3.47 shows that not

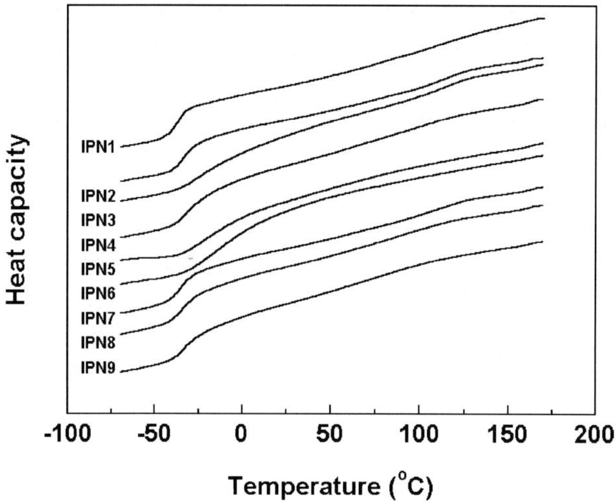

Figure 3.46. Heat capacity versus temperature data for IPN1 to IPN9.

only did the peak location change to higher temperature, but also that the peak decreased in height, and, simultaneously, became broader. The PS transition remained at the same location.

These samples, simultaneous PU/PS IPNs, were synthesised by a one-short route. The IPN topology appears to restrict phase separation, which results in materials with broad transition regions. By variation of the crosslink level in either or both polymer networks, the controlled introduction of inter-network grafting or the incorporation of compatibilisers into the PS network, the compatibility of the two polymer networks can be increased. For simultaneous IPNs, it has been found [129] that the network which is first formed

Table 3.5. Composition of the PU/PS IPN series

Code	PU/PS	Diol/triol	DVB[a]
IPN1	60/40	7:1	5 mol%
IPN2	60/40	3:1	5 mol%
IPN3	60/40	1:1	5 mol%
IPN4	60/40	3:1	5 mol% with 1 wt% of TMI[b]
IPN5	60/40	3:1	5 mol% with 5 wt% of TMI[b]
IPN6	60/40	3:1	5 mol% with 10 wt% of TMI[b]
IPN7	60/40	3:1	5 mol% standard polymerisation
IPN8	60/40	3:1	5 mol% with 10 wt% of compatibiliser[c]
IPN9	60/40	3:1	5 mol% with 2.5 wt% of TMI

[a]DVB: divinybenzene.
[b]TMI: benzene-1-(1-isocyanato-1-methylethyl)-3-(1-methylethenyl).
[c]Compatibiliser: a polyoxypropylene glycol 1025 molecule terminated at both ends with TMI units is incorporated in the PS network.

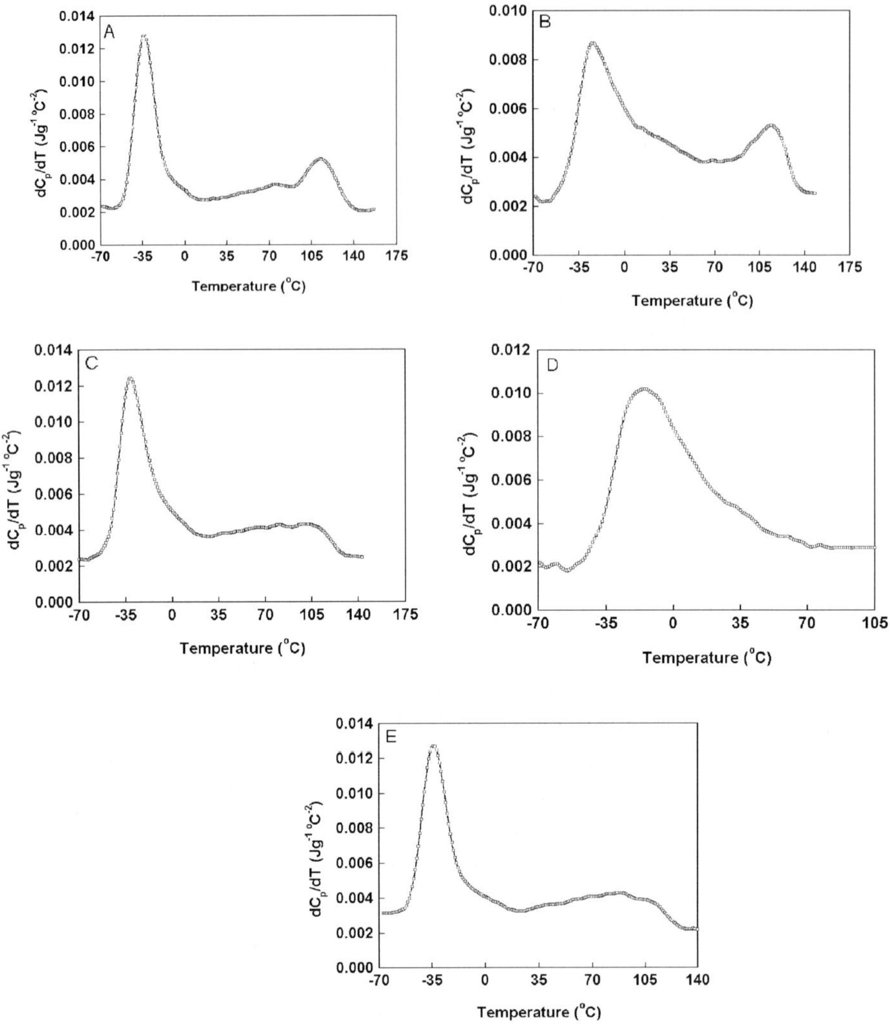

Figure 3.47. dC_p/dT versus temperature data for (A) IPN2, (B) IPN3, (C) IPN4, (D) IPN6 and (E) IPN8.

represents the continuous phase. Hourston and Schafer [133,134] investigated the rate of network formation in the 60:40 PU/PS IPN (IPN7) by means of FTIR spectroscopy coupled with a heated cell unit. The conversion curves of both networks were monitored by following integrated peak areas versus time. This study confirmed that under the given reaction conditions, the PU network formed first. In such a situation, it is believed that several possible morphologies could result. (a) The two networks could be miscible yielding

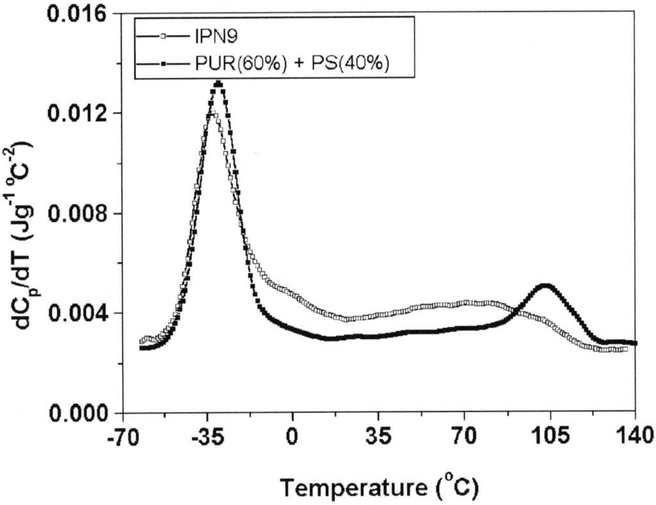

Figure 3.48. dC_p/dT versus temperature data for the 40% PS + 60% PUR physical blend and for IPN9.

a homogeneous material. (b) The first-formed network could be uniformly distributed in space, but with the second-formed network heterogeneously distributed. (c) Both networks could be heterogeneously distributed in space, but with interfacial zones containing a mixture of the two networks. For the first situation, a single glass transition would be obtained. For the second situation, the glass transition temperatures could be shifted somewhat. For the third situation, the glass transition region will broaden.

4.4.2 Model Experiment

The aim is to establish a quantitative analysis method applicable to IPNs. A spectrum can be synthesised by using an analogue method to sum a series of functions representing individual peaks in order to produce a final function that closely represents the experimental spectrum.

Table 3.6. Crosslink level and glass transition temperatures

		T_g (°C)[a]	
Diol/triol	**DVB**	**PU-rich phase**	**PS-rich phase**
7:1(IPN1)	5 mol%	−38	113
3:1(IPN2	5 mol%	−33	113
1:1(IPN3)	5 mol%	−24	113

[a]The T_g values were obtained by the multi-peak resolution technique (133,134).

For an IPN, we may consider dC_p/dT as a multiple Gaussian function in the glass transition region.

$$
\begin{aligned}
dC_p/dT &= B + f(T) \\
f(T) &= \sum_i f_i(T, T_{gi}, \omega_{di}, \Delta C_{pi}) \\
&= \Delta C_{p1}/[\omega_{d1}(\pi/2)^{1/2}]\exp\left[-2(T - T_{g1})^2/\omega_{d1}^2\right] \\
&\quad + \Delta C_{p2}/[\omega_{d2}(\pi/2)^{1/2}]\exp\left[-2(T - T_{g2})^2/\omega_{d2}^2\right] \\
&\quad + \Delta C_{p3}/[\omega_{d3}(\pi/2)^{1/2}]\exp\left[-2(T - T_{g3})^2/\omega_{d3}^2\right] \\
&\quad + \cdots
\end{aligned}
\tag{42}
$$

where $f_i(T)$ is related to the ith phase of the multi-phase system.

To evaluate this model, an experiment with a four-component system was conducted This system was a poly(methyl acrylate)/poly(vinyl acetate) (PMA/PVAc) physical blend, or mixture, consisting of four individual blends (PMA/PVAc (80/20) + PMA/PVAc (60/40) + PMA/PVAc (40/60) + PMA/PVAc (20/80)). PMA is miscible with PVAc. The open squares in Figure 3.49 are the experimental dC_p/dT data. The difference between glass transition temperatures of PMA and PVAc is about 33°C. In the glass transition region, the four-component mixture showed an acceptable fit to the experimental data, see Figure 3.49. The solid lines shown in Figure 3.49

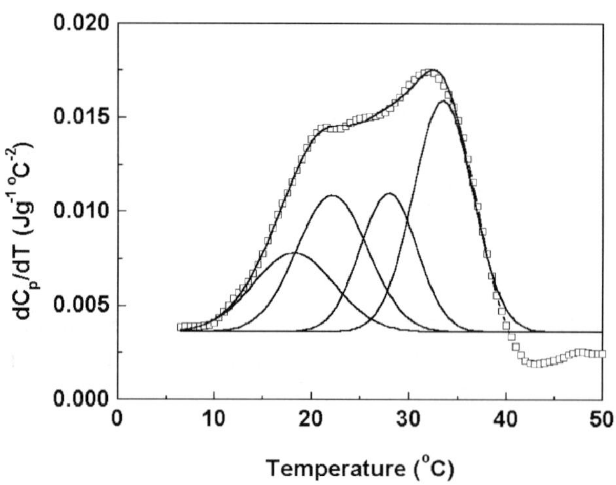

Figure 3.49. Comparison of experimental data with peak resolution results for a four-component model system.

Table 3.7. Comparison of known weight fraction with the calculated value

System	Known	Calculated
PMA-20	30	27.8
PMA-40	23	21.1
PMA-60	25	22.9
PMA-80	22	19.7

are the fitting and peak resolution results. The conditions for the fitting and peak resolution are as follows.

1: ΔC_p (fitting) = ΔC_p (experimental).

2: $\Delta T_g = w_1 \Delta T_{g1} + w_2 \Delta T_{g2}$.

ΔT_g is the transition width and ΔT_{g1} and ΔT_{g2} are the glass transition widths for pure polymer 1 and polymer 2, respectively. Table 3.7 shows the comparison of the known and calculated results. The average difference is about 8%.

Curve fitting of this type assumes that a particular peak profile is uniquely characterised once its peak width at half maximum has been fixed, and cannot be resolved into sub-components. In most practical situations, a Gaussian profile is unique and curve fitting may be undertaken [135].

4.4.3 *Analysis of Phase Structure of IPNs*

Consider that there exist interfacial phases in IPNs. The dC_p/dT signal may then be divided into three parts by the peak resolution method. These are related to the PU-rich, PS-rich and the interfacial phases. The phase that has the lowest T_g is considered as a PU-rich phase and the phase with the highest T_g is considered as being PS-rich. Other phases located between the PU-rich and PS-rich phases are considered as being interfacial.

As examples, Figure 3.50 shows the peak resolution results [131] for the IPN1, IPN2, IPN7, IPN8 and IPN9 materials discussed above.

For IPN1, three transition peaks were obtained. For IPN2, IPN7, IPN8 and IPN9, four transition peaks were involved. DMTA measurements [134] showed that the glass transition temperatures of the PS-rich phase in the IPN1, IPN2 and IPN3 were the same, 133°C. The original MTDSC data for IPN1, IPN2 and IPN3 showed that the glass transition temperatures of the PS-rich phase were different. However, the peak resolution results give the same glass transition temperature, 113°C, for the PS-rich phase in the IPN1, IPN2 and IPN3 materials. The difference may result from the effect of the interface, which results in the shift and broadening of the dC_p/dT peak.

Table 3.8 gives the results of this analysis for the IPN1, IPN2, IPN8 and IPN9 materials.

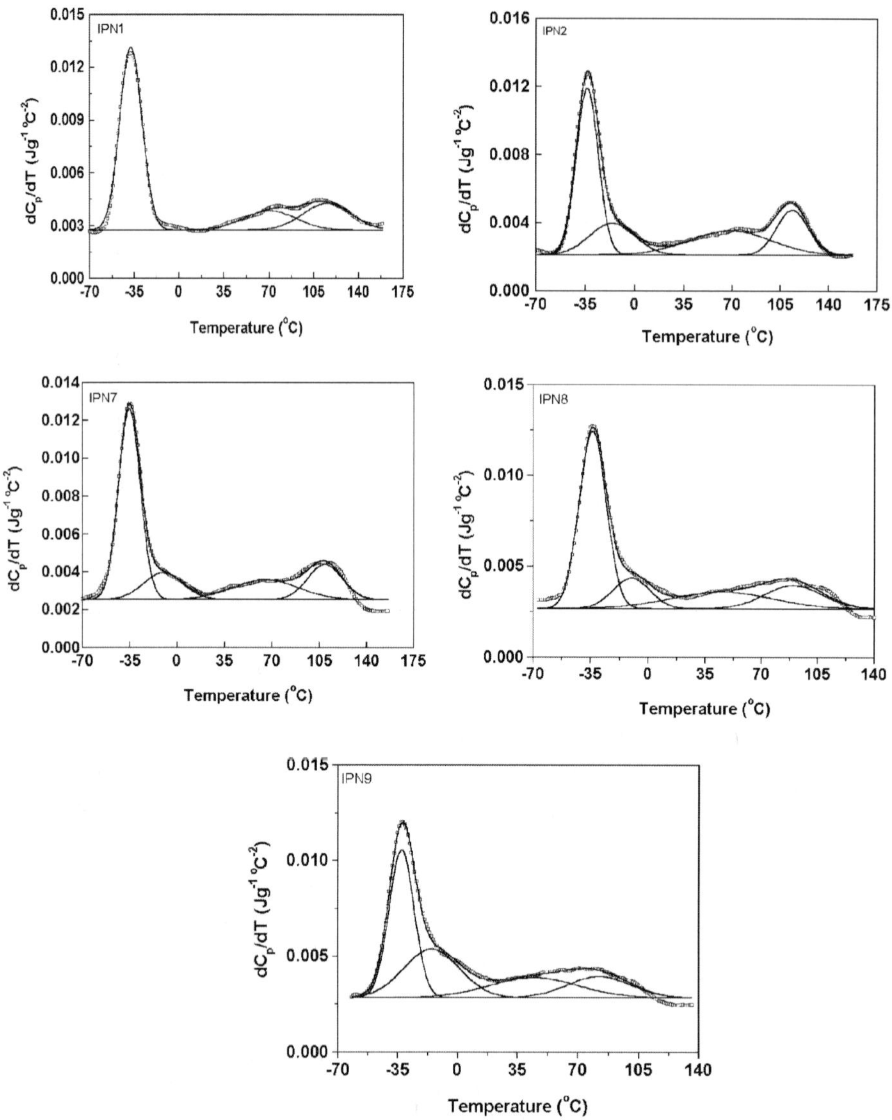

Figure 3.50. Comparison of experimental dC_p/dT data with peak resolution results for the IPN1, IPN2, IPN7, IPN8 and IPN9 materials.

From Table 3.8, it can be seen that there are several different levels of network compatibility. For IPN8, there are PU-rich phases whose T_gs are located at about −33, and −10°C, and PS-rich phases whose T_gs are located at about 46 and 88°C, respectively. For IPN9, there are PU-rich phases at about −33 and −17°C, and PS-rich phases at about 50 and 90°C, respectively. The total interface content in IPN8 and IPN9 is high. This

Table 3.8. T_g *and weight fraction values for the IPNs [131]*

System	T_g (°C)	Weight fraction (%)
IPN1	−38	58
	70	14 interface
	113	30
IPN2	−33	48
	−15	16 interface
	70	15 interface
	113	20
IPN8	−33	43
	−10	17 interface
	46	20 interface
	88	21
IPN9	−33	32
	−17	23 interface
	50	30 interface
	90	19

correlates well the high loss peak in the DMTA data [134]. For IPN2, there are PU-rich phases whose T_g is located at about −15°C and PS-rich phases whose T_g is at about 70°C. In the PU-rich phases, the weight fraction of PS is about 18% and in the PS-rich phases, the weight fraction of PU is about 24%.

By combining the TEM and MTDSC techniques, a clearer understanding of the morphology of IPNs may be obtained. From TEM measurements, phase domain size and shape and connectivity can be determined. From MTDSC measurements, the weight fraction of interphase regions can be obtained. So, the relationships between mechanical properties and IPN morphology can now, in practice, be more comprehensively investigated.

5 Conclusions

It has been shown in this chapter that the MTDSC technique is a very useful tool in the study of several aspects of polymer blends and related materials including structured latexes and interpenetrating polymer networks. It is important to note that the dC_p/dT versus temperature signal may be used not only qualitatively as a sensitive detector of transitions impossible to spot by other thermal techniques such as conventional DSC and DMTA, but it may also be used to significant advantage in a quantitative way. It has been shown that it is sensitive to the diffuse interface between phases. Thus, from dC_p/dT versus temperature signals, the weight fraction of the diffuse interface can be quantified. There are many situations where this will prove to be very valuable.

References

[1] B.J. Hunt and M.I. James, Polymer Characterisation, Blackie Academic & Professional, London, (1993).
[2] L.H. Sperling, Introduction to Physical Polymer Science, John Wiley & Sons, Inc, NY (1992).
[3] S.L. Cooper and G.M. Estes, Multiphase Polymers, American Chemical Society, Washington, DC (1979).
[4] J.A. Manson and L.H. Sperling, Polymer Blends and Composites, Plenum Press, NY (1976).
[5] O. Olabisi, L. Robeson and M.T. Shaw, Polymer–Polymer Miscibility, Academic Press, NY (1979).
[6] D.R. Paul and S. Newman, Polymer Blends, Academic Press, London (1978).
[7] G.D. Spathis, E.P. Sideridis and P.S. Theocaris, Int. J. Adhes. Adhes., 1 (1981) 195.
[8] P.S. Theocaris and G.D. Spathis, J. Appl. Polym. Sci., 27 (1982) 3019.
[9] J. Koberstein, T. Morra and R.S. Stein, J. Appl. Crystallogr., 13 (1980) 34.
[10] F. Annighofer and W. Gronski, Makromol. Chem., 185 (1984) 2231.
[11] R.J. Roe, J. Appl. Crystallogr., 15 (1982) 18.
[12] F. Annighofer and W. Gronski, Colloid Polym. Sci., 261 (1983) 15.
[13] A.C. Fernandes, J.W. Barlow and D.R. Paul, J. Appl. Polym. Sci., 32 (1986) 6073.
[14] A.C. Fernandes, Ph.D. Dissertation, University of Texas, Austin (1986).
[15] M.J. Folkes and P.S. Hope, Polymer Blends and Alloys, Chapman and Hall, London, (1993).
[16] S. Singh, H.L. Frisch and H. Ghiradella, Macromolecules, 23 (1990) 375.
[17] M. Annakutty and P.C. Deb, J. Appl. Polym. Sci., 45 (1992) 2145.
[18] H.R. Allcock, K.B.Visscher and I. Manners, Chem. Mater., 4 (1992) 1188.
[19] Y. Lipatov, V.F. Rosovizky, P.V. Datsko and Y. Maslak, J. Appl. Polym. Sci., 36 (1988) 1143.
[20] D. Vesely, In Polymer Blends and Alloys, M.J. Folkes and P.S. Hope, Eds., Blackie Academic & Professional, London (1993).
[21] K. Kato, Polym. Eng. Sci., 1 (1967) 38.
[22] D.A. Thomas, In Advances in Preparation and Characterisation of Multi-polymer Systems, R.J. Ambrose and S.L. Aggarwal, Eds., John Wiley and Sons, NY (1978).
[23] W. Ruland, Macromolecules, 20 (1987) 87.
[24] T. Nishi, T.T. Wang and T.K. Kwei, Macromolecules, 8 (1975) 277.
[25] M. Song, H. Liang, Y. Chen and B. Jiang, Acta Phys. Chem. Sin., 5 (1991) 513.
[26] V. Nelliappan, M.S. El-Aasser, A. Kilein, E.S. Daniels and J.E. Roberts, J. Appl. Polym. Sci., 58 (1995) 323.
[27] M. Hidalgo, J. Guillot, M.F. Llauro and H Waton, J. Chem. Phys., 89 (1992) 505.
[28] G. Krause and K.W. Rollmann, J. Polym. Sci. Polym. Phys., 14 (1976) 1133.
[29] B. McGarey, In Advances in Interpenetrating Polymer Networks, D. Klempner, K.C. Frisch, Eds. Technomic Publishing Co., Lancaster (1989) p. 69.
[30] M. Song, D.J. Hourston, F.-U. Schafer, H.M. Pollock and A. Hammiche, Thermochim. Acta, 315 (1998) 25.
[31] D.J. Hourston, M. Song, H.M. Pollock and A. Hammiche, J. Thermal Anal., 49 (1997) 209.
[32] D.J. Hourston, M. Song, A. Hammiche, H.M. Pollock and M. Reading, Polymer, 38 (1997) 1.
[33] R.F. Boyer, J. Macromol. Sci. B, 7 (1973) 487.

[34] M.J. Richardson and N.G. Savill, Polymer, 16 (1975) 753.
[35] B. Wunderlich, Thermal Analysis, Academic Press, Boston (1990).
[36] B. Wunderlich and L.D. Jones, J. Macromol. Sci. B, 3 (1969) 67.
[37] B. Wunderlich, Polymer Handbook, Published by John Wiley & Sons New York, Second Edition Section V-55 (1992).
[38] M. Song, A. Hammiche, H.M. Pollock, D.J. Hourston and M. Reading, Polymer, 37 (1996) 5661.
[39] M. Song, A. Hammiche, H.M. Pollock, D.J. Hourston and M. Reading, Polymer, 36 (1995) 3313.
[40] A.K. Nandi, B.M. Mandal and S.N. Bhattacharyya, Macromolecules, 18 (1985) 1454.
[41] P. Perrin and R.E. Prud'homme, Polymer, 32 (1991) 1468.
[42] X. Lu and R.A. Weiss, Macromolecules, 25 (1992) 3242.
[43] J.N. Clark, J.S. Higgins, C.K. Kim and D.R. Paul, Polymer, 33 (1992) 3137.
[44] M. Song, H.M. Pollock, A. Hammiche, D.J. Hourston and M. Reading, Polymer, 38 (1997) 503.
[45] S. Wu, Polymer Interfaces and Adhesion, Dekker, NY (1982).
[46] G. Helfand and Y. Tagami, J. Chem. Phys., 56 (1972) 3592.
[47] K.M. Hong and J. Noolandi, Macromolecules, 15 (1982) 482.
[48] C.I. Poser and I.C. Sanchez, Macromolecules, 24 (1984) 79.
[49] N.H. Sung, A. Kaul, I. Chin and C.S.P. Sung, Polym. Eng. Sci., 22 (1982) 637.
[50] K. Binder and H. Sillescu, In Encyclopedia of Polymer Science and Engineering, Vol. 20, Wiley, NY (1989) p. 297.
[51] H.H. Kausch and M. Tirrell, Ann. Rev. Mater. Sci., 19 (1989) 341.
[52] T.K. Kwei and T.T. Wang, In Polymer Blends, Vol. 1, D.R. Paul and S. Newman, Eds., Academic, NY (1978) pp. 141–185.
[53] H. van Oene, In Polymer Blends, Vol. 1, D.R. Paul and S. Newman, Eds., Academic, NY (1978).
[54] S. Wu, presented at 16th Europhysics Conference on Macromolecules, Polymer Alloys: Structure and Properties, Brugge, Belgium (1984).
[55] Y.H. Kim and R.P. Wool, Macromolecules, 16 (1983) 1115.
[56] K. Judd, H.H. Kausch and J.W. Williams, J. Mater. Sci., 16 (1981) 204.
[57] C.D. Han, Multiphase Flow in Polymer Processing, Academic Press, NY (1981).
[58] P.T. Gilmore, R. Falabella and R.L. Laurence, Macromolecules 13 (1980) 880.
[59] G. Fytas, Macromolecules, 20 (1987) 1430.
[60] S.S. Voyutskii, Adhesion, 3 (1971) 69.
[61] J.K. Kim and C.D. Han, Polym. Eng. Sci., 31 (1991) 258.
[62] P.F. Green and B.L. Doyle, Macromolecules, 20 (1987) 2471.
[63] F. Brochard-Wyart and P.G. de Gennes, Makromol. Chem. Makromol. Symp., 40 (1990) 167.
[64] F. Brochard-Wyart, Proc. Toyota Conf. Stud. Polym. Sci., 2 (1988) 249.
[65] A. Brochard-Wyart and P. Pincus, C. R. Acad. Sci. Ser. II, 314 (1992) 131.
[66] E. Jabbari and N.A. Peppas, Macromolecules, 26 (1993) 2175.
[67] P.G. de Gennes, C. R. Acad. Sci. Ser. II, 292 (1981) 1505.
[68] F. Brochard-Wyart, J. Jouffroy and P. Levinson, Macromolecules, 16 (1983) 1638.
[69] P.G. de Gennes, J. Chem. Phys., 72 (1980) 4656.
[70] S.F. Tead and E.J. Kramer, Macromolecules, 21 (1988) 1513.
[71] P.F. Green, C.J. Palmstrom, J.W. Mayer and E.J. Kramer, Macromolecules, 18 (1985) 501.
[72] E.J. Kramer, P.F. Green and C.J. Palmstrom, Polymer, 25 (1984) 473.
[73] H. Sillescu, Makromol. Chem. Rapid Commun., 8 (1987) 393.

[74] A.Z. Akcasu, G. Nagele and R. Klein, Macromolecules, 24 (1990) 4408.
[75] J.V. Seggern, S. Klotz and H.J. Cantow, Macromolecules, 22 (1989) 3328.
[76] E.A. Jordan, R.C. Ball, A.M. Donald, L.J. Fetters, R.A.L. Jones and J. Klein, Macromolecules, 21 (1988) 235.
[77] G. Reiter, S. Huttenbach, M. Foster and M. Stamm, Macromolecules, 24 (1991) 1179.
[78] S. Wu, H.K. Chuang and C.D. Han, J. Polym. Sci. Polym. Phys., 24 (1986) 143.
[79] E. Kim, E.J. Kramer, W.C. Wu and P.D. Garrett, Polymer, 35 (1994) 5706.
[80] P.G. de Gennes, Scaling Concepts in Polymer Physics, Cornell University Press, Ithaca, NY (1979).
[81] S. Prager and M.J. Tirrell, Chem. Phys., 75 (1981) 5194.
[82] S. Prager, D. Adolf and M. Tirrell, J. Chem. Phys., 78 (1983) 7015.
[83] D.B. Adolf, Macromolecules, 17 (1984) 1248.
[84] J. Crank, The Mathematics of Diffusion, Oxford University Press, London (1975).
[85] U. Murschall, E.W. Fischer, C.H. Herkt-Maetzky and G. Fytas, J. Polym. Sci. Polym. Lett., 24 (1986) 193.
[86] P.T. Gilmore, R. Falabella and R.L. Laurence, Macromolecules, 13 (1980) 880.
[87] M.A. Winnik, Y. Wang and F. Haley, J. Coating Technol., 64 (1992) 51.
[88] M. Chainey, M.C. Wilkinson and J. Hearn, J. Polym. Sci. Polym. Chem., 23 (1985) 2947.
[89] O. Pelcan, Trends Polym. Sci., 2 (1994) 236.
[90] Y. Wang and M.A. Winnik, Macromolecules, 23 (1990) 4731.
[91] M.C. Goh, D. Juhue, O.M. Leung, Y. Wang and M.A.Winnik, Langmuir, 9 (1993) 1319.
[92] M. Song, D.J. Hourston, H.X. Zhang, H.M. Pollock and A. Hammiche, Polymer, 42 (2001) 6299.
[93] D. Juhue and J. Lang, Macromolecules, 28 (1995) 1306.
[94] F. Lin and D.J. Meier, Langmuir, 11 (1995) 2726.
[95] D.J. Hourston, H.X. Zhang, M. Song, H.M. Pollock and A. Hammichc, Thcrmochim. Acta, 294 (1997) 23.
[96] J. Brandrup and E.H. Immergut, Eds., Polymer Handbook, John Wiley & Sons, NY (1966).
[97] M. Song, Y. Huang, G. Cong, H. Liang and B. Jiang, Polymer, 33 (1992) 1293.
[98] L.H. Sperling, Interpenetrating Polymer Networks and Related Materials, Plenum Press, NY (1981).
[99] J.H. Saunders and K.C. Frisch, Polyurethane-Chemistry and Technology, Interscience, New York (1964).
[100] L.H. Sperling, In Advance Interpenetrating Polymer Networks, Vol. 4, D. Klempner and K.C. Frisch, Eds., Technomic, Lancaster, PA (1994) p. 1.
[101] L.H Sperling, Contemp. Top. Polym. Sci., 6 (1989) 665.
[102] Y. Lipatov, S.V.V. Shilov, V.A. Bogdanovic, L.V. Karabanova and L.M. Sergeeva, J. Polym. Sci. Polym. Phys., 25 (1987) 43.
[103] D.J. Hourston and Y. Zia, J. Appl. Polym. Sci., 28 (1983) 3849.
[104] D.J. Hourston and Y. Zia, J. Appl. Polym. Sci., 28 (1983) 3745.
[105] Y. Lipatov, V.F. Rosovizky, P.V. Datsko and Y. Maslak, J. Appl. Polym. Sci., 36 (1988) 1143.
[106] B. Suthar, N. Parikh and N. Patel, Polym. Int., 25 (1991) 173.
[107] D.J. Hourston and Y. Zia, J. Appl. Polym. Sci., 29 (1984) 2963.
[108] D.J. Hourston and Y. Zia, J. Appl. Polym. Sci., 30 (1985) 2157.
[109] D.J. Hourston and S. Decurtins, J. Appl. Polym. Sci., 36 (1988) 365.
[110] L.H. Sperling, J.J. Fay, C.J. Murphy and D.A. Thomas, Makromol. Chem. Makromol. Symp., 38 (1990) 99.

[111] B. Suthar, In Advances in Interpenetrating Polymer Networks, Vol. 2, D. Klempner and K.C. Frisch, Ed., Technomic, Lancaster, PA (1990) 281.
[112] L.H. Sperling and J.J. Fay, Polym. Adv. Technol., 2 (1991) 44.
[113] Y. Suzuki, Nippon Gomit Kyokaishi, 62 (1989) 593.
[114] J. Qin, F. Li, Z. Wu and B. Cian, In Advance Interpenetrating Polymer Networks, Vol. 2, D. Klempner and K.C. Frisch., Eds., Technomic, Lancaster, PA (1990) 205.
[115] T. Akio and H. Mizunachi, In Advances in Interpenetrating Polymer Networks, Vol. 3, D. Klempner and K.C. Frisch, Eds., Technomic, Lancaster, PA (1991) 25.
[116] L.H. Sperling, C.E. Carraher, S.P. Qureshi, J.A. Manson and L.W. Barret, In Biotechnol. Polym. (Proc. Am. Chem. Soc. Symp.), C.G. Gebelein, Ed., Plenum Press, New York (1990) p. 96.
[117] Y. Lipatov, Rev. Macromol. Chem. Phys., C30 (1990) 209.
[118] J.H. An and L.H. Sperling, In Cross-Linked Polymers, R.A. Dickie, S.S. Labana and R.S. Bauer, Ed., ACS Series 376, American Chemical Society, Washington DC (1988).
[119] D. Klempner and H. Berkowski, In Encyclopedia of Polymer Science and Engineering, Vol. 8, H. Mark, N.M. Bikales, C.G. Overberger and G. Menges, Ed., John Wiley & Sons, NY (1988).
[120] H.L. Frisch, D. Klempner, H.K.Yoon and K.C. Frisch, Macromolecules, 13 (1980) 1016.
[121] D.J. Hourston, F.-U. Schafer, J.S. Bates and M.H.S. Gradwell, Polymer, 39 (1998) 3311.
[122] M. Akay and S.N. Rollins, Polymer, 34 (1993) 1865.
[123] S. Ma, Synthesis and Characterisation of Interpenetrating Polymer Networks, Ph.D. Thesis, Jilin University, People's Republic of China (1988).
[124] A. Alig, M. Junker, W. Jenninger, H.L. Frisch and M. Schulz, Morphology of Polymers, Conference lecture, Prague, July 1995.
[125] M.M. Coleman, C.J. Serman and P.C. Painter, Macromolecules, 20 (1987) 226.
[126] B.J. Bauer, R.M. Briber and C.C. Han, Macromolecules, 22 (1989) 940.
[127] B.J. Bauer and R.M. Briber, In Advances in Interpenetrating Polymer Networks, Vol. 4, D. Klempner and K.C. Frisch, Ed., Technomic, Lancaster, PA (1994) p. 45.
[128] N. Parizel, G. Meyer and G. Well, Polymer, 36 (1995) 2323.
[129] A.A. Donatelli, L.H. Sperling and D.A. Thomas, J. Appl. Polym. Sci., 21 (1977) 1189.
[130] Y. Yang, M.K. Winnik, D. Ylitalo and R.J. Devoe, Macromolecules, 29 (1996) 7055.
[131] M. Song, D.J. Hourston, F.-U. Schafer, H.M. Pollock and A. Hammiche, Thermochim. Acta, 305 (1997) 335.
[132] M. Song, D.J. Hourston, F.-U. Schafer, H.M. Pollock and A. Hammiche, Thermochim. Acta, 315 (1998) 25.
[133] D.J. Hourston and F.-U. Schafer, J. Polym. Adv. Technol., 7 (1995) 273 (special edition).
[134] F.-U. Schafer, Ph.D. thesis, Loughborough University (1996).
[135] A. Baruya and W.F. Maddams, Appl. Spectrosc., 32 (1978) 563.

Chapter 4

THE APPLICATION OF MTDSC TO POLYMER MELTING[1]

Bernhard Wunderlich*,¶
*Department of Chemistry, The University of Tennessee Knoxville, TN 37996-1600, USA

¶Chemical Sciences Division, Oak Ridge National Laboratory, Oak Ridge, TN 37831-6197, USA

1 Introduction

The thermal analysis of polymer melting is not an easy or straightforward technique and requires an understanding of both the instrumental and material's problems for its interpretation. Once mastered, however, the analysis allows considerable insight into the structure and properties of the polymers. Attempting the study of equilibrium melting of one-component systems of small molecules, such as pure indium or water, the temperature should be constant from the beginning to completion of melting. A standard differential scanning calorimeter (DSC), however, will produce, because of its instrumental lag, a melting peak of a width of a few kelvins, instead of an infinitely sharp spike. For any data interpretation, this fact must be kept in mind. Going to scanning differential calorimetry with modulated temperature, called MTDSC, or often also abbreviated as TMDSC (temperature-modulated DSC), the instrument lags become even more important and are joined by problems inherent in the analysis method. More details of the complications, which are by now well understood, are explained in Section 3.

[1]The submitted manuscript has been authored by a contractor of the U.S. Government under the contract No. DOE-AC05-00OR22725. Accordingly, the U.S. Government retains a non-exclusive, royalty-free license to publish, reproduce the published form of this contribution or allow others to do so, for U.S. Government purposes.

M. Reading and D. Hourston (eds.), Theory and Practice of Modulated Temperature Differential Scanning Calorimetry, 217–320.

Turning to flexible macromolecules[2] which most often, less precisely, are just called "polymers," one finds that even homopolymers such as polyethylene or polyesters do not crystallise fully, and their melting range may be many Kelvin wide, i.e. their melting range adds considerably to the instrument-caused broadening of a DSC trace. The amount crystallised is measured as crystallinity and has its limitation in strain exerted on the surrounding melt by molecules that pass from the crystals to the melt. This strain interferes with attainment of full crystallisation. The crystals, in addition, are commonly so small that they are best characterised as nanophases, phases whose properties are affected to such a degree by their surfaces, that little or no relaxed bulk material remains in their centre. These effects are at the root of the broad melting range, which may begin 100 K below the equilibrium temperature for the smallest crystals and reach above it when strains prohibit full randomisation during the time available for melting (superheating).

The amorphous phases, areas or defects are similarly small in size as the crystals, i.e. they are also nanophases. The temperature where a solid glass softens and becomes viscous without any heat effect is called the glass transition. The glass transition of amorphous areas in semicrystalline polymers is usually broadened by shifting its upper limit to higher temperatures. This shift of the glass transition region is due to strain exerted by the molecules that emanate from the crystals and continue into the amorphous regions. It is customary to call this strained material an intermediate, third phase. Sometimes, this third phase develops a separate glass transition, which may reach the melting temperature or extend to even higher temperature. The material frozen in this fashion is called the rigid-amorphous fraction (RAF). The RAF can be characterised by thermal analysis of its glass transitions. The change in heat capacity in the broadened low-temperature glass transition and during the glass transition at high temperature allows an assessment of the intermediate phase.

The strain that hinders the motion in the amorphous phases is transmitted by the polymer molecules that traverse large parts of the sample and set up a global, non-equilibrium structure. If these strained molecules were immobilised during mechanical drawing of the sample, as is necessary in the production of films or fibres, the amorphous areas become oriented and reduce the entropy (degree of disorder). Because melting is governed by the entropy of fusion ($\Delta S = \Delta H/T_m$) and the heat of fusion, ΔH, is

[2]H. Staudinger, who received the Nobel Price in Chemistry for 1953 honouring his fundamental work on "Theory of Macromolecules," suggests that a macromolecule or polymer molecule must have more than 1000 atoms. To have the properties known for plastics, the molecules must be flexible, i.e. they must be able to rotate about some or all of the covalent bonds of their backbone.

approximately constant, the melting temperature T_m may temporarily increase until the strain is released, i.e. the DSC curve will show superheating. The amorphous fraction, now oriented, also may assume a "heat of fusion" and have a mesophase structure, similar to a liquid crystal. More about this non-equilibrium melting is discussed in Section 2.

This brief summary reveals a too complex issue to be discussed in all details in a single book chapter. More information is available through the 2879 screens of the computer course "Thermal Analysis of Materials," available by downloading from the Internet [1] and the reference work of 2547 pages on "Thermal Characterisation of Polymeric Materials" [2], or in the treatises on "Thermal Analysis of Polymeric Materials" [3] and "Macromolecular Physics" [4–6]. General information and data can be found in "Calorimetry" [7], the "Encyclopaedia of Polymer Science and Engineering" [8], the "Polymer Handbook" [9] and the ATHAS Data Bank [10]. All these sources should give you access to the information needed for interpretation of the instrumental and polymeric materials problems beyond this chapter.

Section 2 of this chapter contains the basics needed to understand melting and crystallisation, mainly using equilibrium and irreversible thermodynamics and kinetics. Section 3 comprises a summary of the details on instrumentation and data treatment. Both of these sections can be bypassed initially when the main goal is to get started quickly on experiments. As the need arises, the basic material can then be filled in by reading Sections 2 and 3 and consulting the references.

Combining MTDSC and polymer science is a challenge, which when met, yields so much additional information on the subjects that the added effort to understand it is well worthwhile. This chapter can, naturally, only point the way and help in avoiding the most common pitfalls.

2 The Melting and Crystallisation Behaviour of Polymers

2.1 EQUILIBRIUM MELTING

The equilibrium description of melting and crystallisation is a subject of the field of thermodynamics. The basic quantity of calorimetry is the heat capacity, C_p (at constant pressure, in J K^{-1} mol^{-1}), which represents the amount of heat, Q (in joules, J), needed to be added to raise the temperature by 1 K or to be extracted to lower the temperature by 1 K for 1 mol of material. If the material analysed has a mass of 1 g, one calls this quantity the specific heat capacity, c_p (at constant pressure, in J K^{-1} g^{-1}). In the more precise differential notation, one writes for the heat capacity that

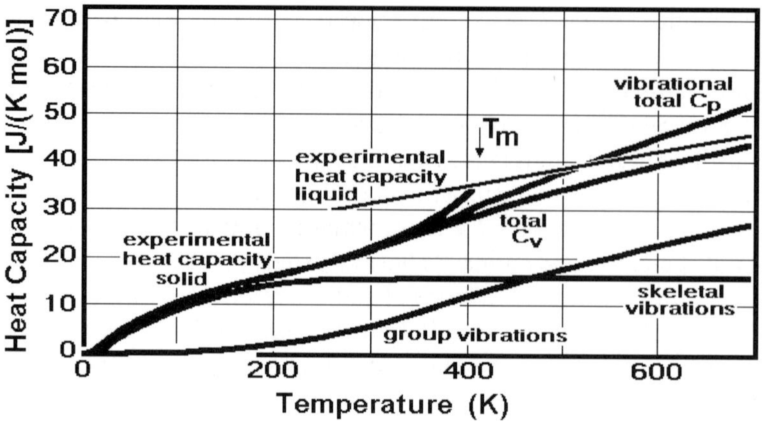

Figure 4.1. Heat capacity of polyethylene. The experimental heat capacity of the solid is caused by skeletal and group vibrations (and needs to be changed from the heat capacity at constant volume, C_v, to C_p).

$C_p = (\partial H/\partial T)_{p,n}$, where H is the enthalpy or heat content, and the subscripts p and n signify that the partial differential is taken at a constant pressure (usually atmospheric pressure) and without changing the amount of material which is expressed in number of moles n, respectively.

The origin of the heat capacity is the vibrational energy, and effects connected with large-amplitude motion, such as translation, rotation and internal rotation. The internal rotation is also called conformational motion since the rotation about C-bonds leads to distinctly different isomeric structures. Figure 4.1 shows the fully analysed heat capacity of polyethylene [11].

At room temperature, the vibrations of the backbone are fully excited and reach a constant contribution to the heat capacity. The group vibrations have higher frequencies and their contribution to the vibrational heat capacity keeps on increasing. The total vibrational heat capacity can be used as a baseline for the interpretation of the large amplitude motion. The liquid has a maximum of large amplitude motion and has at lower temperatures, but above the glass transition, a higher heat capacity than the solid. Below the glass transition temperature, crystals and glasses have similar heat capacities until temperatures are reached that are lower than 50 K. At higher temperatures, usually approaching the melting temperatures of the polymer crystals, the heat capacity of the liquid is less than expected from the vibrations alone. Note that the experimental heat capacity of the solids shows also contributions from defects that raise the heat capacity beyond the total vibrational contributions. A detailed discussion of the correlation of vibrational spectra and heat capacity is given in Ref. [12]. Use of the calculated vibrational heat capacities and the measured heat capacities will be made as baselines for the interpretation of transitions.

Heat, Q, however, can also be exchanged without affecting the temperature of a sample. This occurs during chemical or physical transitions of the material. The heat involved is generally called a latent heat, L (at constant pressure in J mol^{-1}). From heat capacity and latent heats measured from the zero of temperature to the value of interest, it is possible to establish the integral thermal properties:

$$H(T) = H_0 + \int_0^{T_m} C_p\,(\text{crystal})\mathrm{d}T + \Delta H_f + \int_{T_m}^{T} C_p\,(\text{liquid}) \tag{1}$$

where $H(T)$ is the enthalpy at temperature, T; H_0 is the (usually unknown) heat content at absolute zero; ΔH_f is the heat of fusion ($= L$, a latent heat); and T_m is the melting temperature. If there are additional transitions, Eq. (1) must be expanded accordingly.

The other two integral functions are given by the two parts that $H(T)$ can be separated into when equilibrium is maintained and are based on the second law of thermodynamics. The first is the entropic part of the enthalpy, TS, with the entropy $S(T)$ (expressed in J K^{-1} mol^{-1}), representing a measure of the disorder of the system. At 0 K, S_0 of a crystal is 0, a consequence of the third law of thermodynamics (a perfect crystal at 0 K is fully ordered):

$$S(T) = \int_0^{T_m} \frac{C_p\,(\text{crystal})}{T}\,\mathrm{d}T + \Delta S_f + \int_{T_m}^{T} \frac{C_p\,(\text{liquid})}{T}\,\mathrm{d}T \tag{2}$$

where ΔS_f is the entropy of fusion ($= L/T_m$). The connection to the heat content or enthalpy is given through the free enthalpy, $G(T)$, which is a measure of the stability of a system (a lower G represents a more stable system):

$$H(T) = G(T) + TS(T) \tag{3}$$

With these three simple equations, all equilibrium calorimetry can be described, so that measurement of heat capacity and latent heat allows a full thermal characterisation. Figure 4.2 illustrates a typical diagram of the thermal properties of crystalline polyethylene and its melt. The data were obtained by extrapolation of measurements of heat capacities on

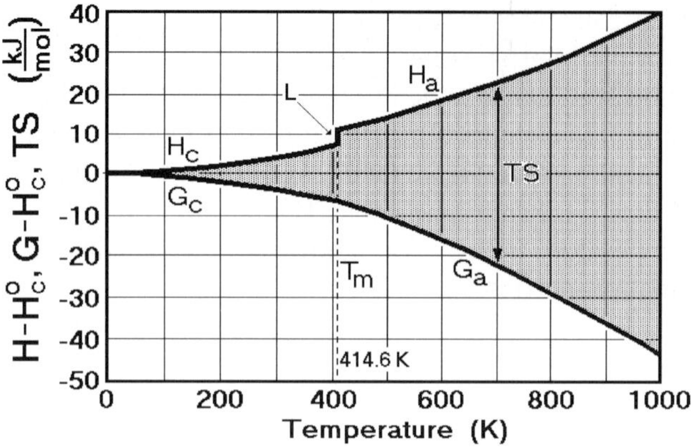

Figure 4.2. Enthalpy, entropy and free enthalpy of polyethylene. The subscripts c and a stand for crystalline and amorphous (liquid) polyethylene, respectively, and *TS* is the positive quantity defined by Eq. (3).

(non-equilibrium) semicrystalline polyethylenes to full crystallinity and of the transition temperatures to equilibrium.

Concentrating on the melting, it is useful to magnify the free enthalpy of the crystal and liquid in the vicinity of the melting temperature, as is done in the schematic of Figure 4.3. The state with the lowest free enthalpy represents the stable equilibrium for the chosen temperature. At low temperature

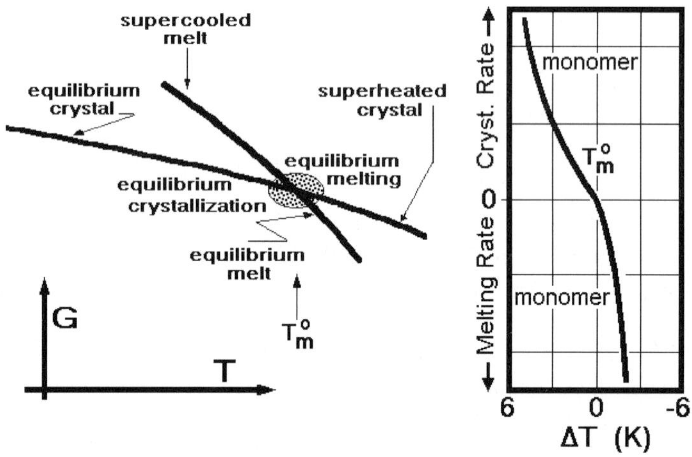

Figure 4.3. A schematic of the free enthalpy in the vicinity of the equilibrium melting temperature (left) and a plot of the linear crystallisation and melting rates of gaseous selenium (Se_2) (right). Selenium crystallizes or sublimes to and from selenium crystals made up of flexible, linear macromolecules. The process can be expressed as: $x Se_2$ (gaseous) ⇔ Se_{2x} (cryst.).

it is the crystal, and at high temperature it is the melt. The point of intersection of the enthalpies of crystal and melt identified the equilibrium melting temperature T_m^0. At T_m^0, the enthalpies of melt and crystal are identical. The dynamic melting point is written as crystal ⇔ melt, and the change of free enthalpy for this process is $\Delta G = G(\text{melt}) - G(\text{crystal}) = 0$. An important conclusion from this simple discussion is that at T_m^0, the following relationship must hold:

$$T_m^0 = \frac{\Delta H_f}{\Delta S_f} \tag{4}$$

Equation (4) is the basic equation needed to describe the melting temperature of materials. The melting temperature sets the temperature limit of use of a material and is thus an important piece of information. Fortunately for a discussion of Eq. (4), one does not need independent information about all enthalpies and entropies contained in the expression. It could be shown that for all flexible polymers, the molecule can be divided into "beads," which mark the basic flexible units. For polyethylene, one CH_2– unit corresponds to a bead, while for polypropylene, the repeating unit (CH_2–$CHCH_3$) consists of two beads, and for polyisoprene (rubber), the repeating unit (CH_2–CCH_3=CH–CH_2) consists of three beads. Each beads contributes 7–12 J K^{-1}(mol of beads)$^{-1}$ to the entropy of fusion [6]. The enthalpies of fusion, in addition, can similarly be estimated from their cohesive energy [13]. A simple rule can be derived from this information. Higher melting temperatures can be attained by decreasing the flexibility of a polymer by introducing fewer beads per unit mass (by increasing the number of atoms per bead), or by increasing the interaction between the beads (by using chemical groups with stronger intermolecular forces) [6]. Typical examples are the melting temperatures of poly(ethylene suberate), $T_m^0 = 348$ K, poly(ethylene terephthalate) (PET), $T_m^0 = 553$ K and nylon-2,6, $T_m^0 \approx 575$ K. Each of these molecules has 14 large atoms of type –C=, –CH=, –CH_2–, –O–, =O or –NH– in its repeating unit. Dividing the structures into rigid beads, however, one finds 10 [(CH_2–)$_2$ (OCO–) (CH_2–)$_6$ (OCO–)], 5 [(CH_2–)$_2$ (OCO–) (C_6H_4–) (OCO–)] and 10 beads [(CH_2–)$_2$ (NHCO–) (CH_2–)$_6$ (OCNH–)], respectively. The poly(ethylene suberate) is very mobile with 10 beads, while PET has four mobile groups connected to a rigid phenylene ring, so that the entropy of fusion is much less. Nylon is more mobile, but the amide groups have a larger interaction due to dipoles and hydrogen bonds. To understand the thermodynamic parameters of polymers, thus, it is a vital step to establish the link to the molecular structure.

Returning to Figure 4.3, one can see from the free-enthalpy curves that the more a melt is supercooled or a crystal is superheated, the higher are the respective driving forces ΔG for crystallisation or melting. The actual rates of crystallisation and melting depend on the detailed mechanism, but MTDSC should be able to measure the kinetics, as it is shown schematically on the right side of Figure 4.3. For a dynamic equilibrium at the melting temperature, the molecular rates of crystallisation and melting are equal, i.e. no macroscopic kinetics is observable. If the curve is continuous through T_m^0, the response to temperature modulation should be symmetric and the kinetics measurable. Section 2.2 shows that for polymers, this is, however, rarely the case. The crystallisation/melting curve becomes discontinuous at T_m^0.

This discussion of the thermodynamics of melting reveals a rather simple theory with good predictive capability for the melting parameters. The heat capacities are also well linked to the underlying molecular motion, and various quantitative baselines can be generated. Only with such quantitative information is it possible to analyse the common deviations from equilibrium. Section 2.2 will expand this discussion to non-equilibrium systems.

2.2 NUCLEATION OF CRYSTALS AND MOLECULES

Supercooled melts and solutions are well known. For small molecules, the major reason for supercooling is the need to overcome the free enthalpy barrier hindering the formation of small crystals. Figure 4.4 illustrates the change of onset temperatures of melting and crystallisation of indium as a function of heating and cooling rates, as measured by DSC. The instrument lag causes the change of the onset of melting and crystallisation with heating

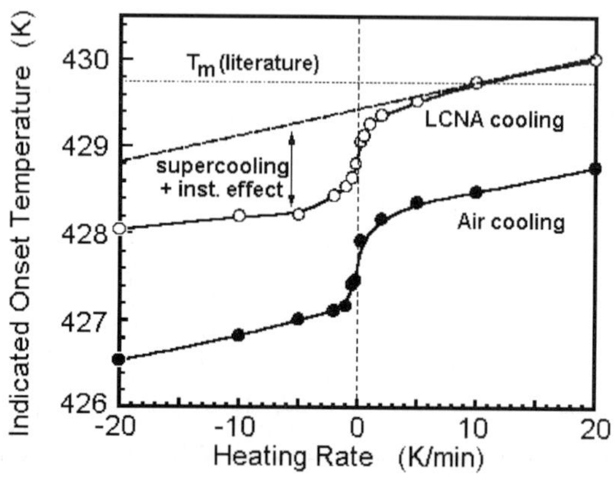

Figure 4.4. Change of onsets of melting and crystallisation as a function of the rate of temperature change. The data were obtained with a TA Instruments 2920 DSC in two different configurations, with air cooling (lower curve) and cooling with the liquid nitrogen accessory (upper curve).

rate, q. For correction, the data taken with the liquid nitrogen accessory (LNCA) are expressed as:

$$T_m \text{ (measured)} = [T_m \text{ (literature)} = 429.75 \text{ K}] - 0.0308(q - 10) \quad (5)$$

where the rate of temperature change, q, is expressed in K min^{-1}, and the calibration is to be made at a heating rate of 10 K min^{-1}. Note that in some DSCs, this correction is included into the data analysis, so that the two linear portions of the curve are close to horizontal and that the heating and cooling branches of the graph do not necessarily have the same slope. Most obvious, however, is the supercooling before crystallisation. For indium, this supercooling is about 1 K. Other molecules may have largely different supercoolings. Seeding with crystals usually can eliminate this supercooling and allows the measurement of the rates of crystallisation, as is shown on the right-hand side of Figure 4.3.

Oligomers, the polymers of low molar mass, and polymers behave drastically differently. Their crystals do not grow even in the presence of nuclei. Figure 4.5 illustrates that the melting and crystallisation rates of polymers and oligomers have a discontinuity at the equilibrium melting temperature, different from the monomer which exhibited a dynamic molecular equilibrium at the melting temperature in Figure 4.3. The horizontal portions of the crystal-growth-rate curves indicate a temperature region of metastability. In this region, crystals cannot melt and melts cannot crystallise, even in the presence of nuclei of the phases. Recent MTDSC studies of paraffins and

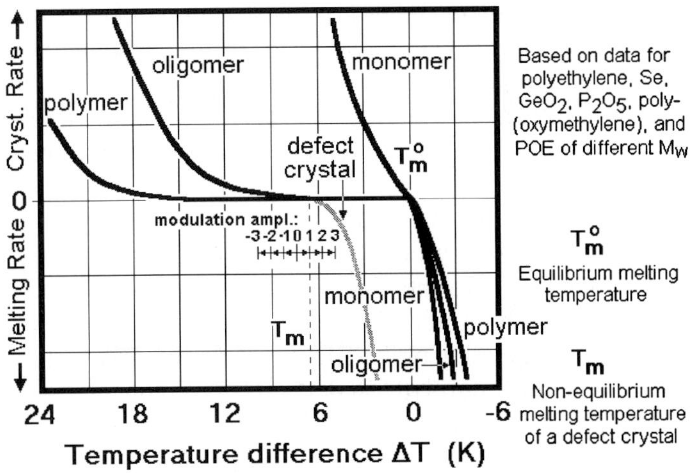

Figure 4.5. Plot of the linear crystallisation and melting rates of polymers, oligomers and monomers. The shaded line indicates the shift that is possible for the melting of defect crystals.

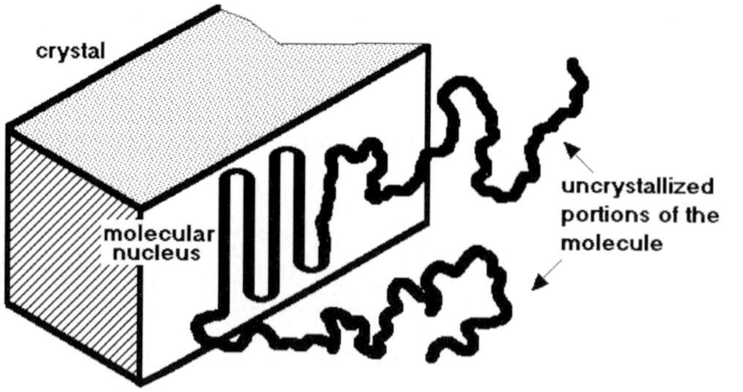

Figure 4.6. Schematic of a molecular nucleus with uncrystallised chain ends.

fractions of low molar mass polyethylenes have shown that this metastability shows for molecules longer than about 10 nm [14].

Figure 4.6 illustrates the type of nucleation that is active for oligomers and polymers longer than the critical length of 10 nm, that of molecular nucleation. The flexible macromolecules are sufficiently large so that they need many molecular steps in the proper sequence to achieve a start on the crystal surface to initiate crystallisation of the specific molecule and produce the linear crystal growth rate shown in Figure 4.5. While there is an easy way by adding seeds of crystals to a melt to avoid supercooling of small molecules, there are no seeds for the macromolecules. Special occasions for faster molecular nucleation exist during mechanical deformation of polymers, as observed during drawing. Not only is the local melting temperature increased, as seen from the thermodynamic equation (4), but also the molecular nucleation barrier is reduced due to the stretching of the molecules, so that during drawing, fibres crystallise faster at higher temperatures. Furthermore, molecular nucleation can also be avoided on partial melting and recrystallisation of molecules during MTDSC. During the cooling cycle, the melted portions of the molecules can recrystallise, as will be discussed in Section 4.

Superheating of crystals, in contrast, is not so commonly observed since the surfaces and corners of crystals are sufficiently rough to serve for the nucleation of the melt [15]. The melting is then just determined by its kinetics with minimal nucleation barriers. For small molecules, the melting is usually so fast that the conduction of the latent heat across the crystal surface is the limiting time factor for growth. Superheating is known for crystals with molecular networks, such as silicates [16] where melting 50–100 K above the melting temperature may take many hours. Superheating of polymer crystals is also observed, but usually only for rather perfect, extended-chain crystals. Polymer molecules at the surface of a crystal can initiate melting only at chain ends or on folds since polymer molecules usually do not break on fusion. If the crystals are extended, these sites for the initiation of melting are rare. This slows the melting rate to make it observable. As will be discussed in Section 2.3, strain within the macromolecules crossing the phase boundary, coupled with orientation in the melt, may cause an increase in melting temperature and appear in DSC similar to superheating. A simple method to distinguish between the two causes is to etch the polymer crystals. The strained molecules are oxidised first and quickly relieve the strain on the remaining parts of the molecule, so that the melting experiment by DSC shows strongly reduced superheating after etching. Extended chain crystals, in contrast, are affected little. They change their superheating character only slowly as the crystals are etched to smaller sizes.

Before one can study the crystallisation behaviour of polymers, it is thus necessary to be informed about their nucleation behaviour. Supercooling by substantial amounts is common, making the crystallisation/melting transition irreversible. The rates of melting in Figure 4.5 are not much affected by nucleation, but show a characteristic kinetics that is different for different molecules. Since crystallisation of polymers is shifted to rather large supercooling, the resulting crystals are different in perfection depending on the crystallisation conditions. Although on heating, perfection and recrystallisation of the imperfect crystals may occur and cause a major difficulty in the analysis of DSC traces, the so perfected crystals are usually still far from equilibrium and melt at a lower temperature, as will be discussed next. The shift in melting kinetics for an imperfect crystal is also indicated in Figure 4.5. Under some conditions, it may approach the crystallisation curve and may be bridged by MTDSC with a sufficiently large temperature amplitude. In this case, MTDSC is an ideal tool to study the non-equilibrium crystallisation and melting.

2.3 IRREVERSIBLE MELTING

Although it may look that polymer melting cannot be analysed with DSC since it can rarely be brought into a state of equilibrium, irreversible thermodynamics is by now well understood [17]. A detailed description of the use of time-dependent heat capacities in MTDSC was published [18], and early efforts to describe the melting of polymers and copolymers were presented some 40 years ago [19]. Overall, it turns out that the information on reversible melting of polymers in equilibrium, as described in Section 2.1, is largely only available by extrapolation from data of systems not in equilibrium and serves mainly as a base for the discussion of irreversible melting. A major importance of the analysis of irreversible processes lies in the possibility of documenting the thermal history and the actual properties

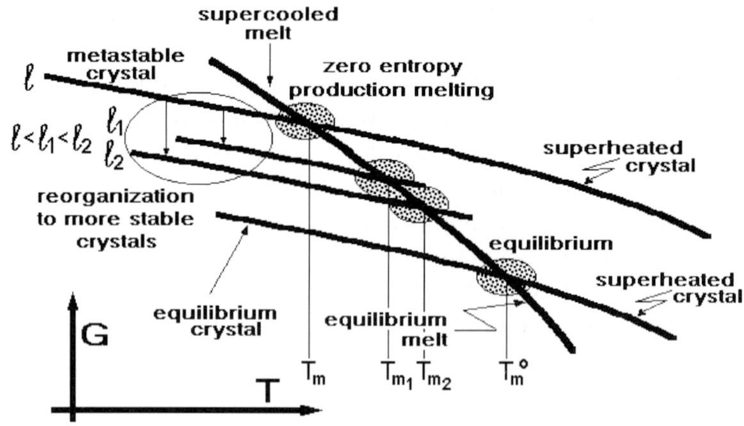

Figure 4.7. Schematic diagram of the free enthalpy as a function of temperature. For the metastable lamellar crystals, their lamellar thickness, *l*, is indicated. Compare to Figure 4.3.

of the sample on hand. By analysing the defect structure of polymer crystals, it is possible, for example, to identify the crystallisation conditions; see the effects of possible annealing, drawing, texturing, etc. Similarly, the glass transition is a good indicator for the thermal history of non-crystalline materials. In semicrystalline samples, the study of the glass transition of the amorphous fraction can supplement the analysis of the melting of the crystals and identify the state of internal strains within the sample.

Figure 4.7 gives a summary of the changes that occur when going from an equilibrium analysis as illustrated in Figure 4.3 to irreversible situations. Plotted is the free enthalpy as obtained from thermal analysis as described in Eq. (3). Besides the extrapolation of the equilibrium crystal and melt into the temperature region of superheating and supercooling, shown already in Figure 4.3, the free enthalpy of a metastable crystal is indicated, as it is typically found in the lamellar crystals of a semicrystalline polymer. Since the major difference of such crystals from equilibrium is the specific surface free energy, γ, in J cm^{-2}, one can write for the representation of the free enthalpy of such a crystal:

$$G\ (\text{lamellar crystal}) = G\ (\text{equilibrium crystal}) + \frac{2m_c\gamma}{\rho l} \qquad (6)$$

where m_c is the mass of the lamellar crystal, ρ is its density and l is the lamellar thickness, usually from 0.5 to 50 nm. The expression $2m_c\gamma/\rho l$ represents simply the area of the upper and lower lamellar surfaces. The side surfaces of the lamellae are so much smaller that their effect can be neglected. Similarly, the effect of internal defects has been neglected in Eq. (6). Both effects could, however, be easily added to the equation if their free enthalpy contributions are known. Each of these defects increases the free enthalpy of the crystal above that of the equilibrium crystal, as is shown in Figure 4.7 for lamellae of thickness l. On heating, the free enthalpy of the non-equilibrium crystal remains largely parallel to that of the equilibrium crystal as long as the crystal dimensions remain constant, i.e. the crystal is metastable.

At the point of intersection of the free enthalpy of the metastable, lamellar crystal with the free enthalpy of the supercooled melt, at T_m, the two phases have the same degree of metastability as expressed by the distance from equilibrium melting at T_m^0. On heating a defect crystal, this is the point where most often melting is observed. Since such non-equilibrium melting goes without change in free enthalpy, just as the case of the equilibrium in Figure 4.3, this is called the zero-entropy-production melting. It does not mean that there is no entropy of fusion, but it means that the entropy flow from the surroundings, although it occurs at the lower non-equilibrium melting temperature, T_m, and is larger than at the equilibrium melting temperature, is compensated exactly by the entropy of fusion of the defect crystal and yields no excess entropy, i.e. no production of entropy. The excess entropy of fusion due to the lower melting temperature is exactly compensated by the defect contribution of the elimination of the crystal surfaces:

$$G(\text{melt}) - G(\text{lamellar crystal}) = G(\text{equilibrium crystal}) - \frac{2m_c\gamma}{\rho l} = 0 \qquad (7)$$

where the melt is assumed to be a large phase without surface effects, and the free enthalpy of fusion of the equilibrium crystal, ΔG_f, at the lower than equilibrium melting temperature is positive and approximated by ΔH_f $(T_m^0 - T_m)/T_m$. From this consideration, the zero-entropy-production melting temperature of lamellar crystals can be written as the Gibbs–Thomson equation:

$$\Delta T_m = T_m^0 - T_m = \frac{2\gamma T_m^0}{\Delta h_f \rho l} \qquad (8)$$

where Δh_f is the specific heat of fusion in J g^{-1}. The quality of the fit of the Gibbs–Thomson equation is seen in Figure 4.8 on polyethylene lamellae of different thicknesses. The lamellar thickness was obtained in these examples from low-angle X-ray diffraction, electron microscopy or interference microscopy. The extrapolation of the best fit marked in Figure 4.8 intersects the ordinate at the equilibrium melting temperature, found in this case also experimentally (414.6 K).

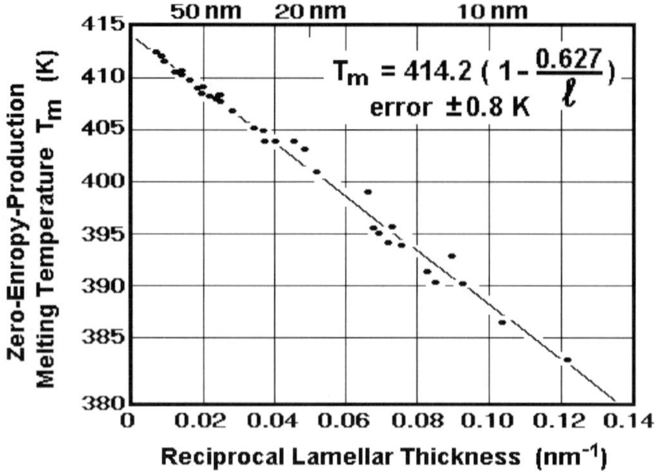

Figure 4.8. Zero-entropy-production melting temperatures of lamellar crystals of polyethylene [6].

All changes in Figure 4.7, which involve a downward movement (decrease) in free enthalpy in going from one state to the other, are spontaneous, irreversible processes with an entropy production. All processes that would involve an increase in free enthalpy in going from one state to the other are forbidden by the second law of thermodynamics. In the upper left corner of Figure 4.7, the changes in free enthalpy on annealing or reorganisation of the lamellar crystals of thickness l to the larger thicknesses l_1 and l_2 are illustrated. On further heating, the thicker lamellae would melt at higher zero-entropy-production melting temperatures. Figure 4.9 shows a thermal analysis result that documents a decreasing melting temperature

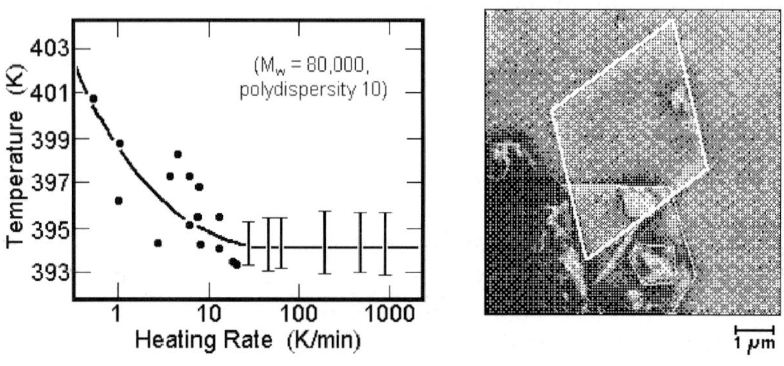

Figure 4.9. Decrease of melting temperature of lamellar crystals of polyethylene when heating with increasing rates. The right figure is a typical electron micrograph of a lamella of such polyethylene. DSC data from Ref. [20].

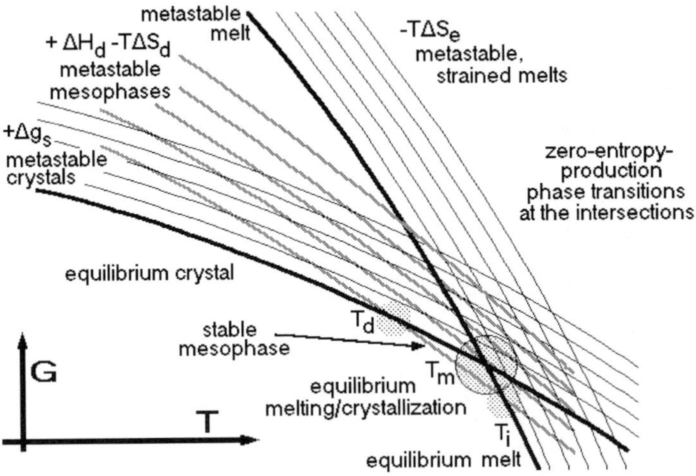

Figure 4.10. Schematic diagram of the free enthalpy as a function of temperature, illustrating the effect of orientation in the amorphous regions and partial disordering of the crystals to a mesophase. Compare to Figures 4.3 and 4.7.

of lamellar crystals of polyethylene of initially about 13 nm thickness, as shown on the right, when the heating rate increases. As the heating rate increases, the thickening of the crystals is reduced, and finally, on heating faster than 50 K min^{-1}, the zero-entropy-production melting of the initial crystals is reached. This example shows how important it is to make sure that a thermal analysis trace gives results that refer to the initial sample that was intended to be analysed. Naturally, changes during heating can also be analysed quantitatively and can serve to characterise samples.

Two examples of other complications that may occur in analyses of one-component systems are given by the free enthalpy schematic of Figure 4.10. Besides the supercooled, superheated and equilibrium crystals of Figure 4.3 and the metastable crystals of Figure 4.7, two additional sets of curves are drawn. The almost parallel curves to the free enthalpy of the melt are produced by straining the melt by mechanical drawing, which results in crystals embedded in oriented melt of certain degrees of orientation. The orientation in the melt reduces the entropy and accordingly increases the zero-entropy-production melting temperature as long as the higher degree of order is maintained. Figure 4.11 illustrates a thermal analysis of such a sample in the form of a PET fibre. The unrestrained fibre shows some reorganisation with decreasing heating rates as discussed with help of Figure 4.9 for polyethylene. Only at heating rates above about 40 K min^{-1} is the zero-entropy-production melting realised, and the observed melting temperature can be used to discuss the crystals in the initial sample. Keeping the fibres

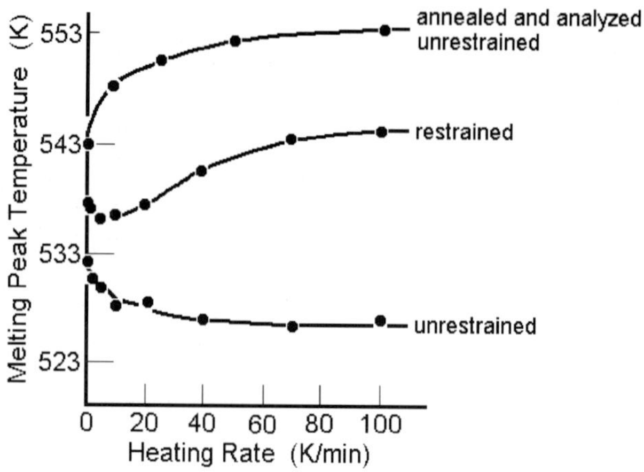

Figure 4.11. Melting peak temperatures of PET fibres under different conditions [21].

restrained at constant length increases the observed melting temperature since the amorphous fraction is oriented and can relax less at higher heating rates, yielding the higher melting temperatures. Annealing the fibres before analysis increases the melting temperatures even further because of crystal perfection at the annealing temperature, which further increases the strain on the amorphous fraction.

The final set of curves in Figure 4.10 refers to a possible mesophase of different lamellar thickness, i.e. a crystal phase that shows an intermediate degree of order. Most common in polymer crystals are the conformationally disordered crystals or condis crystals [22]. Since the entropy of the mesophase is intermediate between the crystal and melt, the slope of the free enthalpy curve, which is given by the expression $\partial G/\partial T = \Delta S$, is also intermediate. The illustrated case has the proper enthalpy level, so that the mesophase has a small temperature range of stability.

A typical example is represented by the plot of the apparent heat capacity of *trans*-1,4-poly(butadiene) in Figure 4.12. The melting occurs in two steps: a disordering transition, T_d, and a final melting, the isotropisation transition, T_i. Since the sample is only semicrystalline, it shows an additional glass transition temperature T_g. The large-amplitude conformational motion is proven by the line-width of the proton NMR signal, given by its second moment. Below the glass transition all parts are rigid. At the glass transition, the amorphous polymer becomes mobile and leads to a decrease of the line width. On disordering at T_d, further mobility is introduced into the crystals which are now a mesophase with higher mobility. The final narrowing occurs on full isotropisation to the melt at T_i. This example of a mesophase, along with the possibility of orientation in the sample as shown in Figure 4.11,

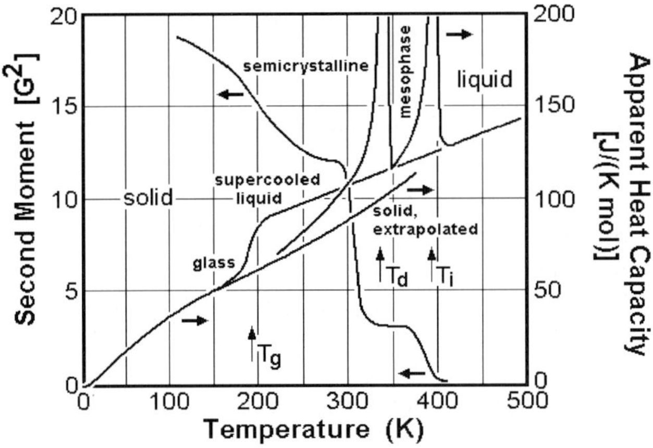

Figure 4.12. Line width of proton NMR signal and apparent heat capacity from DSC of *trans*-1,4-polybutadiene [22].

summarises the large range of results that need to be understood for the interpretation of thermal analyses of semicrystalline polymers.

Addition of temperature modulation to the analysis brings the need to interpret quantitatively the transition kinetics, as is indicated in Figure 4.5. Although this enormous breadth of possible effects may look overwhelming on first study, it also illustrates the enormous potential information on thermal analysis. A much broader study of these effects is offered with many examples in the earlier-mentioned computer course [1]. A final, general topic is given in Section 2.4, where local equilibria are discussed as they were discovered in semicrystalline samples.

2.4 LOCAL EQUILIBRIA

The structure of semicrystalline polymers can best be described as a nanophase aggregate of two or more phases such as crystalline, amorphous and intermediate, to be described later. Within this global, metastable phase structure, local areas may be contained which may have a melting and crystallisation equilibrium that can be detected by MTDSC, as will be described in Section 4.5.

The main reason behind the local equilibria is the rather strong interaction between the phases in the schematic of Figure 4.13, caused by the molecules that traverse various phases. Two interesting facts need to be considered in the interpretation of these local equilibria. First, small flexible, linear molecules like paraffins seem to need little or no supercooling for crystallisation. Second, chain segments coupled by as few as 4–6 flexible

Nanophases

A phase is a macroscopic aggregate of molecules, usually with well defined boundaries. To be able to describe the phase thermodynamically, one assumes that all molecules are inside the boundaries. As the phase volume approaches in at least one direction micrometre-dimensions, the surface properties affect the phase, now called a microphase. Microphases were first observed in the form of colloids. More recently, nanophases are being studied that have at least one dimension in the nanometre range.

Macromolecules exceed by far the nanophase dimension and can, thus, enter and leave many different nanophases. Understanding of the nanophase interactions is at the root of the understanding of partially crystallized or ordered macromolecules.

Schematic of a semi-crystalline polymer with nanophase separated crystalline and amorphous phases

Figure 4.13. Description and schematic of polymeric nanophases [1].

CH_2- groups to a macromolecule behave largely as the small molecule by itself with respect to crystallisation, melting and disordering transitions [6]. One can, thus, understand that partially melted polymer chains, which are still attached to the polymer crystal by at least a molecular nucleus, as shown in Figure 4.6, may crystallise and melt reversibly, as does the corresponding paraffin, as long as the molecular nucleus is hindered to melt itself. Similarly, segments of sufficient length within the amorphous nanophases may crystallise and melt like corresponding short-chain molecules.

Since oligomers with chemically identical repeating units have practically the same heats of fusion per oligomer (except for a change with temperature), and the melting point changes smoothly, quantitative analysis of such local equilibria can be attempted. For example, the melting temperature of paraffins up to polyethylene is well represented by a semiempirical expression given by Broadhurst [23]:

$$T_{\mathrm{m}}^{0} = 414.3\frac{x - 1.5}{x + 5.0} \text{ (in Kelvins)} \qquad (9)$$

where x represents the number of CH_2 groups. While polyethylene has an equilibrium melting temperature of about 414.6 K (and crystallises only with a typical supercooling of 10–20 K), segments of about 20 CH_2- groups that are sufficiently mobile in a structure as shown in Figure 4.13 may melt and crystallise practically reversibly at their equilibrium melting temperature of 305.8 K, not far from room temperature.

2.5 SUMMARY OF THERMAL EFFECTS

There are six different thermodynamic contributions to the apparent heat capacity in the melting and crystallisation region of the analysed polymer [24]. The first three can be truly reversible, and the second three are increasingly irreversible.

(1) The first and largest contribution to the thermodynamic heat capacity of polymers is always vibrational, as shown in Figure 4.1. The vibrational heat capacity has been calculated from the density of states of the skeletal and group vibrations as determined from normal mode calculations and matching to low-temperature, experimental heat capacities. The skeletal vibrations contribute most of the low-temperature heat capacity and for polyethylene level to their limit of $2R$ at about 300 K (R is the gas constant of 8.314 J K^{-1} mol^{-1}). As shown in Figure 4.1, the group vibrations start contributing at about 150 K, reach $1.4R$ by about 400 K and have a limit of $7R$ which would be approached only far above the decomposition temperature of the polyethylene.

(2) The second contribution originates from the emergence of dynamical changes between conformational isomers. In polyethylene, the low-energy *trans*-conformation can reach a local equilibrium with its two, higher-energy *gauche*-conformations. For the glassy polymer, this process starts above 100 K, and for the crystalline polymer, above 250 K (see Figure 4.1). At the beginning of the glass transition (237 K), this process contributes about 3.4 J K^{-1} (mol of CH_2)$^{-1}$ to the heat capacity of the glass; at the equilibrium melting temperature (414.6 K), it contributes about 5.0 J K^{-1} (mol CH_2)$^{-1}$ to the heat capacity of the orthorhombic crystal. For the amorphous polyethylene, the local *trans–gauche* processes change at the glass transition from a local to a global equilibrium involving a co-operative process that extends over a small volume of, perhaps, 1 nm in diameter. In the glass-transition range, this co-operative process is slower than the calorimetry and may not be fully reversible until the temperature is sufficiently high so that the co-operative kinetics is faster than the rate of measurement. This reversibility is reached at the end of the glass transition. The analysis by MTDSC allows the study of this co-operative kinetics of the glass transition [25]. The *trans–gauche* exchange is an internal rotation (conformational motion) between

states of different potential energy. Calorimetrically, the change in conformation is most easily correlated to the change in potential energy. A change from torsional oscillation to internal rotation between conformational isomers of equal potential energy would change the heat capacity only negligibly. At both, the disordering transition from the orthorhombic to the pseudo-hexagonal phase and to the melt, this increase in potential energy is a significant part of the latent heat of transition. The change in *trans* to *gauche* ratio also causes much of the difference between the solid and liquid heat capacities of polyethylene [26]. At 250 K, the liquid heat capacity is higher by 10.3 J K^{-1} (mol of CH_2)$^{-1}$ than for the orthorhombic crystals; at about 400 K, as shown in Figure 4.1, the orthorhombic and liquid heat capacities are about equal.

(3) The third contribution is the reversibly melting fraction discussed in Section 2.4. It has been observed some time ago that not all melting in polymers removes entire molecules from the crystals [27]. Rather, molecules may melt partially and then cannot be extractable from the remaining semicrystalline, higher-melting fraction, but can recrystallise at lower temperature, or crystallise and melt reversibly. Not only must there be a crystal to obviate primary and secondary nucleation for a reversible crystallisation and melting, but also there must be a molecular nucleus to initiate the reversible process. It is of interest to note that linear paraffins up to (at least) $C_{50}H_{102}$ melt practically reversibly [28], in contrast to most other small molecules which need at least crystal nucleation (see Section 2.2, Figure 4.4).

(4) The fourth contribution involves crystal perfection. Typically, the perfected crystals melt 5–20 K above the annealing temperature as demonstrated by the common annealing peaks [6]. A continuous annealing during heating is illustrated in Figure 4.9. The faster one goes through the range of temperatures where annealing can occur, the closer one approaches the zero-entropy-production melting temperature, explained in Section 2.3.

(5, 6) The fifth contribution is identified as secondary crystallisation, i.e. it involves less perfect crystal growth, occurring later than the initial crystallisation. The sixth contribution is the well-studied initial crystallisation with kinetics as demonstrated in Figure 4.5. The analysis of the thermodynamic stability of both primary and secondary crystals is complicated by crystal perfection, even when growth occurs isothermally. The interpretation of the irreversibility of the secondary crystallisation is obscured by the existence of a global network of the primary crystals, as seen in Figure 4.13. The fifth contribution may also contribute to the annealing peak. For homopolymers, the sixth contribution, the primary crystallisation, yields commonly the biggest latent heat effect. Overall, the ordering of polymers may go to fully ordered crystals, but mesophases are also possible, such as the condis crystals shown schematically in Figure 4.12 and discussed in Ref. [22].

Research on poly(ethylene-*co*-octene) (PEcoO) shows all six caloric effects just summarised [24,29]. The described processes in this chapter have rather broad applicability to polymer melting, and their understanding will permit a better link between structure, properties and processing. Examples are given in Section 4 after the discussion of instrumental and analysis problems in Section 3 that arise from MTDSC in the transition region of polymers due to the different degrees of reversibility and changes of temperature gradients within the samples.

3 Instrument and Deconvolution Problems

3.1 EVALUATION OF HEAT CAPACITY BY MTDSC AS A BASELINE FOR THE STUDY OF MELTING

The basic theory and practice of MTDSC is given in Chapter 1. In this section, a brief review is given of the changing analysis methods that are needed when the stringent restrictions are relaxed, which apply for measurements at the steady state and with negligible temperature gradients within the sample. In addition, the nomenclature is changed somewhat to adjust to common symbols found in the literature and the ones used in our research papers, books and courses found in the references.

The standard, non-modulated DSC can easily be described as long as one assumes a negligible temperature gradient within the sample and steady state during the period of measurement [30,31]. Under such idealised conditions, the heat-flow rates of the sample calorimeter, consisting of pan and sample, and the reference calorimeter, usually only an empty pan, are governed solely by the heating rate, q (in K min^{-1}), and the heat capacities, written as $C_s = (mc_p + C'_p)$ and $C_r = C'_p$, where m is the sample mass, c_p is the specific heat capacity of the sample and C'_p is the heat capacity of the empty calorimeter. Figure 4.14 illustrates such a DSC experiment, started at time 0 with a linear increase of the temperature of the heater, T_b. After about 100 s, the reference and sample temperatures reach steady state, i.e. both increase with the same q as T_b, and ΔT becomes constant. The data in Figure 4.14 are calculated by assuming Newton's law constant, K, to be

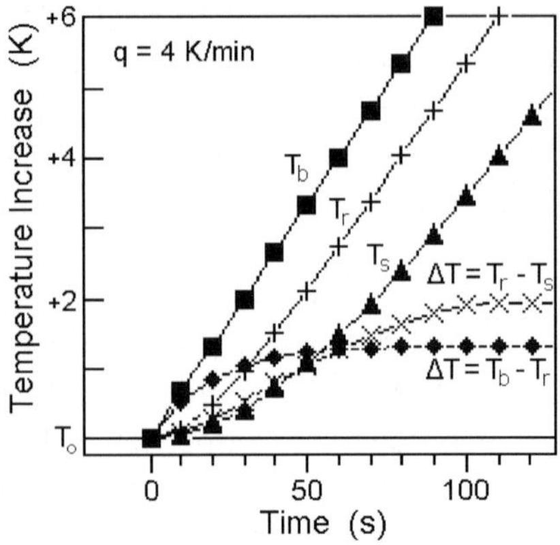

Figure 4.14. Temperatures in a standard DSC during the start of an experiment with heating rate 4 K min^{-1}. The differential heat-flow rate is proportional to $\Delta T = T_r - T_s$.

$C'_p/20$ in J K^{-1} s^{-1}, and that the Fourier equation of heat flow is valid. At steady state, the heat capacity is:

$$mc_p = K\frac{\Delta T}{q} + C_s\left(\frac{d\Delta T}{dT_s}\right) \tag{10}$$

where the second term on the right-hand side is a small correction term with C_s representing the total heat capacity of the sample calorimeter (pan + sample). The correction is needed since the sample and reference calorimeters change their heat capacities with temperature, i.e. T_r and T_s in Figure 4.14 are not strictly parallel to T_b. This correction needs no further measurement and is typically of the order of magnitude of 1% as long as mc_p is a substantial portion of C_s. Similarly, the "negligible temperature gradient within the sample" is not a stringent condition as long as steady state is kept. A substantial temperature gradient of 2.0 K across a sample of crystalline polyethylene will cause an error in the measurement of the magnitude of dc_p/dT of about 0.3% at 300 K.

The simplicity of the data analysis for a standard DSC is extended to modulated temperatures, as long as the condition of steady state and negligible temperature gradients within the sample can be maintained throughout the modulation. In Figure 4.15, the curves illustrate sinusoidal modulation.

For simplicity, the heater temperature is modulated. An immediate observation is that the sinusoidal modulation reaches a constant average level after a few cycles and that the sliding averages over one modulation period $<T_b>$, $<T_s>$ and $<\Delta T>$ yield the same curves as seen in Figure 4.14 for standard DSC. The contribution of the modulation is usually called the

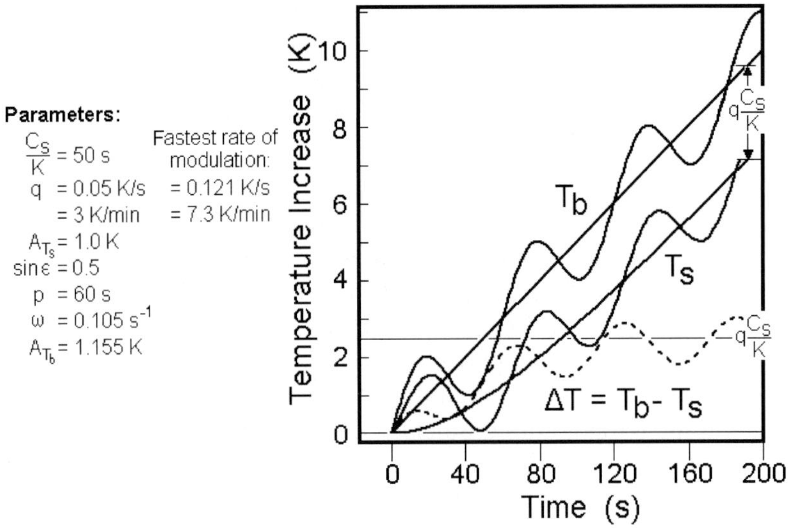

Figure 4.15. Temperatures in an MTDSC experiment with sinusoidal modulation.

reversing signal and is obtained by simple subtraction of the averages from their instantaneous values. The term "reversible" contribution is reserved for cases which have been established to be truly thermodynamically reversible. After the initial 200 s, the deconvolution of the two responses is thus quite simple.

An easy analysis of the reversing heat capacity is possible as long as $<T_b>$, $<T_s>$ and $<T_r>$ change linearly with time. At any time t, $<T_s(t)> - T(t)$ is identical to the result expected from a quasi-isothermal experiment ($<q> = 0$, see Section 3.3) and is called pseudo-isothermal. The quasi-isothermal analysis has been described in detail and yields for the heat capacity the following expression which also holds for the pseudo-isothermal case [31] where A_Δ is the modulation amplitude of ΔT which is proportional to the heat-flow rate, HF ($A_\Delta \propto A_{HF}$):

$$(C_s - C_r) = \frac{A_\Delta K}{A_{T_s}\omega}\sqrt{1 + \left(\frac{C_r\omega}{K}\right)^2} \tag{11}$$

Furthermore, A_{T_s} is the modulation amplitude of T_s. The frequency ω is given in rad s^{-1}. The similarity of Eqs. (10) and (11) becomes obvious if one uses an empty reference calorimeter. Then, C_r is equal to C' and $C_s - C_r = mc_p$, and $A_{T_s}\omega$ represents the amplitude of the modulation of the instantaneous heating rate $q(t) - <q> = dT_s(t)/dt$. The square root accounts, as in Eq. (10), for the difference between the modulation of reference and sample calorimeter. Only, if there is an empty reference position (no pan, $C_r = 0$) does Eq. (11) change into:

$$C_s = mc_p + C' = \frac{A_\Delta K}{A_{T_s}\omega} \tag{12}$$

an equation often erroneously also used with a reference calorimeter ($C_r \neq 0$). Note, however, that if calibration and measurement are done at the same frequency and the reference pans do not change between runs, all differences between Eqs. (11) and (12) can be eliminated by calibration. In Eq. (12), K is then $K \times$ [the square root part of Eq. (11)].

Equations (11) and (12) hold only if steady state is not lost during modulation, and the temperature gradient within the calorimeters is negligible. This condition is kept more stringent than for the standard DSC, because if even a small temperature gradient is set up within the sample during the modulation, each modulation cycle has smaller positive and negative heat flows which depend on the unknown thermal conductivities. A negligible temperature gradient within the sample requires, thus, that the sample calorimeter oscillates in its entirety as shown in the graph of Figure 4.15. It also requires a negligible thermal resistance between thermometer and pan, and the pan and sample calorimeter. The phase lag ε between heater and sample must, in this case, be entirely due to the thermal diffusivity of the Constantan disc [$T_s(t) = A_{T_s} \sin(\omega t - \varepsilon)$]. Typical conditions that have been used for measuring C_p of polymers with sinusoidal modulation are masses of about 10 mg, amplitudes of modulation of 1.0 K and modulation periods of $\geq$60 s.

Turning to the sawtooth modulation displayed in Figure 4.16, the input parameters for the calculation are the same as used for the standard DSC shown in Figure 4.14, just that at the time $t_1 = 100$ s, the heating rate is changed into cooling. The equations derived for the description of this calculation are listed in Figure 4.16. As long as the Fourier equation of heat flow holds, the solutions for different events in the DSC are additive. Beyond time t_1, for example, one can describe the temperature changes of T_s and T_r by assuming the heating is terminated at t_1, resulting in an approach to an isotherm at +6.67 K, and simultaneously a new experiment of cooling is initiated. The sum of these two events yields the bottom equation in Figure 4.16 and is plotted in the curves beyond t_1. The top equation describes the temperatures T_s and T_r for times from 0 to t_1. Steady state is lost at the sharp change of $q(t)$, at t_1 and on continuation of the sawtooth every time the rate of temperature change reverses. In the chosen calculation condition, the heat-flow rate of the sawtooth is not in steady-state during about half of the modulation period, p, of 400 s, i.e. there should be no way

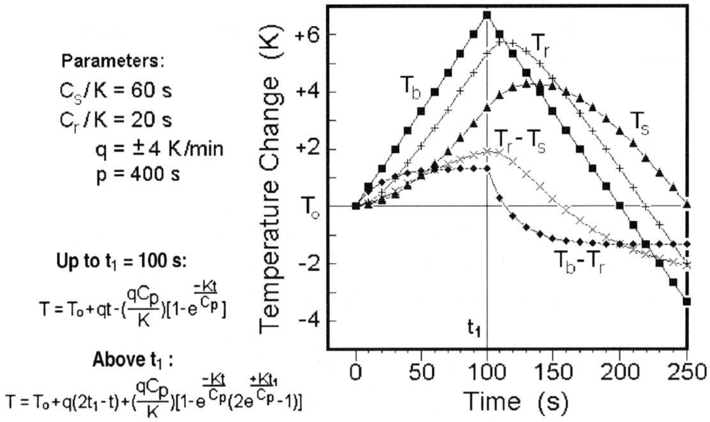

Figure 4.16. Temperatures in the beginning of MTDSC experiments modulated with a quasi-isothermal sawtooth.

to measure heat capacities using Eq. (11) or (12). Attempts were made to use the maxima and minima of the modulation response. This corresponds to a use of a standard DSC to test for the reversing nature of the sample. Naturally, this simple measurement is only correct if steady state is reached in each half cycle. The advantages over the standard DSC with separate heating and cooling cycles are discussed in detail by Ref. [32].

An extensive analysis of the sawtooth modulation brought a number of interesting results. Mathematically, it could be shown that if there were no temperature gradients within the sample and if all other lags and gradients could be assessed with the Fourier heat-flow equation, Eq. (11) does allow the calculation of the precise heat capacities [33]. Temperature gradients are, however, almost impossible to avoid. Especially in the power-compensated calorimeter, the temperature sensor is much closer to the heater than the sample and cannot avoid gradients. The empirical solution to this problem was to modify Eq. (11) as follows [34]:

$$(C_{\mathrm{s}} - C_{\mathrm{r}}) = \frac{A_{\Delta} K}{A_{T_{\mathrm{s}}} \omega} \sqrt{1 + (\tau\omega)^2} \qquad (13)$$

where τ, which has the dimension s rad^{-1}, is an adjustable time constant to be determined by measurements at different frequencies. It depends not only on the heat capacity of the reference calorimeter and the Newton's-law constant, as one would expect from Eq. (11), but also on the mass and thermal conductivity of the sample, as well as on all of the involved thermal contact resistances and, depending on the calorimeter type, possibly also on cross-flow between sample and reference calorimeters. Modelling of such complicated situations has been attempted by Hatta and Katayama [35], Höhne [36] and others (see also Chapter 1), but it is difficult to evaluate the various constants such treatments generate. As a result, we decided to use Eq. (13) as a tool to study τ empirically. As long as only mass and thermal conductivity of the sample affect τ (in addition to C_r and K), a plot of the squared inverse of the uncorrected heat capacity of Eq. (12) versus the square of the frequency should be linear, as it was indeed found for the Perkin–Elmer DSC with long modulation periods. With modulation periods beyond about 250 s, the frequency dependence becomes finally negligible.

To describe the sample temperature, $T_s(t)$ and the reversing heat-flow rate response, HF(t) which is proportional to $\Delta T(t)$, one uses the amplitude of the first harmonic of the Fourier representations. For the computation of the amplitude of the heat-flow rate, A_{HF} ($\propto A_{HF}$), one obtains:

$$\mathrm{HF}(t) = \langle\mathrm{HF}\rangle + \sum_{\nu=1}^{\infty} [A_\nu \sin(\nu\omega t) + B_\nu \cos(\nu\omega t)] \quad (14)$$

where A_ν and B_ν are amplitudes that must be determined in the usual manner, and ν is an integer. An analogous equation is used for the sample temperature. As long as the modulation is symmetric about $\langle q\rangle t$, and begins at time $t = 0$, it is centrosymmetric and all B_νs are 0, i.e. the series contains only the sinusoidal harmonics. For a linear response of the sample to sinusoidal modulation, no higher harmonics are generated in heat-flow rate, i.e. $A_{HF} = A_1$ of Eq. (14). A centrosymmetric sawtooth modulation also simplifies the Fourier representation: it shows only odd, sinusoidal harmonics with $\nu = 1, 3, 5, \ldots$, as is illustrated in Figure 4.17 for $\Delta T(t)$ of a

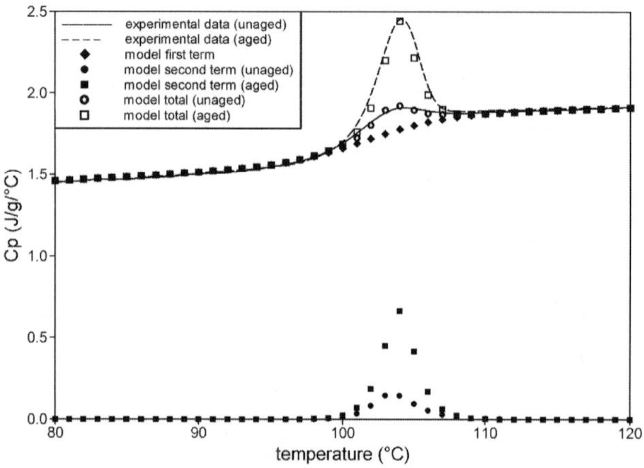

Figure 4.17. Heat flow rate in MTDSC as a function of time for a quasi-isothermal sawtooth modulation, with indicated first, third and fifth harmonic Fourier components.

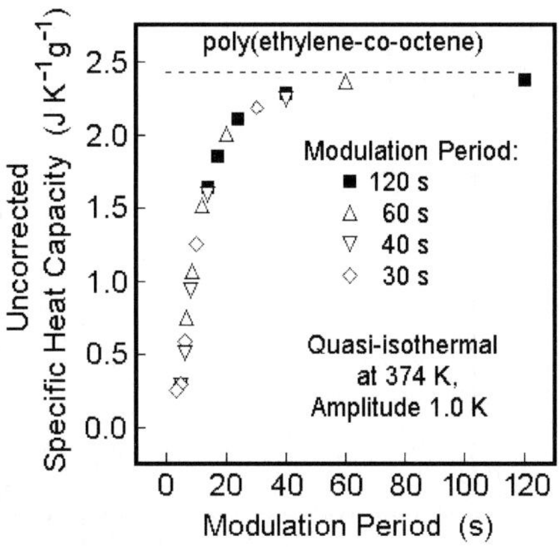

Figure 4.18. Uncorrected specific heat capacity. Calculated from Eq. (12) for different modulation frequencies, generated by four sawtooth runs of different frequencies at 374 K. The dashed line at $c_p = 2.43$ J g^{-1} K^{-1} is the expected specific heat capacity of the amorphous PEcoO in the liquid state.

centrosymmetric sawtooth modulation. In this case, $\Delta T(t)$ reaches steady state after about half of every heating and cooling segment. If Eq. (14) describes the MTDSC, each sinusoidal harmonic can separately be used to compute the heat capacity. Although the amplitudes of the higher harmonics decrease quickly, up to the 11th harmonics could be used to establish τ. Figure 4.18 shows the results for a typical copolymer, analysed with a Perkin–Elmer calorimeter [37]. Several runs with different modulation periods were used, and the uncorrected heat capacities are plotted as calculated from Eq. (12). Figure 4.19 illustrates the evaluation of τ using Eq. (13).

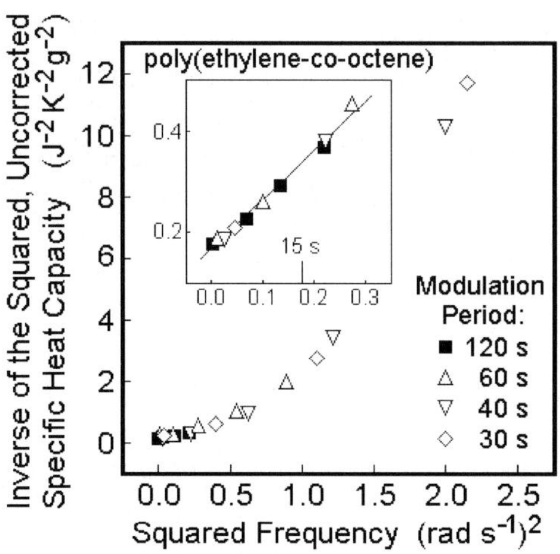

Figure 4.19. Evaluation of τ from the data of Figure 4.18, as suggested by Eq. (13).

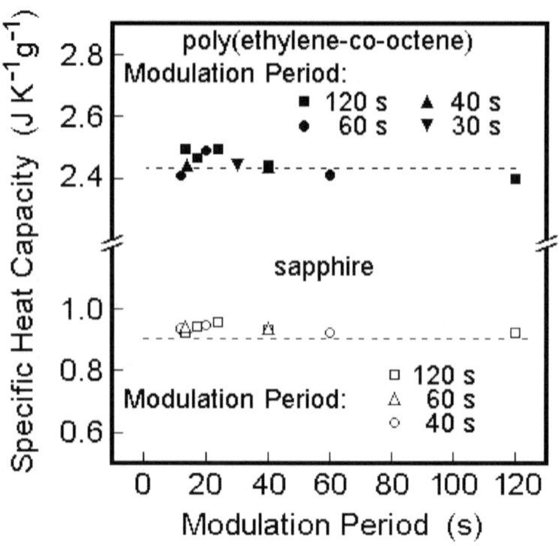

Figure 4.20. Corrected specific heat capacities of liquid PEcoO and the reference material sapphire. Analysed as shown in Figure 4.19 with values of τ of 2.40 and 2.24 s rad^{-1}, respectively. All data of periods >10 s were used for the evaluation of τ.

The corrected values from Eq. (13) are plotted in Figure 4.20 for measurements that show a frequency-independent τ for higher harmonics of periods longer than 10–15 s ($\tau_{\text{polymer}} = 2.40$ s rad^{-1} and $\tau_{\text{sapphire}} = 2.24$ s rad^{-1}). The dashed lines indicate the expected value of the heat capacity. Note that the calibration run with sapphire, which is also shown in Figure 4.20, needs a different τ and can thus only be used after evaluation of its separate τ value and extrapolation to zero frequency. It could also be shown that the common practice of subtracting a baseline of a run with two empty calorimeters to correct for the asymmetry of the calorimeter is not mathematically sound, but for highest precision must similarly be converted into a heat capacity contribution at zero frequency.

The final step in the analysis is to eliminate the problem that arises for the use of multiple frequencies from the quickly decreasing amplitudes of the higher harmonics in Eq. (14). This was accomplished by replacing a simple sawtooth with one designed to have similar amplitudes for the 1st, 3rd, 5th and 7th harmonics:

$$T(t) - T_0 = A[0.378 \sin \omega t + 0.251 \sin 3\omega t + 0.217 \sin 5\omega t + 0.348 \sin 7\omega t - 0.067 \sin 9\omega t \cdots] \qquad (15)$$

Heat capacity with a precision approaching 0.1% could be measured with this complex sawtooth using the usual sequence of a calibration run with sapphire, an asymmetry calibration with two empty pans and the measurement. A single run gives then enough data at different frequencies

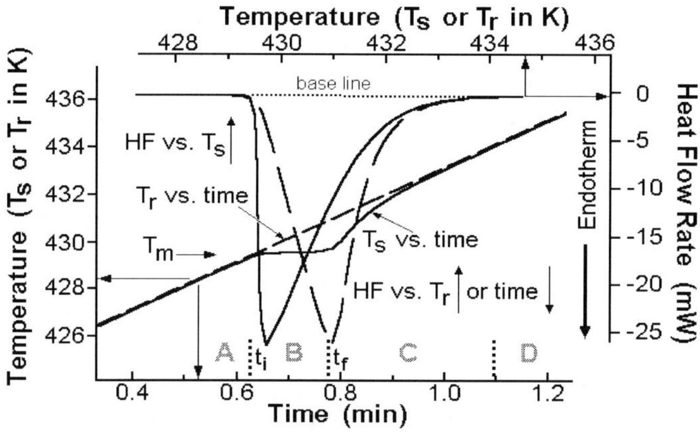

Figure 4.21. Changes of temperature and heat flow rate in a Mettler-Toledo DSC on melting of 7.584 mg of indium at a heating rate of 10 K min^{-1} [41].

to analyse the data, as shown in Figures 4.18–4.20. Detailed discussions of the techniques for the three most common scanning calorimeters are described by Kwon *et al.* [38], Pyda *et al.* [39], and Pak and Wunderlich [40]. The quantitatively determined heat capacity of a sample produces the baseline needed for the analysis of the latent heat effects as seen on melting and crystallisation.

3.2 MELTING AND CRYSTALLISATION BY STANDARD DSC

To develop a feel for the study of melting and crystallization, it is useful to observe these processes first by standard DSC. Figure 4.21 illustrates the changes in heat-flow rate on melting of indium of the reference and sample temperatures. Indium is known from quasi-isothermal MTDSC to melt within ±0.05 K or less. As a consequence, the melting peak in the heat-flow rate, HF (proportional to ΔT), is not a true record of melting, but is broadened considerably by instrument lag and changes in steady state within the calorimeter. From A to B, the baseline trace, which is due to the heat capacity of the solid indium, changes to a new steady state due to the melting of indium at a constant temperature of 429.75 K. Then, after melting is complete (C), a new baseline is approached, now due to the heat capacity of liquid indium. The width of the melting peak for the shown mass, calorimeter and heating rate is about 2.0 K instead of the expected upper limit of 0.05 K.

The plot of HF versus sample temperature, T_s, represented by the left peak of Figure 4.21, illustrates the initial, rounded approach to the steady

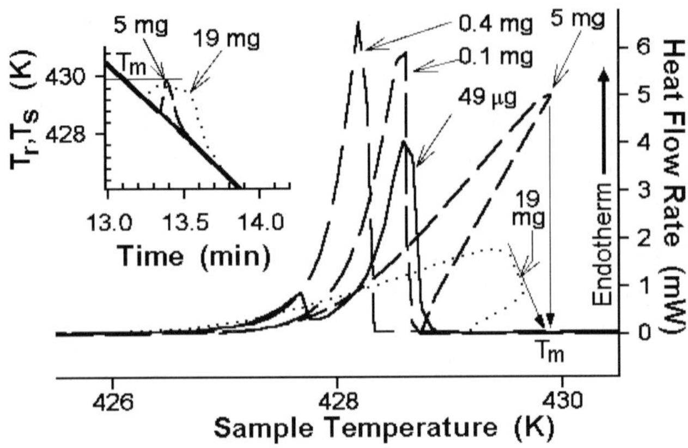

Figure 4.22. Changes of temperature and heat flow rates in a Mettler-Toledo DSC on crystallisation of different masses of indium on cooling at 5.0 K min^{-1} [41].

state of melting, followed by an almost vertical recording of the melting temperature. The remaining slight slope of the leading edge of the melting peak is used to establish the lag of the calorimeter (sometimes called the τ-correction). Substances of different thermal conductivity, however, will have somewhat different slopes. By extrapolation of the linear portion of the leading edge of the melting peak back to the baseline, one determines the onset temperature of melting for calibration purposes and for measurement of the melting temperature of sharp-melting substances.

The plot of HF versus reference temperature, T_r, on the right of Figure 4.21, approximates the heating rate, q, as the slope for the leading edge of its melting peak. The area under the baseline of this peak in the time recording is a measure of the heat of fusion. An exact derivation can be found in Figure 4.23 (below) and Screens TAM20-31 to 34 of Ref. [1].

The recording of T_r and T_s versus time reveals that in the Mettler-Toledo DSC, the heating rate is controlled close to the heater, so that the T_r is not affected by the melting of the sample. The two changes of steady state from heating of the solid to melting at constant temperature and then to continued heating of the liquid indium are clearly seen in Figure 4.21. Indium is an ideal calibrant for temperature and heat-flow rate because of its sharp melting peaks and the horizontal baselines. The latter is caused by an almost identical heat capacity of the liquid and solid indium.

Figure 4.22 represents an analogous recording of the heat-flow rate versus T_s on cooling at 5.0 K min^{-1}. Extrapolating the leading edge of the crystallisation temperature to the baseline for the determination of the onset of crystallisation illustrates a supercooling of about 1 K (see Section 2.2). For larger masses, the slope of HF is retrograde and may, with

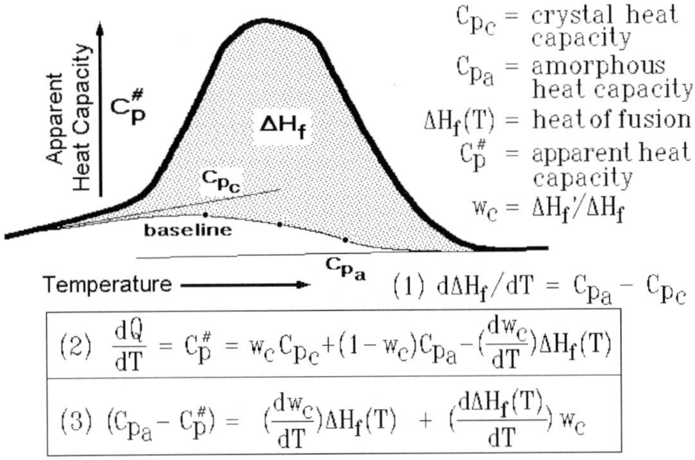

Figure 4.23. Evaluation of the heat of fusion for a sample melting over a broad temperature interval. A summary is given of the method of estimating a proper baseline and of the computation of crystallinity based on the measured apparent heat capacity and the baseline of the liquid.

appropriate mass (and cooling rate), reach back up to the melting temperature. Much interesting information about the performance of the DSC can thus be extracted from analyses as given in Figures 4.21 and 4.22. Temperatures of the onset of melting and crystallisation can be measured, estimations of the instrument lags can be made and heats of fusion and crystallisation can be obtained. What cannot be determined directly is the kinetics of the phase transitions because of the over-riding lags. Special methods have been developed for kinetics analyses using isothermal and lag-corrected procedures. Both are described in the general literature (see also the literature on "purity measurement").

An additional comment is needed for the determination of the heat of fusion. As discussed above, the integral of the heat-flow rate (HF $= \mathrm{d}Q/\mathrm{d}t$) should extend over time, not sample temperature. Since T_r is not affected by the melting, the recording of HF versus t and T_r is similar ($\mathrm{d}T_r/\mathrm{d}t \approx q$), and either area can be used to determine the heat of transition. From Figure 4.21, it is obvious that even HF versus T_s has a similar, but not identical, area.

Polymers have much broader melting ranges and do not crystallise fully, as outlined in Section 2. An important quantity is, thus, the determination of the mass fraction of crystallinity, w_c, and its change with temperature on heating and cooling. One assumes that:

$$w_c = \frac{\Delta H_f}{\Delta H_f^0} \tag{16}$$

where ΔH_f is the measured heat of fusion, and ΔH_f^0 is the heat of fusion of the 100% crystalline sample. If the measurements on heating occur under zero-entropy-production conditions, as described in Section 2.3, they can be used to characterise the initial sample. Measuring on heating, or cooling, with entropy production, some information on the kinetics of the transition can be gained. In either case, the baselines must be evaluated. An empirical construction is shown in Figure 4.23. The baseline of the semicrystalline sample is guessed-at from the low temperature baseline due to heat capacity, C_{p_c}, only. The baseline of the liquid sample due to the heat capacity, C_{p_a}, is measured after melting is complete. The three indicated points are marked by an estimate (or calculation) where the amount of melting reaches 25%, 50% and 75%, and the baseline is drawn (or calculated) accordingly for the proper integration.

The mathematical representation of the apparent heat capacity, $C_p^{\#}$, which contains both heat capacity and latent heat contributions, is given as method (2) in Figure 4.23. Over the wide melting ranges of polymers which may exceed 100 K, the temperature dependence of both heat capacities and also the heat of fusion must be considered. Information for many polymers is available in the ATHAS Data Bank [11,42]. Noting that the change of the heat of fusion with temperature is known through Eq. (1) in Figure 4.23, it is easy to eliminate one of the variables in method (2). Since C_{p_a} is easily available in form of the (calibrated) baseline after melting and is usually linear, it is best to eliminate C_{p_c}, as shown in method (3). In this way, the crystallinity can be generated out of the measured heat-flow rate curve alone [43].

Figure 4.24 illustrates a rather complicated example of a poly (oligoamide-12-*alt*-oligooxytetramethylene) copolymer which has its two components separated into different phase areas. At low temperature, both crystals, those of oligotetramethylene and the oligoamide, are present. First, the glass transition of the oligotetramethylene is seen, followed by the melting of its crystals. Next, the broad glass transition of the oligoamide shows, and the final melting peak accounts for the oligoamide crystals. After establishing the two melt baselines with method (3) of Figure 4.23, the crystallinities are computed, as shown in Figure 4.25. The crystallinities are based on the fraction within their corresponding component (11% and 17% of the 50% components of the total polymer).

Besides the lags during melting, which are complicated by temperature modulation, one must choose the proper modulation type for the measurement. Furthermore, the deconvolution of the signal in MTDSC discussed in Section 3.1 is in need of special attention before one is prepared to study the melting (and crystallisation) of polymers. These two topics are addressed

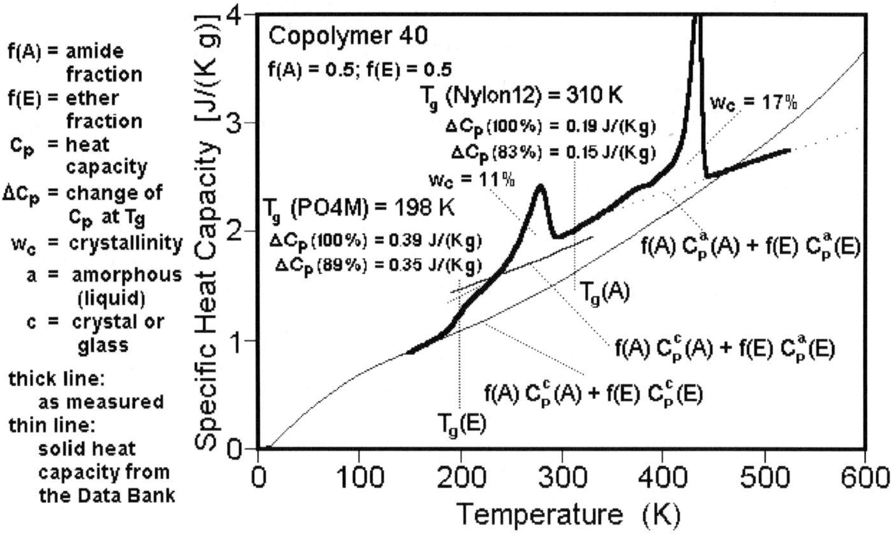

Figure 4.24. Apparent specific heat capacity of a poly(oligoamide-12-*alt*-oligooxytetramethylene) copolymer. Copolymer is analysed with a quantitative baseline based on heat capacity.

next. It will become obvious that several analyses, such as the determination of the onset of melting and crystallisation, as well as the measurement of the total heat of fusion and the change in crystallinity under zero-entropy-production limits, are best determined by the standard DSC methods, as discussed in this section.

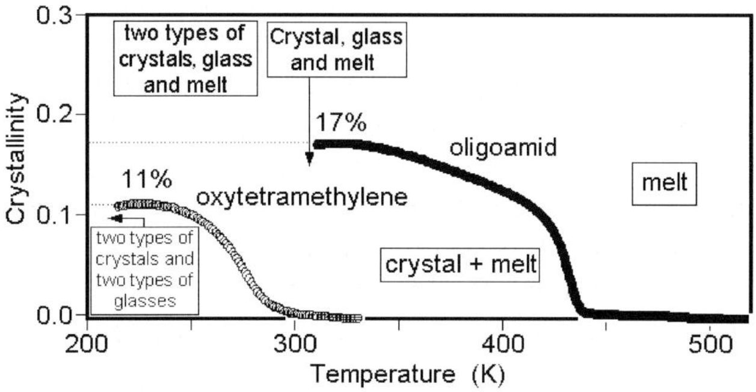

Figure 4.25. Example of the calculation of the crystallinity of the 50/50 copolymer of Figure 4.24 using method (3) of Figure 4.23 as modified by the copolymer concentrations and the baselines shown in Figure 4.24.

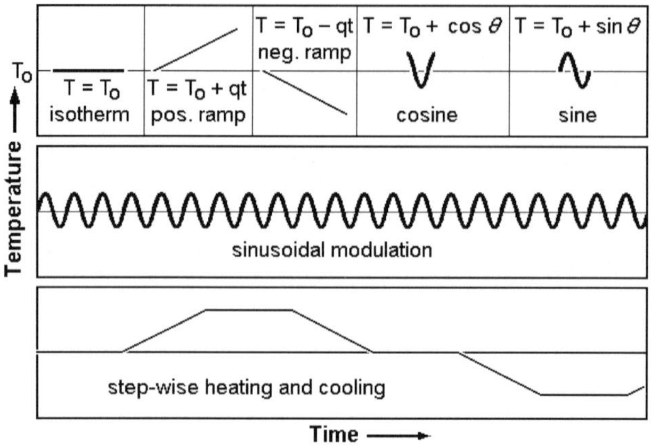

Figure 4.26. Temperature profiles for MTDSC experiments involving sinusoidal and stepwise heating and cooling.

3.3 TYPES OF MODULATION OF TEMPERATURE

The modulation of temperature for MTDSC can take on many forms. Figure 4.26 contains a number of segments in the upper row, which can, when properly linked, give a large number of modulation types. The most common modulation is shown in the second row, which is the sinusoidal modulation. As drawn, it represents only the modulation part, i.e. it is quasi-isothermal about the average temperature, T_0. The modulation can then be added to a linear temperature increase, qt, for the common MTDSC. For analysis, the two components are separated again, as discussed in Section 3.1. The advantage of the sinusoidal modulation is that it can be described by only one term of the Fourier series of Eq. (14), simplifying the analysis.

The step-wise heating and cooling shown at the bottom of Figure 4.26 are actually slight modifications of the standard DSC. An initial isotherm is followed by a heating ramp and concluded with another isotherm. In MTDSC, the heating ramp is continued by a cooling ramp. Naturally, if the segments of temperature change are sufficiently long, a standard DSC analysis is possible for its steady state portions [32]. Equation (10) can then be used for analysis, as was first proposed as dynamic differential thermal analysis [44]. In addition, the periodic function can also be characterised by the first harmonic of the Fourier series of Eq. (14). As long as the analysed segment is centrosymmetric, only sinusoidal terms exist in the series, and the first harmonic can be extracted as the reversing signal. Since practically all common DSCs are linear in response, the ratio of the amplitudes of the first harmonic of the temperature and temperature difference can then be inserted into Eq. (11) for analysis, omitting the higher harmonics. If the

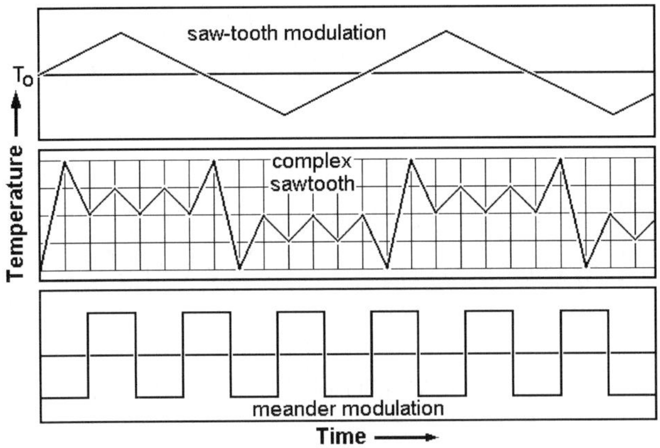

Figure 4.27. Modes of MTDSC involving sawtooth and meander-type modulation.

higher harmonics have a sufficiently high amplitude, they can be used as well for the analysis with a higher frequency of modulation, as is illustrated, for example, in Figure 4.17. Then, the frequency $\nu \times \omega$ must replace the frequency ω in Eq. (11).

Figure 4.27 illustrates in its top sketch the simple sawtooth modulation, discussed in Section 3.1, and the response to a sawtooth modulation is shown in Figure 4.17. The amplitudes of the Fourier series of the sawtooth modulation decrease with $1/\tau^2$, so that the precision of the analysis of higher harmonics decreases rapidly. To overcome this difficulty, the complex sawtooth shown in the center sketch of Figure 4.27, as well as given by the series of Eq. (15), was proposed [45]. Its first four Fourier terms describe practically all the variation shown in Figure 4.17. An overall modulation repeat of 210 s yields almost equal temperature amplitudes with periods of 210, 70, 42 and 23.3 s.

A similarly useful complex modulation is the meander modulation at the bottom of Figure 4.27. In this case, the temperature amplitudes of the Fourier series decrease linearly with the order of the harmonics, but the derivative $dA_{T_s}/dt = A_{T_s} \times T$ which actually enters into Eq. (10) is constant. This method is particularly easy to program for any standard DSC.

The combination of modulations, as shown in Figures 4.26 and 4.27, with an underlying heating rate, $<q>$, yields an MTDSC experiment as illustrated in Figure 4.15. For the study of melting and crystallisation of polymers, special combinations of underlying rates and modulations are of interest. Figure 4.28 illustrates several rates of temperature change as can be generated by sinusoidal modulation or the corresponding Fourier components of a more complicated modulation. The top examples represent the common MTDSC with overall heating or cooling. The centre examples

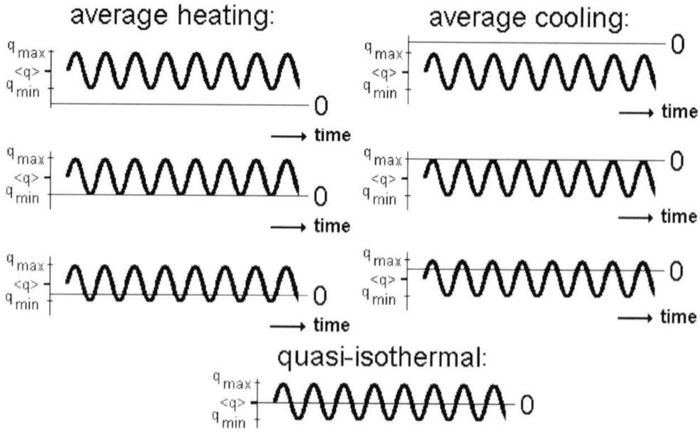

Figure 4.28. Types of sinusoidal temperature modulation with underlying heating rates. Plotted is the instantaneous heating rate: $q(t) = \mathrm{d}T/\mathrm{d}t = A_{T_s}\omega\cos(\omega t + \gamma)$.

result in heating or cooling only. The lowest, respectively, highest rates of temperature-change are 0, so that on heating, no recrystallisation can occur, and similarly in a crystallisation experiment, the once crystallised material cannot remelt. In both cases, an easier interpretation of the data is possible. Only reorganisation and crystal perfection can be superimposed on the simple melting and crystallisation processes of such experiments.

In many cases, fully reversible melting and crystallisation and reorganisation and crystal perfection processes can be further studied by using the bottom modulation of Figure 4.28: the quasi-isothermal case. Reference to this quasi-isothermal analysis is also made in Section 3.1. A fully reversible process, as discussed in Sections 2.1 and 2.4, yields a response that is perfectly sinusoidal with a constant amplitude over many modulation periods. Irreversible processes, such as crystallisation with supercooling, melting with superheating, reorganisation and crystal perfection, as discussed in Section 2.3, change with time. For fast kinetics, the process deforms the sinusoidal response, as will be analysed in Section 3.4. With slow kinetics, the irreversible processes can be followed over many hours, and their kinetics can be evaluated from the decrease of the amplitude of HF(t). This section has illustrated choices for the analysis with MTDSC and shown that a good understanding of the melting of polymers is necessary, as well as mastery of the available DSC.

3.4 DECONVOLUTION OF THE REVERSING HEAT CAPACITY

The standard deconvolution is described in Chapter 1 and summarised in Section 3.1. It gives only a proper answer as long as sample response and

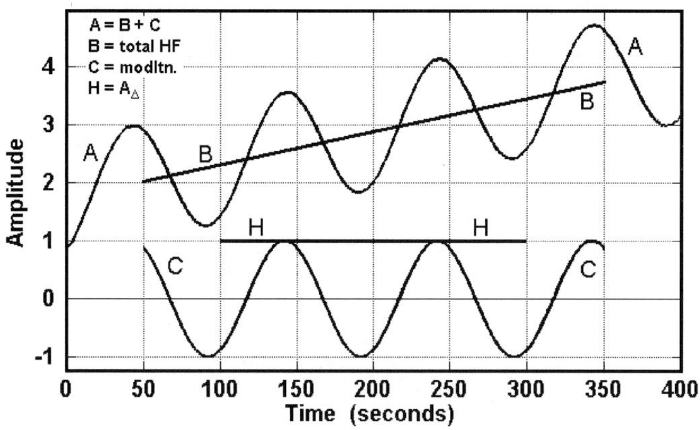

Figure 4.29. Analysis of the heat-flow rate for MTDSC. A: measured response, HF(t) or ΔT; B: total heat flow rate <HF(t)> or <$\Delta T(t)$>; C: modulation response HF(t) − <HF(t)> or $\Delta T(t)$ − <$\Delta T(t)$>; H: amplitude from the Fourier transformation, A_{HF} or A_Δ.

instrument performance are linear, i.e. for a sinusoidal excitation of temperature, the response in HF(t) is similarly sinusoidal (with a phase lag determined by instrument and sample properties). The same applies to the Fourier components of any other modulation. Figure 4.29 illustrates the deconvolution of HF(t), which is the response in an MTDSC experiment (A). The total heat flow rate, <HF(t)>, is given by (B) as the sliding average over one modulation cycle, <>. As long as (B) has a constant slope, (C) can be deconvoluted by forming <HF(t)> − HF(t), as described in Section 3.1. Because of the averaging, the curves (B) and (C) are not available for the first and last half cycles. Next, the amplitude of curve (C) is determined as the first Fourier amplitude in Eq. (14). It yields curve H. Again, another half cycle is lost by this procedure. The negligible changes in the slope of curve (B) are called the stationarity condition, which needs careful attention in the presence of latent heats [46].

By modelling the changes in the analysis when adding irreversible thermal events to the modulation, one can assume a quasi-isothermal analysis, <$q(t)$> = 0, and then carry out the same analysis. Examples of this process with worked-out spreadsheets are in the literature [47]. Figure 4.30 illustrates the influence of a small spike in the modulation, as one may see in very small transitions, inadvertent heat losses, electronic interferences, or mechanical disturbances. Clearly, the Fourier transformation broadens the defect over the averaging range of one cycle, so that it barely shows up in the computed reversing heat capacity using Eq. (11) or (13). This is an advantage in case the spike is caused by an error, but if this effect is to be measured, only curve (C) is a reasonable representation. The heat-flow amplitude (H) is a poor rendition of the effect. The irreversible effect, which

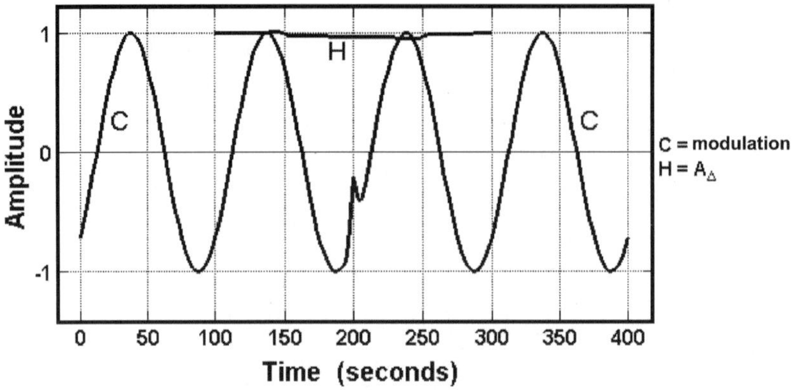

Figure 4.30. Modelling of quasi-isothermal MTDSC with a small transition of a peak height of 50% occurring at 200 s.

is calculated from the total heat-flow rate, <HF(t)>, by subtracting a sinusoidal curve corresponding to the changing amplitude of (H), is similarly displaced. Although averaging curve (C) gives the proper <HF(t)>, it does not occur at the proper temperature.

Figure 4.31 represents the effect of a sudden addition of a substantial, irreversible heat flow. Only the initial sharp increase in heat-flow rate has a small Fourier component of frequency ω. After two cycles, no further reversing component occurs, and the irreversible heat effect is perfectly separated. Figure 4.32 shows similar effects of linearly (top) and exponentially (bottom) increasing irreversible heat-flow rates. Again a good separation is possible. In general, linearly changing or constant heat-flow rates

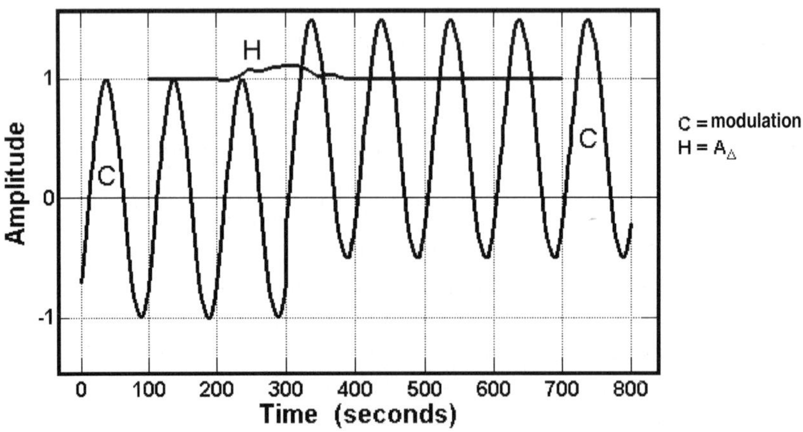

Figure 4.31. Modelling of quasi-isothermal MTDSC with an addition of a constant 50% change in heat-flow rate at 300 s.

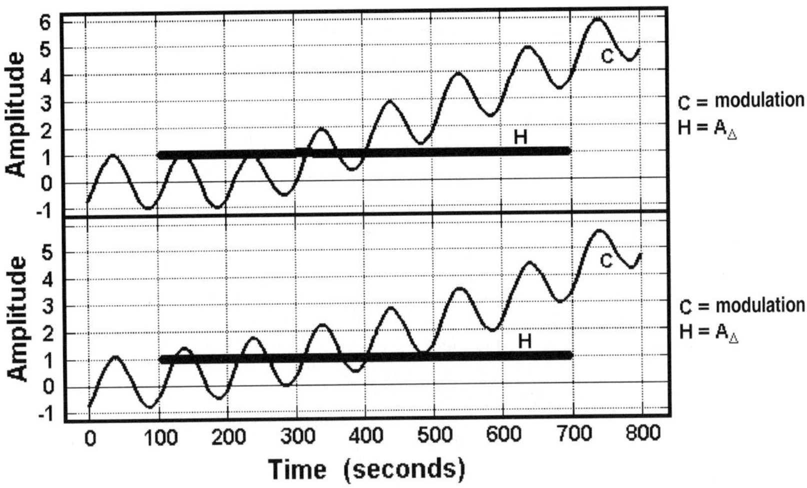

Figure 4.32. Modelling of quasi-isothermal MTDSC with an additional linear increase of the heat flow rate, beginning at 250 s (upper curves), and an additional exponential increase of the heat flow rate starting at the beginning of modulation (lower curves).

are fully rejected, but small effects appear at the beginning and end of such irreversible processes and must be considered [48].

Next, an approximately reversible transition is added to the modulation, as displayed in Figure 4.33. Although the assumed process is reversible, the Fourier analysis used in finding the reversing signal is not able to give a proper representation, as is seen from curve (H). Even the smoothing of the output, as often used in MTDSC, does not improve the signal. It is spread

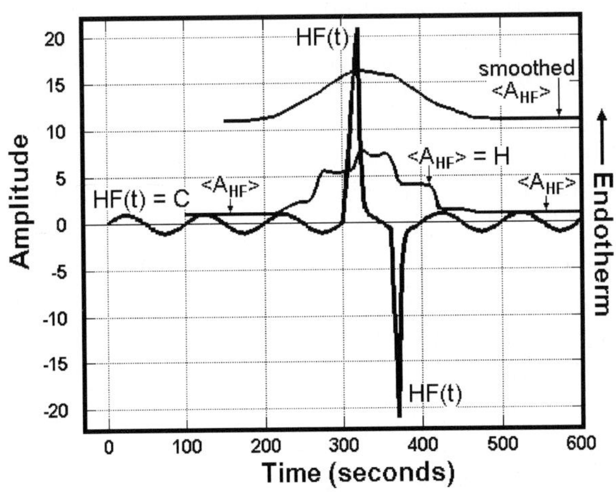

Figure 4.33. Calculation of a reversing melting and crystallisation in the quasi-isothermal modelling of MTDSC.

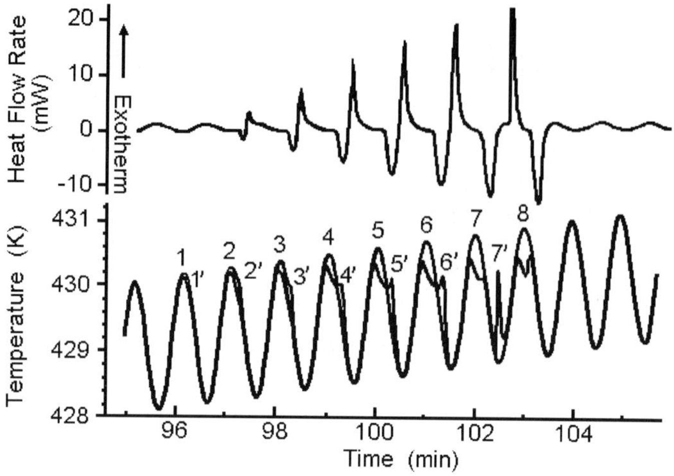

Figure 4.34. Multiple melting and crystallisation peaks of indium on MTDSC, 5.87 mg at an underlying heating rate $<q>$ of 0.1 K min^{-1}. The undistorted sinusoidal sample temperatures are also indicated. The two modulation parameters are as follows: $A = 1.0$ K, $p = 60$ s, Mettler-Toledo DSC 820, sinusoidally modulated, controlled close to the heater [41].

over the usual two modulation periods and improperly deconvoluted into the total and reversing signals. As before, the total signal is representing a proper average, but it is similarly broadened and cannot be used to assess the proper progress of the reversible endothermic and exothermic transitions. For a quantitative analysis of reversing latent heats, it must be possible to treat the added heat effects as approximately constant (stationary) over the sliding analysis window of one modulation cycle <>.

Figure 4.34 illustrates an actual MTDSC result from the melting of indium. The deviations of the sample temperature from the sinusoidal modulation due to instrument lag on melting and crystallisation are obvious and can be easily interpreted in terms of Figures 4.21 and 4.22. The increasing amount of melting in the modulation cycles 1–8 illustrates the lag of the calorimeter. There is not enough time for complete melting of indium in peaks 1–8 within the limits of the programmed temperature (see also Figure 4.21). This lack of time can also be concluded from crystallisation without supercooling (2′–6′) and with some supercooling (7′). As soon as the temperature of the heating cycle remains long enough above the melting temperature, all indium melts, as seen in heating cycle 8. Crystallisation then becomes impossible because of the missing crystal nuclei (see Section 2.2). Figure 4.34, thus, demonstrates clearly the instrumental and nucleation problems of melting of indium.

Figure 4.35 shows the further problems that arise from the Fourier analysis of the already flawed data of Figure 4.34. A similar shape of the amplitude

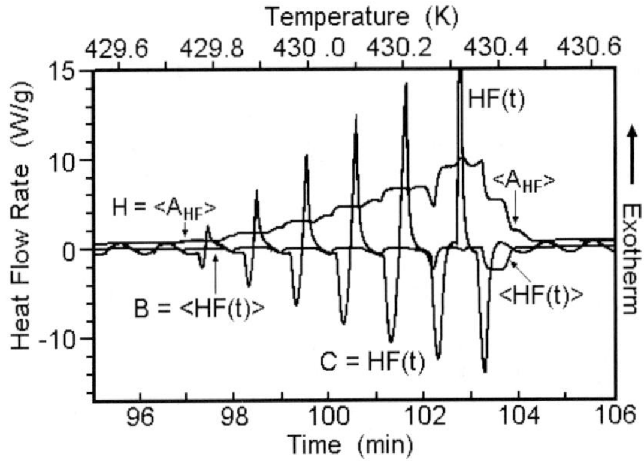

Figure 4.35. First harmonic contribution to the heat flow amplitude $<A_{\mathrm{HF}}>$ of the Fourier transformation of HF(t) of the data of Figure 4.34. Also shown is the total heat flow, <HF(t)>, that is often considered equal to the standard DSC heat flow (compare to Figure 4.21 to see the large differences) [41].

of the reversing heat flow signal (H), as modelled for Figure 4.33, but of increasing size, is superimposed for each successive modulation cycle of Figure 4.34. The total heat-flow rate (B) is also divided into increasing, broadened peaks, which mark the imbalance between melting and crystallisation in the sliding averaging window, with the total area representing ultimately the heat of fusion that corresponds to the initial sample. All recrystallisation in the cooling cycles is ultimately reversed. If one would integrate an apparent heat capacity from curve (H) and the corresponding temperature amplitudes using any of the Eqs. (11)–(13), the reversing heat of fusion would exceed the total heat of fusion by a factor that would depend on the number of times the indium is melting over cycles 1–8 and would represent a rather useless quantity.

A useful analysis is a direct integration of curve C, the measured heat-flow rate in the time domain. Figure 4.36 shows the resulting integral analysis

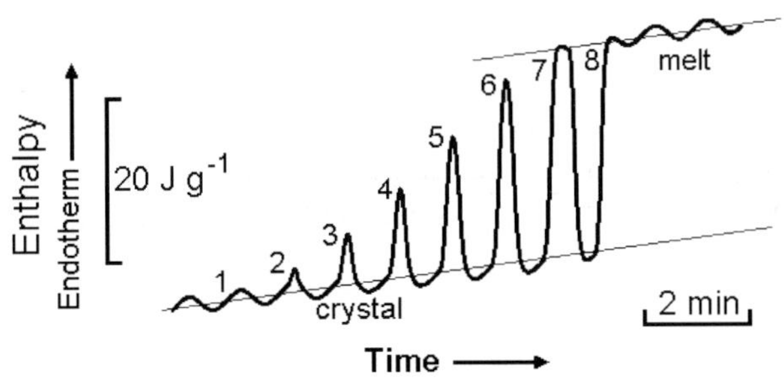

Figure 4.36. Integral analysis of the enthalpy in the melting and crystallisation region for the MTDSC run of Figure 4.34 [41].

[41]. Before cycle 1 and after cycle 8, the respective sinusoidal changes of the enthalpy of the solid and liquid indium are represented. The increasingly higher peaks illustrate the step-wise approach to the liquid and indicate that all cycles before cycle 8 reach practically full crystallisation.

4 Applications of MTDSC to Polymer Melting

Only with a grasp of the discussion of Section 3 one can attempt even a qualitative interpretation of the melting of polymers with MTDSC. Despite the fact that the study of the first part of this chapter seems to indicate that little useful analysis may come from MTDSC of melting transitions due to instrument lags (Figures 4.21 and 4.22) and deconvolution problems (Figure 4.33), experiments since 1994 have revealed many useful and unique qualitative and quantitative insights into polymer melting. In the following parts, examples are displayed which introduce the new analysis techniques. First, qualitative separations of glass transitions, enthalpy relaxations, recrystallisation, irreversible crystallisation and melting and reversible melting processes are described. These are, at present, still the most common applications. Also possible are quantitative analyses of apparent heat capacities in the glass-transition regions, which lead to a description of the relaxation kinetics (see Section 2). The quantitative analysis of the various latent heats similarly allows an evaluation of melting and crystallisation kinetics, as is shown in Section 3. Most quantitative data arise by quasi-isothermal analyses (see Figure 4.28) and have most recently led to the identification of a small reversible fraction of polymer melting, as is discussed in Section 4.

4.1 QUALITATIVE AND SEMI-QUANTITATIVE ANALYSIS OF POLYMER MELTING BY MTDSC

An MTDSC heating trace of amorphous (quenched) PET is reproduced in Figure 4.37. It is a partial copy of the example displayed in the original patent about MTDSC [49]. The insert shows the response of the heat-flow rate to the modulation of the temperature [curve (A) in Figure 4.29]. The main plot reveals the deconvoluted total heat flow rate [curve (B) in Figure 4.29] and the amplitude of the reversing heat flow rate [curve (H) in Figures 4.29–4.33]. The also shown amplitude of the non-reversing heat flow rate derives from the difference between total and reversing heat flow rates [$\langle HF(t)\rangle - A_{HF}$, curves (B)–(H) in Figure 4.29, see also Eq. (14)]. In the glass transition region, the increase in heat capacity appears in the reversing

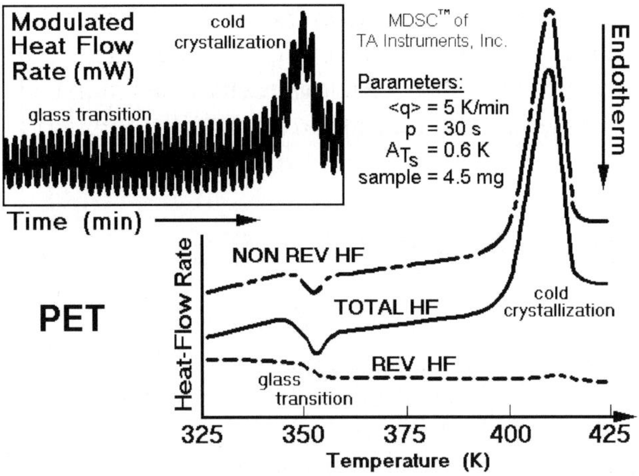

Figure 4.37. Total, reversing and non-reversing heat flow rates of PET using sinusoidal temperature modulation [49].

heat flow rate, and the endothermic enthalpy relaxation, caused by an annealing of the glass below its transition temperature before the analysis, is seen in the non-reversing heat flow rate, while the total heat flow rate is similar to a standard DSC trace. More details on the analysis of this glass transition are given in Section 4.3.

Above the glass transition temperature, the PET shows what is called a cold crystallisation, a crystallisation that contrasts the more common crystallisation by cooling from the melt [50]. This process is fully irreversible, and only a minor deviation appears in the reversing heat flow rate due to incomplete deconvolution, as also seen in Figure 4.30 (see also Ref. [48]). Although the deconvolution of fully non-reversing and partially non-reversing components of melting is not fully quantitative, as will become obvious later in this section, with some care, the components can be identified qualitatively and often, as in this case, analysed semi-quantitatively. The MTDSC yields, thus, important additional information on crystallisation and melting. In addition, special methods have been developed to get quantitative results, as is described in Sections 4.2–4.5. Of particular importance is the good separation of the heat flow rate outside of the transition regions, which allows a precise determination of the heat capacity. Note that the slope of the total heat flow rate of Figure 4.37 changes to the expected, gradual increase caused by the heat capacity with temperature [see also Figure 4.29, curve (H)].

The melting range of polymers is much more complicated, as is shown on a more recent example, the analysis of the poly(trimethylene terephthalate) (PTT). Figure 4.38 illustrates the heat-flow rate of a quenched PTT

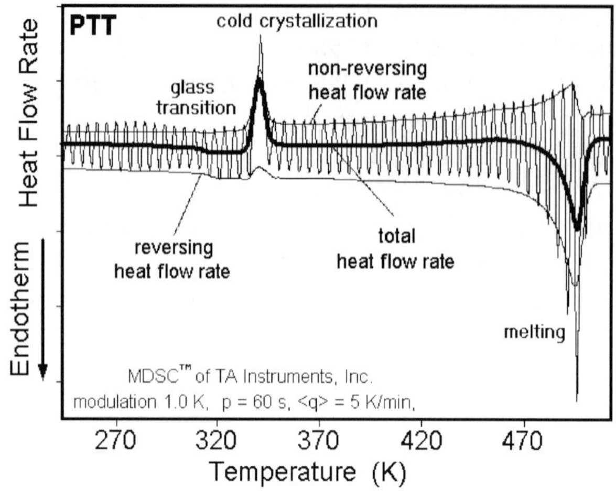

Figure 4.38. Total, reversing and non-reversing heat flow rates of PTT (sinusoidal modulation) [51].

without annealing below the glass transition. In this case, the glass transition displays no enthalpy relaxation. The cold crystallisation is less well separated from the reversing heat flow than for PET in Figure 4.37 since fewer modulation cycles cover the transition. In general, one finds that five or more cycles of modulation are needed in the crystallisation region to approach stationarity during the transition and cause only negligible errors in the deconvolution (see also Ref. [48]). After cold crystallisation, a slowly increasing non-reversing, exothermic contribution signals further crystallisation or reorganisation. In this temperature region, from 350 to 470 K, one still expects rather small instrument lags and a semi-quantitative analysis is possible. The processes of secondary crystallisation, pre-melting, recrystallisation and crystal perfection, however, can be so slow, that they lag due to their kinetics and not due to an instrument effect [6]. Quantitative analysis then needs a separation from the heat capacity and a study over long periods of time, as will be detailed in Sections 4.3–4.5. Qualitatively, one can separate such non-reversing exothermic processes from the reversing processes in this pre-melting temperature region.

Beyond 470 K, the major melting region can be seen in Figure 4.38. It displays a small exothermic, and a larger, endothermic contribution. On averaging, the heat flow rate, HF(t), yields the total endotherm, again similar to the melting peak of standard DSC, but smaller in size than the reversing contribution to the melting. It is quite clear by the strong deviation of the modulation from a sinusoidal response that the first harmonic, which yields the reversing contribution, cannot give quantitative information about the heat involved [see Figures 4.29–4.33 and Eq. (14)]. Higher harmonics would have to be added or a direct analysis in the time domain would have to be done. A further complication for the melting of polymers is seen by inspection of Figure 4.5. One would expect that polymer crystals do not recrystallise after melting on the small amount of cooling of the magnitude of the modulation amplitude. This would leave only crystal perfection as the exothermic process. The MTDSC trace indicates, thus, a change in the crystal perfection due to heating during the analysis, a most important observation since melting is used to characterise the polymer for use at low temperature and any change during analysis itself must, naturally, be considered. In Section 4.5, an additional, surprising, reversible melting will be described that was discovered by MTDSC and seems to be common for all polymers.

A special analysis method for the identification of a non-reversing exotherm hidden in the total heat-flow rate of melting is displayed in Figure 4.39 for PET. The special method involves a modulation without ever reversing the temperature to lower values, as indicated in the left centre curve of Figure 4.28. The heating rate changes between 0 and 10 K min^{-1}. Since there is no cooling, it is not possible to supercool the once-melted crystals, and any remaining exotherm cannot be an instrument effect, but must be due to crystal reorganisation or recrystallisation to a higher degree of perfection.

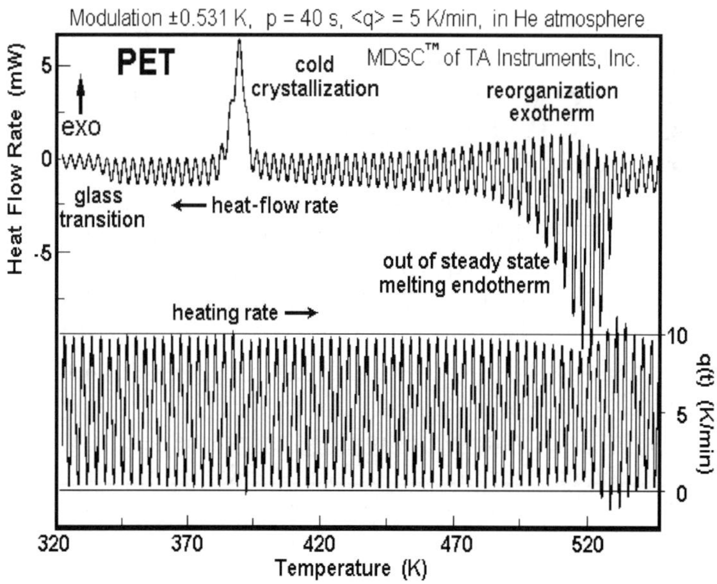

Figure 4.39. Heat flow rate and heating rate as a function of time of PET (TA Instruments, Application Notes).

Figure 4.39 shows, in addition, that the instrument control is lost at about 500 K. The programmed heating rate cannot be reached because of the large heat of fusion.

To gain further insight into the melting behaviour and the effect of instrument lag, a series of different materials is discussed next. First, the reversible melting of the metal indium with a high heat of fusion and high thermal conductivity is analysed further (see also Figures 4.34–4.36). This is followed by the reversible isotropisation of azoxyanisole, a liquid crystal with a low heat of transition and low thermal conductivity. The third example is the reversible melting of the paraffin *n*-pentacontane. It has a high heat of fusion and low thermal conductivity. These three low molar mass examples of reversible transitions involving 100% crystallinity are followed by the description of a semicrystalline poly(oxyethylene) that could be crystallised to an extended-chain macro-conformation and melted almost completely irreversibly.

Figure 4.40 illustrates the change of sample temperature (dashed line) and heat flow rate (heavy solid line) for a quasi-isothermal MTDSC experiment, as suggested by the bottom heating rate program of Figure 4.28 ($<q> = 0$). This analysis should be compared to Figure 4.34, which was

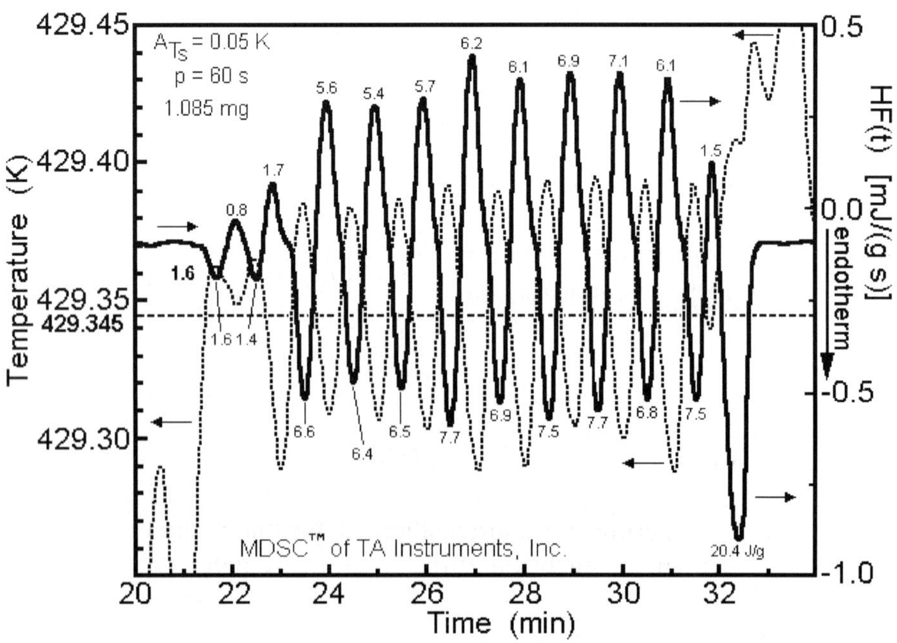

Figure 4.40. Quasi-isothermal melting of indium using a heat-flux calorimeter with control of the modulation at the sample temperature (sinusoidal modulation, the indicated sample temperatures are uncorrected) [52].

taken with an underlying heating rate. The quasi-isothermal method allows to extend the experiment to times when all non-reversing effects have decreased to zero and, thus, eliminates the kinetics of slow processes within the sample. Figure 4.40 shows on the far-left and -right sides the last and first modulation cycles before and after melting. The minimal heat flow rates are caused by the low heat capacity of the solid indium. At 21 min, T_0 is increased to 429.35 K and after about three cycles, repeatability is reached. The heat flow rate has increased enormously, indicative of melting followed in each cooling cycle by crystallisation. The numerals indicate the heats of transition, estimated by subtracting the modulated baseline. The observed 6–7 J g^{-1} corresponds only to 1/3 of the heat of fusion and crystallisation. No indication of supercooling exists because the melting is partial and keeps sufficient nuclei for instantaneous crystallisation. In each segment, the amount of melting is set by the thermal resistance of the calorimeter (see Figures 4.21 and 4.22 for the lag of a standard DSC, but with larger indium mass and faster heating rate). At about 32 min, the temperature is increased by 0.1 K, leading to a full melting peak of 20.4 J g^{-1}, followed by the low heat flow rate of the liquid indium. Complete melting between 329.35 and 329.45 K leaves no nuclei. The quasi-isothermal temperature could have been corrected with this experiment (literature value 329.75 K), and the experiment proved that indium melts within <0.1 K. Practically, all of the width of the standard DSC melting peaks in Figures 4.21 and 4.22 are due to lag of the instrument.

The calorimeter used in Figure 4.40 is controlling the sample temperature at the position of measurement of the sample temperature, i.e. the heating and cooling applied during the transition increases dramatically. Using a calorimeter that is modulated at the heater or reference temperature changes some details of the experimental result, as shown in Figure 4.41. The first modulation sequence indicates practically no melting and crystallisation. The central one indicates a small amount of incomplete reversible melting. The third run, however, saturates the possible heat flow rate. Each heating cycle melts a certain amount of indium. The following cooling cycle recrystallises only a portion of the just-melted indium, so that after seven melting steps, all crystals are melted and no crystal nuclei are left and the only calorimeter response comes from the heat capacity of the melt. Note that in this experiment, a short isotherm at the maximum temperature is inserted at the end of each run at the maximum temperature. The first run shown decays quickly to a baseline (at a). The second run indicates a small amount of continuing melting (at b). The point b must, thus, be in the melting range of the sample. These experiments with the two calorimeters suggest that each instrument type should be analysed and may need a different approach to MTDSC.

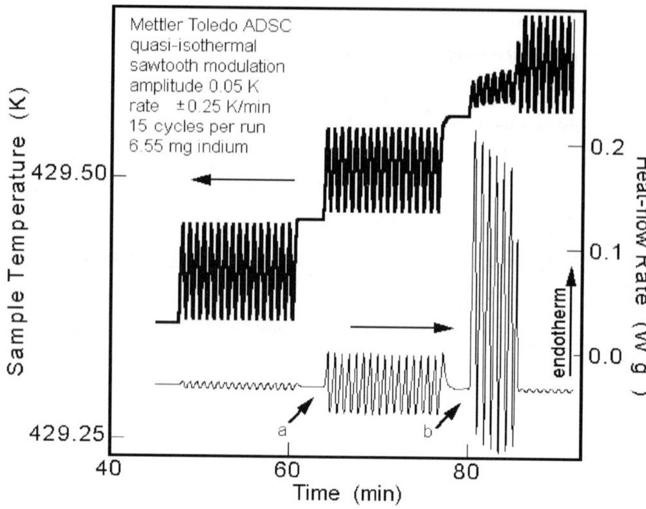

Figure 4.41. Quasi-isothermal melting of indium using a heat-flux calorimeter with control of the modulation at the heater temperature (sawtooth modulation, the indicated sample temperatures are uncorrected) [53].

The just-described reversible melting and crystallisation of indium are characterised by a high heat of transition (28.62 J g^{-1}) and high thermal conductivity (0.008 W m^{-1} K^{-1}). The liquid crystal azoxyanisole shown in Figure 4.42 is also known to have a reversible transition to the isotropic melt, but with a heat of transition of only 2.56 J g^{-1} coupled with a thermal

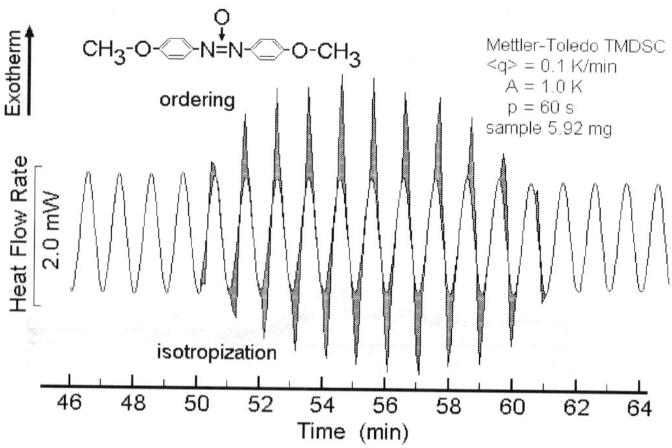

Figure 4.42. Heat flow rate as a function of time in the transition temperature range of isotropisation of 4,4′-azoxyanisole on MTDSC (calorimeter with modulation control at the heater temperature, sinusoidal modulation) [54].

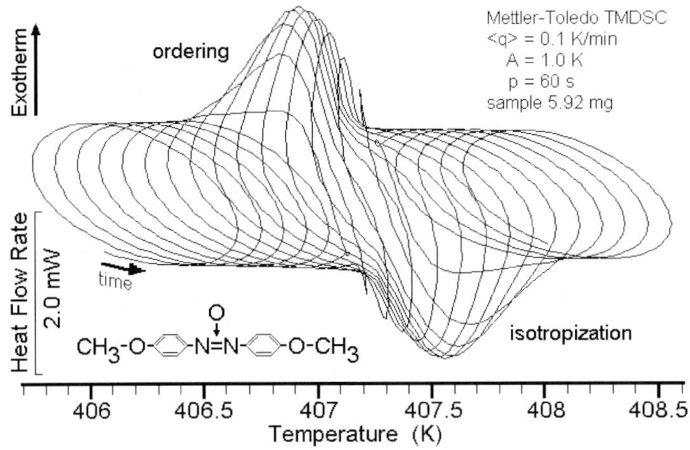

Figure 4.43. Heat flow rate as a function of temperature through the isotropisation range of 4,4′-azoxyanisole on MTDSC as in Figure 4.42 (Lissajous figure) [54].

conductivity 1000 times less than that of indium. The shaded areas of heat flow rate curve in Figure 4.42 delineate the small transition on heating. The plot of heat flow versus temperature gives the Lissajous figure, depicted in Figure 4.43. Figure 4.44 compares the results of standard DSC, MTDSC on heating and cooling and quasi-isothermal MTDSC. The quasi-isothermal result reveals a small broadening of the transition, but a sharp, <0.2 K wide main transition. All other broadening is caused by the lag of the calorimeter

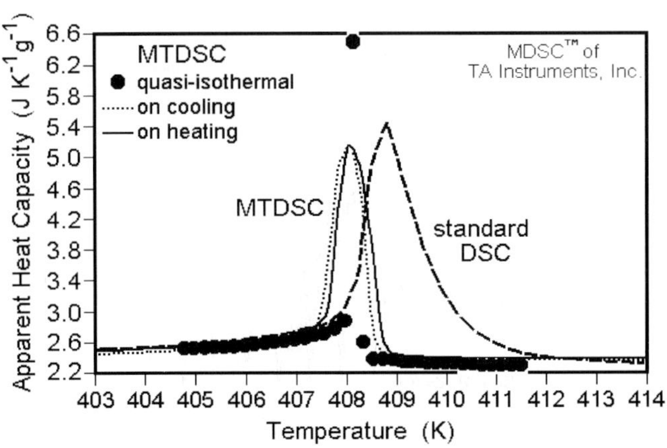

Figure 4.44. Apparent heat capacity in the temperature range of isotropisation of 4,4′-azoxyanisole. Analysis of 5.01 mg. Standard DSC (heating rate 10 K min^{-1}), MTDSC on heating and cooling ($A = 0.5$ K, $p = 60$ s; $<q> = 0.2$ K min^{-1}) and quasi-isothermal MTDSC ($A = 0.1$ K, $<q> = 0$ and $\Delta T_0 = 0.2$ K) [54].

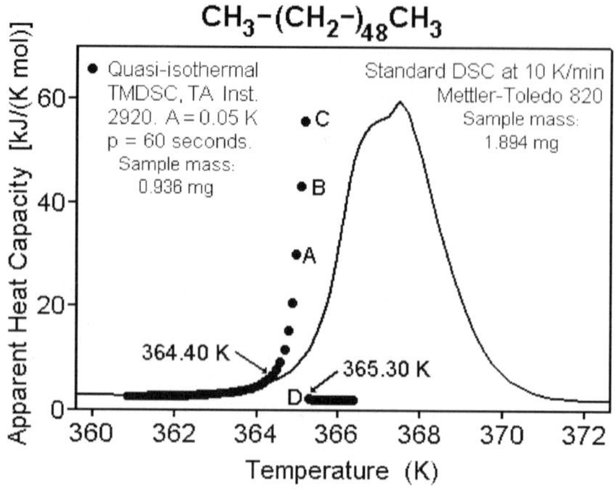

Figure 4.45. Apparent heat capacity in the melting range of 0.936 mg of *n*-pentacontane by standard DSC (heating rate 10 K min^{-1}) and quasi-isothermal MTDSC ($A = 0.05$ K, $p = 60$ s and $\Delta T_0 = 0.1$ K) [28].

and the slow conduction of the heat of transition into or out-of the sample. The fast standard DSC shows the largest lag. The MTDSC on heating and cooling at only ± 0.2 K min^{-1} is still broadened to over 1 K, and there is a difference between heating and cooling, mainly due to the averaging over the modulation period for the deconvolution.

Figures 4.45 and 4.46 illustrate the increase in lag by going to a paraffin which has a high heat of fusion (*n*-pentacontane, $C_{50}H_{102}$, 224.87 J g^{-1}) and the low thermal conductivity of an organic material, but for analysis, a lower sample mass was used to compensate some of the additional lag.

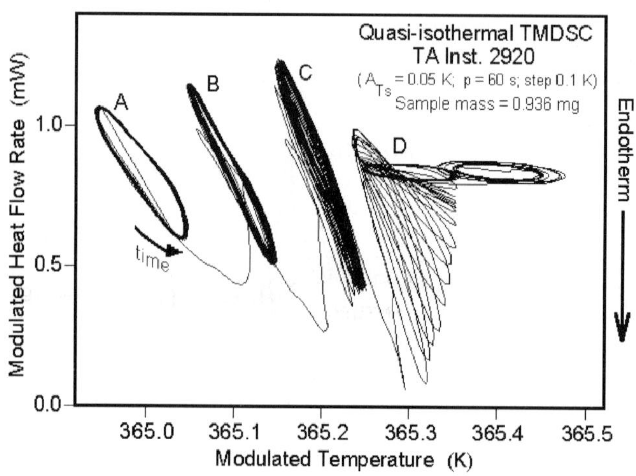

Figure 4.46. Lissajous figures for the quasi-isothermal MTDSC of Figure 4.45 [28].

Figure 4.45 shows a broadened standard DSC melting peak of about 5 K. On quasi-isothermal MTDSC, most of the melting occurred on going from 365.25 to 365.35 K, as can be seen from the Lissajous figures reproduced in Figure 4.46. The heat flow is limited so that the melting cannot be completed in one cycle. It takes about 15 cycles at 365.30 K before melting is complete, and an ellipse is obtained, which is characteristic for the constant heat capacity of the melt. The small exotherms of crystallisation seen in the curves labelled D seem to arise from the inability of the modulation to follow the temperature program until most of the melting is completed. As soon as the temperature does not go below 325.25 K, crystallisation stops. There seems to be little need of crystal or molecular nucleation in this experiment. The Lissajous figures A–C deviate from the expected elliptical shape due to crystallisation and melting and, thus, cannot be deconvoluted quantitatively into a reversing and a non-reversing component using the first harmonic of Eq. (14), as outlined in Section 3, but it is possible to follow the melting in the time domain. Considering the various lags, one can follow the reversing processes and eliminate the irreversible processes by longer-time quasi-isothermal measurements.

The irreversible melting of a polymer is demonstrated with Figure 4.47 on the example of melt-crystallised poly(ethylene oxide) (PEO) of molar mass 5000 Da (POE5000). This low molar mass polymer is known to crystallise mainly in the extended-chain conformation, i.e. close to equilibrium [57,58]. The standard DSC trace shows the large and lag-broadened melting peak with the indicated heat of fusion and crystallinity. The quasi-isothermal MTDSC, in contrast, shows practically no contribution from a reversible melting process, a result one would expect from the diagram of crystallisation and melting rates summarised in Figure 4.5. All melting

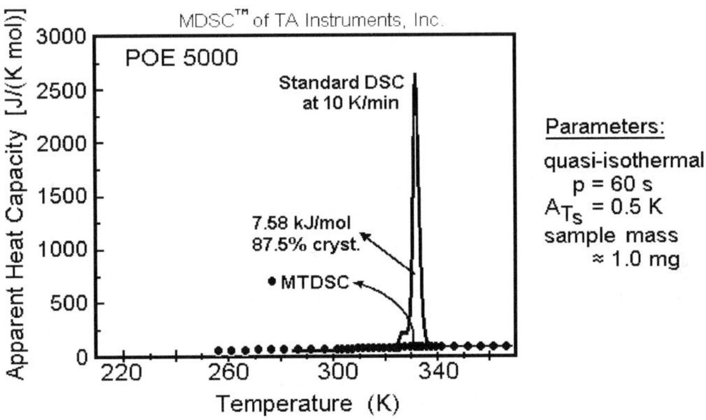

Figure 4.47. Apparent heat capacity measured by DSC and MTDSC for POE5000 crystallised at 320 K [59].

occurs during the initial increase of temperature to the average temperature, T_0, of the quasi-isothermal run and the approach to steady state. Measurements are done only after this steady state is reached. Under these conditions, no melting effect can be seen at the given scale of plotting. Expanding the scale and comparing the measured to the expected heat capacity of the crystal, one finds a too large apparent heat capacity, as will be discussed in Section 4.5 considering heat capacities of solids and melts, heats of transitions, kinetics, heats of annealing and reorganisation. A summary of each sub-section is given to permit a quick search for the experiment of interest.

4.1.1 Summary

The glass transition is represented by the reversing heat flow rate (Figures 4.37–4.39). The enthalpy relaxation appears almost completely in the non-reversing heat-flow rate (see Figure 4.37). The exotherm of cold crystallisation is fully non-reversing (Figure 4.37), but may not be fully deconvoluted (Figure 4.38). Exothermic and endothermic contributions to the heat flow rate may be found in the pre-melting range due to secondary crystallisation, pre-melting, recrystallisation and crystal perfection (Figure 4.38). In the melting range, the separation of revering and non-reversing effects is only qualitative and hindered by instrument lag (Figures 4.38 and 4.39).

Reversible melting of the metal indium with a high heat of fusion and high thermal conductivity is shown in Figures 4.34–4.36, 4.40 and 4.41. These experiments reveal a narrow melting range (<0.1 K), the need of crystal nucleation and the characteristic lag of the calorimeters. The difficulties in maintaining the programmed temperature are seen in Figures 4.34 and 4.41, and the complications in interpretation are analysed in Figure 4.35. The analysis of indium is followed by a study of the reversible isotropisation of azoxyanisole, a liquid crystal with a low heat of transition and low thermal conductivity. It shows a larger broadening of the transition peak due to the low thermal conductivity of the sample (Figures 4.42 and 4.43). The reversible melting of the paraffin *n*-pentacontane is analysed in Figures 4.44 and 4.45 as an example of high heat of fusion and low thermal conductivity, as in polymers, but without the difficulty of partial crystallisation.

Finally, a semicrystalline extended-chain poly(oxyethylene) which melts almost completely irreversibly is shown in Figure 4.47. It is a specially perfect polymer, and more quantitative analyses of the less perfect polymers are detailed in Sections 4.2–4.5. This concludes Section 4.1 and sets the basis for a more quantitative analysis of the melting and crystallisation region of polymers.

4.2 DETERMINATION OF HEAT CAPACITY OF SOLIDS AND MELTS

Knowledge of the heat capacity of solids and melts is of importance not only for its own sake, but also for the discussion of the various latent heat effects in polymers which occur commonly between the glass transition temperature and the melting point and will be analysed in more detail in Section 4.3. Furthermore, in the glass transition region, the heat capacity may become time dependent without the presence of a latent heat effect. Both of these topics will be discussed in this section.

A straightforward MTDSC experiment for the non-isothermal measurement of heat capacity of sapphire (Al_2O_3) over a wide temperature range is characterised in Figure 4.48 by its Lissajous figure. It takes about two to three modulation cycles until an approximate steady state is reached, but this is followed by some instability of the measurement at the lowest temperatures. Once steady state is reached, the ellipse increases in size as the heat capacity increases and slightly tilts its orientation angle. Samples of polystyrene and quartz were run in the same fashion and all three runs were baseline-corrected for the asymmetry of the calorimeter and elimination of the effect of the aluminium pans by an extra run with two empty pans. The correction factor K in Eq. (11) was established from the sapphire run shown in Figure 4.48. The corrections and calibrations were done at each temperature. Typical average and standard deviations of the heat capacities from 320 to 550 K were 1.4 $\pm$ 0.8% for polystyrene and 0.02 $\pm$ 1.5%

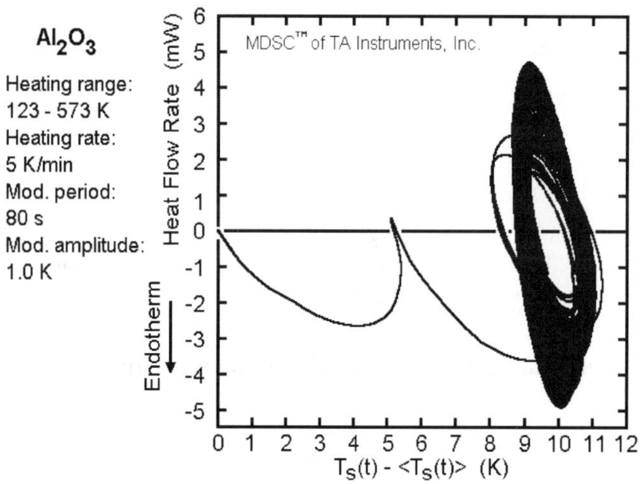

Figure 4.48. Lissajous figure for the heat capacity measurement of sapphire by MTDSC under the given conditions. Data from Ref. [60].

for quartz [60]. Similarly, a quasi-isothermal analysis was made for polystyrene, quartz, sodium chloride, selenium and aluminium [61]. The average and standard deviations for these quasi-isothermal measurements in the temperature range from 300 to 600 K were $-0.9 \pm 1.0\%$, with possible improvements suggested for optimisation of the variables which were not quite identified at that time. Finally, Figures 4.18–4.20 illustrate how effects such as thermal resistance and changes in thermal conductivity for the different samples can be eliminated by empirically calibrating the constant τ in Eq. (13). Using the complex sawtooth of Figure 4.27, it is possible to measure with five frequencies simultaneously and reach standard deviations as little as $\pm 0.1\%$ [40] and establish the enthalpy, entropy and free enthalpy data for polymers, as shown in Figure 4.2. To be able to evaluate Eqs. (1)–(3), the transition enthalpies and entropies, their equilibrium values must be known. The measurement of these data is discussed in Section 4.3. The extrapolation of the observed melting temperatures to equilibrium temperatures is developed for many polymers [1,3,11,12].

Figure 4.49 shows the results of adiabatic calorimetry, standard DSC and quasi-isothermal MTDSC for poly-*p*-dioxanone ($(-CH_2-CH_2-O-CH_2-COO-)_x$, (PPDX). The ordinate is labelled as apparent heat capacity since in the transition region, latent heat contributions may increase the heat capacity. Up to 250 K, the heat capacity is practically fully vibrational as is typical for glassy and crystalline solids. The skeletal and group vibrational contributions are then extrapolated to higher temperature, as is discussed with Figure 4.1 for polyethylene. The sample analysed with

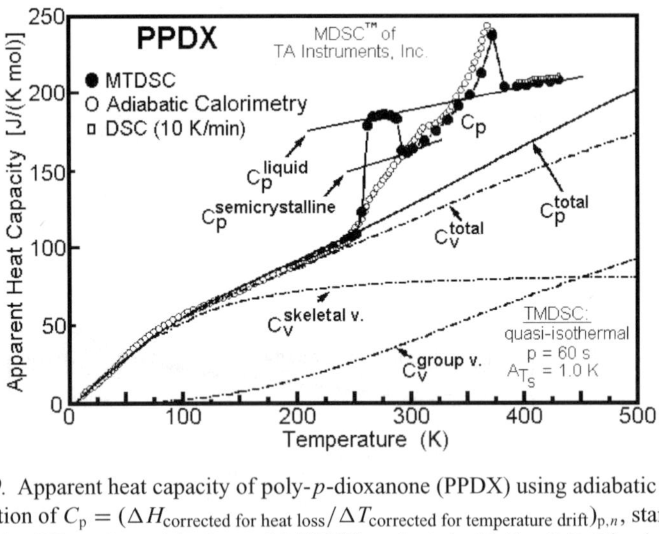

Figure 4.49. Apparent heat capacity of poly-*p*-dioxanone (PPDX) using adiabatic calorimetry by calculation of $C_p = (\Delta H_{\text{corrected for heat loss}}/\Delta T_{\text{corrected for temperature drift}})_{p,n}$, standard DSC using Eq. (10), and quasi-isothermal MTDSC evaluated with Eq. (11). The data were measured in Ref. [62].

adiabatic calorimetry (○) was semicrystalline and shows a reduced, broadened glass transition, typical for the presence of small polymer crystals with tie molecules to the amorphous fraction. Melting begins immediately after the glass transition. The largely irreversible melting peak that shows in the adiabatic calorimetry is outside of the limits of the graph. The MTDSC data (•) were taken on an initially amorphous, glassy material. The glass transition reaches up to the level of the heat capacity of the melt. This supercooled melt crystallises, and at about 290 K, the sample reaches the level of the heat capacity of a semicrystalline polymer, as expected from the heat of fusion measured using the methods discussed with Figure 4.23 or by dilatometry, X-ray diffraction, or infrared spectroscopy [4]. With the four heat capacity baselines, the extrapolated vibrational heat capacity, C_p^{total}, $C_p^{semicrystalline}$ and C_p^{liquid}, the quantitative discussion of glass transition and melting transition is possible, even if the transitions stretch over wide temperature ranges. Surprising is the small reversing melting peak seen in MTDSC which will need further analysis in terms of its latent heat contributions (see Section 4.3). Annealing and reorganisation are treated in Section 4.4. A small truly reversible contribution hidden under the reversing melting peak is discussed in Section 4.5. Despite the fact that cold crystallisation and melting of perfect crystals of polymers were shown in Section 4.1 to be irreversible (Figures 4.37 and 4.47), some reversing, apparent heat-capacity contributions are seen in the broad melting range of PPDX.

A comparison of quasi-isothermal MTDSC and standard DSC of the melting of PET crystallised from the melt is shown in Figure 4.50. This graph was the first proof that there is an apparent, reversing heat-capacity contribution to the melting [63]. The higher melting temperature of the

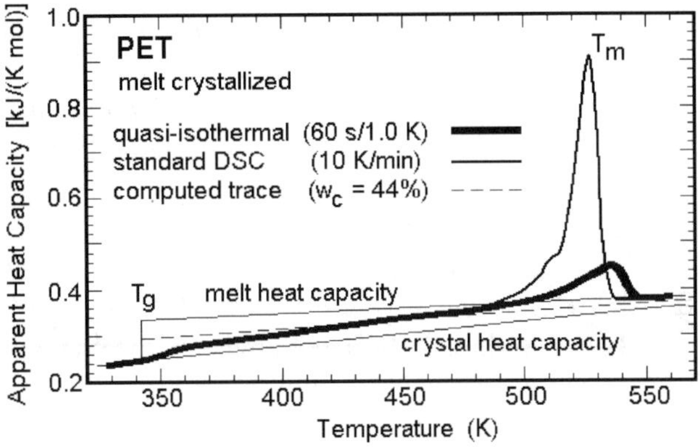

Figure 4.50. Apparent heat capacity of melt-crystallised PET as measured by standard DSC and quasi-isothermal MTDSC, compared to the baselines of the heat capacity of the melt, solid (vibrational contributions only) and semicrystalline polymer [63].

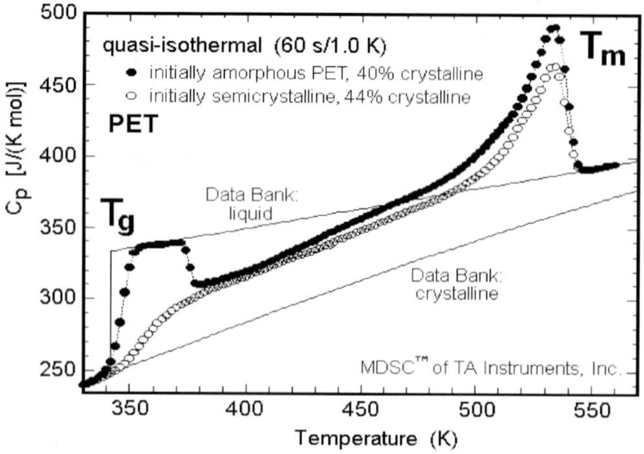

Figure 4.51. Apparent heat capacity of amorphous PET by quasi-isothermal MTDSC, compared to data on semicrystalline, melt-crystallised PET as shown in Figure 4.50 [63]. Compare also to Figure 4.39.

quasi-isothermal analysis when compared with the standard DSC trace is caused by the continuing annealing, which occurs during the long modulation experiments. The glass transition of the semicrystalline sample is smaller than expected from the crystallinity up to almost 450 K. This points to a RAF in the polymer that does not contribute to the glass transition [64].

An enlargement of the quasi-isothermal MTDSC results of Figure 4.50 is plotted in Figure 4.51 (○) and compared to an identical analysis of a quenched, amorphous glass of PET (•). Glassy PET was also the polymer analysed qualitatively in the plot of Figures 4.37 and 4.39. From the comparison in Figure 4.51, one deduces that after cold crystallisation, PET has a larger reversing contribution than the melt crystallised PET of slightly higher crystallinity (44% versus 40%). The quasi-isothermal data in Figure 4.51 are free of instrument-lag, in contrast to the data with an underlying heating rate illustrated in Figure 4.39, and thus are more suited for a quantitative analysis.

The effects due to very slow processes in the sample, however, may not be removed even in the quasi-isothermal MTDSC with data collection after 10 min. The analysis of the slow response of the sample in the glass transition region will be treated in Section 4.3. The decrease in the heat capacity due to cold crystallisation can easily be converted into a plot of the crystallisation kinetics. Additional points for the kinetics plot can be generated at shorter and longer analysis times of the quasi-isothermal runs. The time-scale can easily be adjusted to modulations from 1 min to many hours, limited only by the patience of the operator and the stability of the calorimeter.

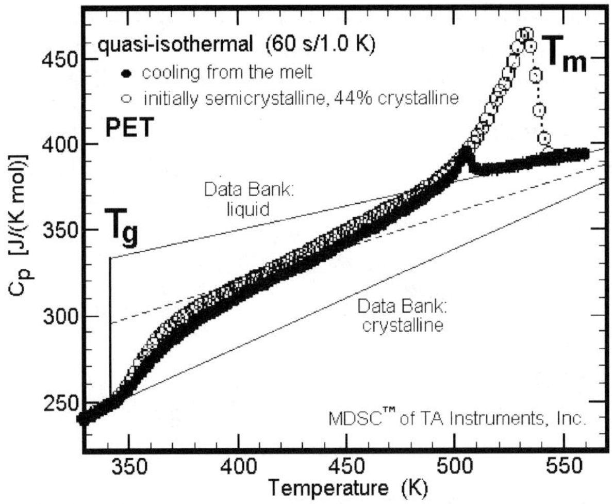

Figure 4.52. Apparent heat capacity of PET on cooling by quasi-isothermal MTDSC compared to data on semicrystalline, melt-crystallised PET measured on heating, as shown in Figure 4.50 [63].

On cooling of the melt, one sees in Figure 4.52 the expected large super-cooling, but after crystallisation at about 500 K, the same increased apparent heat capacity is observed as for a sample that was reheated after cooling from the melt. On step-wise further cooling, the quasi-isothermal MTDSC yields a slightly higher crystallinity (49%).

A biaxially drawn film is shown in Figure 4.53. It has a 42% crystallinity and, again, is different from the melt-crystallised sample. Its glass transition

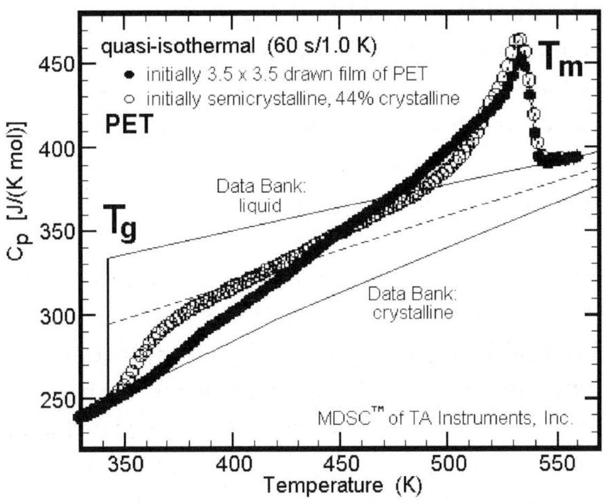

Figure 4.53. Apparent heat capacity of drawn PET film by quasi-isothermal MTDSC, compared to data on semicrystalline, melt-crystallised PET as shown in Figure 4.50 [63].

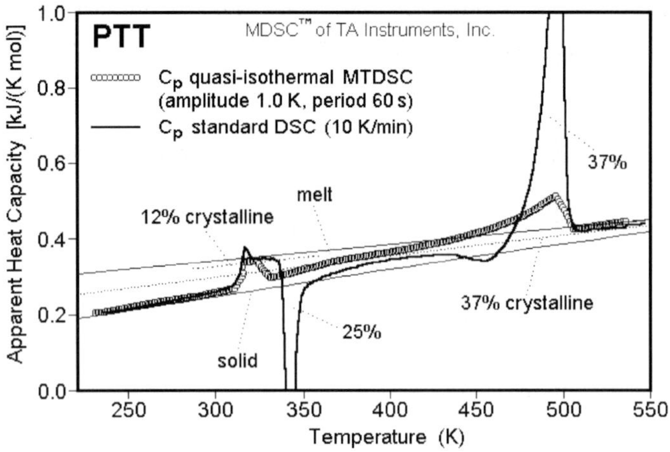

Figure 4.54. Apparent heat capacity of low crystallinity PTT by quasi-isothermal MTDSC compared to data from standard DSC measurements [51,65]. See, for review, Figure 4.38.

is much broader and shifted to higher temperature, the RAF is larger and the distribution of the reversing latent heat contribution is different. Overall, in this example of quasi-isothermal analysis, four samples of different process history, which yields diverse morphologies, produce different behaviour on analysis by MTDSC in the temperature region between the glass and melting transitions. Since the morphology is also the main determinant for the physical properties, MTDSC can be used to characterise the structure-processing-property triangle.

Figure 4.54 is a quantitative quasi-isothermal MTDSC trace for quenched, poorly crystallised PTT. The corresponding semiquantitative MTDSC is depicted in Figure 4.38. The cold crystallisation at about 325 K, the recrystallisation, 450 K, and the small enthalpy relaxation at 320 K are seen to be fully irreversible, and as in PET, the kinetics of the glass transition and the cold crystallisation can be further analysed quantitatively making use of the reversing heat capacity. It is also clear that during the standard DSC measurement, the cold crystallisation never stops completely between the two peaks and considerable errors in the crystallinity may result from choosing a baseline without MTDSC data.

4.2.1 Summary

The heat capacity of solids and melts can be determined by adiabatic calorimetry and various methods of DSC and MTDSC, where with modern methods, the latter may reach a precision equal or better than the adiabatic measurement, so that curves like in Figures 4.1 and 4.2 are available for an increasing number of polymers. A comparison of the heat capacity measured with standard DSC and with MTDSC shows agreement for the temperatures outside of the transition regions. Within the transition regions, differences arise. They permit a characterisation of the processing and thermal history as well as determination of transition kinetics. More quantitative studies of these added capabilities of MTDSC are shown in Sections 4.3–5, but even empirical links between structure, property and processing variables are possible, as is seen particularly well in a comparison of Figures 4.50–4.53.

4.3 DETERMINATION OF HEAT OF FUSION, CRYSTALLINITY AND KINETICS

4.3.1 Heat of Fusion and Crystallinity

The determination of the heat of fusion of the pure crystals of polymers, ΔH_f^0, involves always a coupling of measurement of the heat of fusion of semicrystalline samples, ΔH_f, and their weight-fraction crystallinity, w_c, as expressed by:

$$w_c = \frac{\Delta H_f}{\Delta H_f^0} \tag{17}$$

Either, one uses an independent method for the crystallinity determination, such as dilatometry, X-ray diffraction or infrared spectroscopy for the determination of w_c [4], or one tries to determine the amorphous fraction, w_a, from the measured increase of the heat capacity at the glass transition temperature, ΔC_p, and the same quantity for the fully amorphous sample, ΔC_p^0:

$$w_a = \frac{\Delta C_p}{\Delta C_p^0} = (1 - w_c) \tag{18}$$

Figure 4.55 shows a plot of various heat capacity data of PTT as shown in Figure 4.54 [65]. The fully amorphous point was calculated from the heat capacity of the glass and the melt, both extrapolated to the glass transition temperature. The heat of fusion of the 100% crystalline sample agrees also with a discussion of the entropies expected from similar polymers. The data points with somewhat lower ΔC_p are most likely due to a small amount of RAF [64], frozen at the crystal interface, as indicated by the thin line.

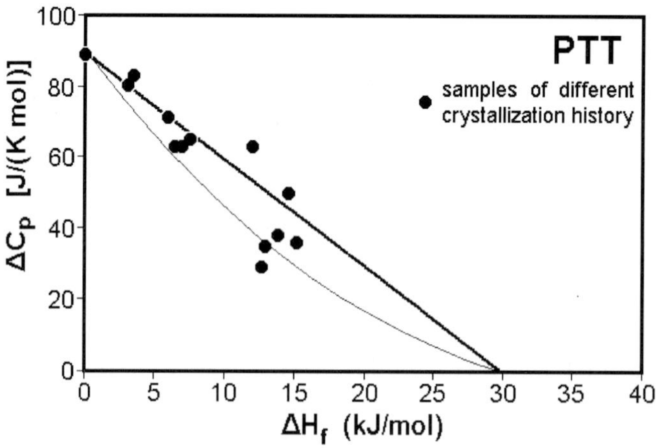

Figure 4.55. Plot of the change of the heat capacity at the glass transition as a function of the heat of fusion of the material [65].

4.3.2 Baseline Fits

The discussion of Section 4.2 has shown that MTDSC can produce a reliable baseline when coupled with the analysis of the baseline of vibrational motion in the solid state and the heat capacity of the liquid. Between the glass transition and the melting transition, one finds, however, sufficient latent heat contributions, that a detailed analysis of cold crystallisation and reorganisation is needed to be able to separate the heat flow rate due to heat capacity from the latent heat contributions. Fortunately, the method (3) in Figure 4.23 allows an analysis of the total crystallinity and its change on heating without knowledge of the true semicrystalline heat capacity, as long as the temperature dependence of the heat of fusion is known. The ATHAS Data Bank contains information for many polymers for such analyses [42]. Figure 4.56 illustrates the changes observed when analysing an initially melt-crystallised sample by DSC and MTDSC. Using the apparent heat capacity curve of the DSC run or the total heat flow rate of the MTDSC yields a curve as reproduced in Figure 4.57. Net melting begins at 390 K. Recrystallisation reverses this trend at about 470 K, to be followed by the major melting peak. A similar analysis is shown in Figure 4.25 for a copolymer. The MTDSC must now separate any possible compensating effects of melting and recrystallisation or secondary crystallisation, as well as reorganisation and reversible melting and crystallisation, as will be discussed in Sections 4.4 and 5.

4.3.3 Quasi-Isothermal Kinetics of the Glass Transition

The kinetics in the glass transition and melting regions are of interest for a better understanding of the materials, prediction of their performance and analysis of their history. To describe the glass transition, following the

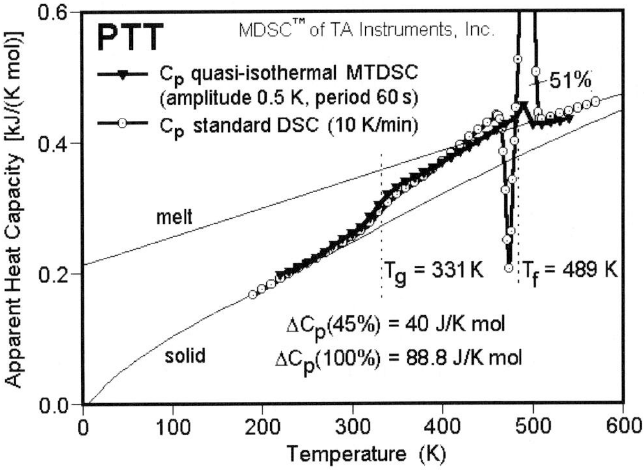

Figure 4.56. Apparent heat capacity of high crystallinity PTT by quasi-isothermal MTDSC, compared to data measured with standard DSC [51,65] (see also Figures 4.38 and 4.54).

summary of Ref. [66], the heat capacities of liquids can be divided into a fast-responding part due to the vibrations, C_{p_0}, as shown in Figure 4.1 for the solid, and a slow, co-operative part due to large-amplitude, conformational motion common for the liquid:

$$C_p\,(\text{liquid}) = C_{p_0} + \varepsilon_h \left(\frac{dN^*}{dT}\right) = C_p + \varepsilon_h \alpha \tag{19}$$

where an equilibrium number, N^*, of configurations of energy ε_h govern the extra contribution.

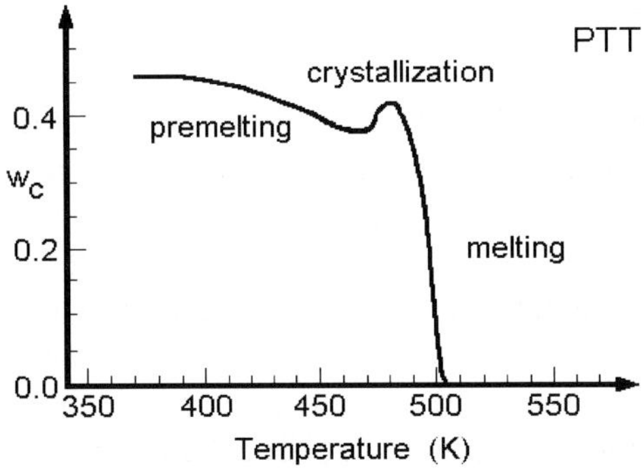

Figure 4.57. Crystallinity of PTT from standard DSC data of Figure 4.54 using the method of Figure 4.23. See also Figure 4.25 [51,65].

In the glass transition region, the approach to equilibrium may be approximated by a first-order kinetics as long as the distance from equilibrium is not too large. The instantaneous number of high-energy configurations is then represented by N and the relaxation time by τ:

$$\left(\frac{\mathrm{d}N}{\mathrm{d}t}\right) = \frac{1}{\tau}(N^* - N) \tag{20}$$

Under quasi-isothermal conditions and at steady state, the solution of Eq. (20) can be written with constants $A_N = A\alpha/N_0$ and $A_\tau = A\varepsilon_j/(RT_0^2)$, where A is the amplitude of temperature modulation.

$$\begin{aligned}\frac{N - N_0^*}{N_0^*} &= \frac{A_N A_\tau}{2} + (A_N A_\tau)\cos\gamma\,\sin(\omega t - \gamma)\\ &\quad - \frac{A_N A_\tau}{2}\cos[2\beta]\cos[2(\omega t - \beta)]\end{aligned} \tag{21}$$

and ε_j is the activation energy for the formation of the high-energy configurations, assumed to be describable by an Arrhenius expression and written as $\tau = B \exp \varepsilon_j/(RT)$. From Figure 4.29, it is obvious that the reversing heat capacity of Eq. (11) makes use only of the middle term on the right-hand side of Eq. (21) since only it has a frequency of ω. The first term is constant with time and contributes only to the total heat capacity, the last is a second harmonic and is rejected in the calculation of the first harmonic, i.e. it also only occurs averaged in the total heat capacity. The phase shift γ is linked to the relaxation time τ at T_0 via $\tan\gamma = \omega\tau$, and the apparent heat capacity, which is measured as the reversing C_p, is now equal to:

$$C_\mathrm{p}^{\#}(\text{liquid}) = C_{\mathrm{p}_0} + N_0\varepsilon_\mathrm{h}\left[\frac{(A_N + A_\tau)}{A}\right]\cos\gamma \tag{22}$$

From Eq. (22), one sees that the large-amplitude motion contributes fully in the liquid state where $\gamma = 0$, and not at all in the glassy state where $\gamma = \pi/2$. The glass transition temperature, defined at the temperature of half-vitrification or devitrification, occurs at $\gamma = \pi/3$.

An example of the data treatment for PET is shown in Figure 4.58. The experimental data of ΔC_p [$= C_\mathrm{p}^{\#}(\text{liquid}) - C_{\mathrm{p}_0}$] were first normalised to the equilibrium difference of the liquid and vibrational heat capacities. The curves [A] represent ΔC_p. Equation (22) allows then the transformation to γ, shown in the curves [B]. Using the values of γ, a plot of $\ln\tau'$ versus $1/T_0$ can be drawn $\{\cos\gamma = (1/\tau)/[(1/\tau)^2 + \omega^2]^{1/2}\}$. The different modulation

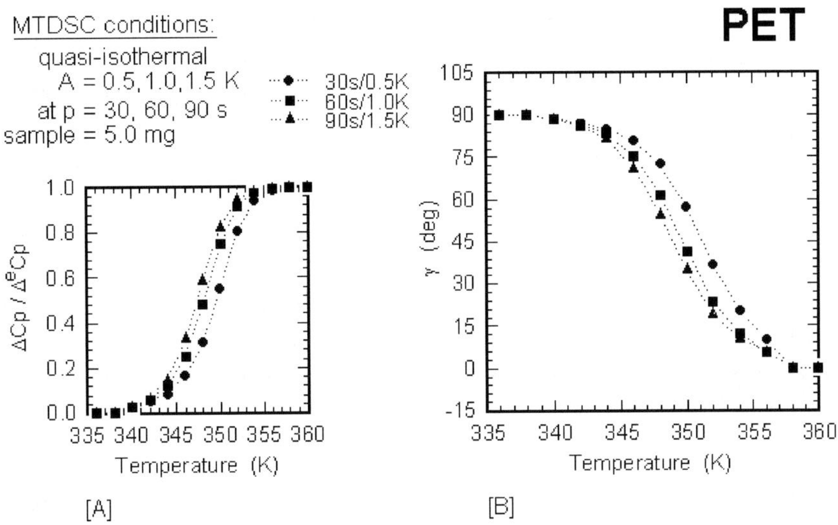

Figure 4.58. Quasi-isothermal analysis of the glass transition of amorphous PET. Data as shown in the glass transition region of Figure 4.51. [A] Normalisation to the ΔC_p of Figure 4.51. [B] Conversion as suggested by Eq. (22).

amplitudes, however, give different average values of τ and different activation energies, as is seen from curves [A] of Figure 4.59.

This observation points to an important difference between MTDSC and dynamic mechanical analysis (DMA). In DMA, the stress or strain is modulated, keeping the temperature, and with it the relaxation time, constant (as long as the strain is sufficiently small to keep the sample in the range of

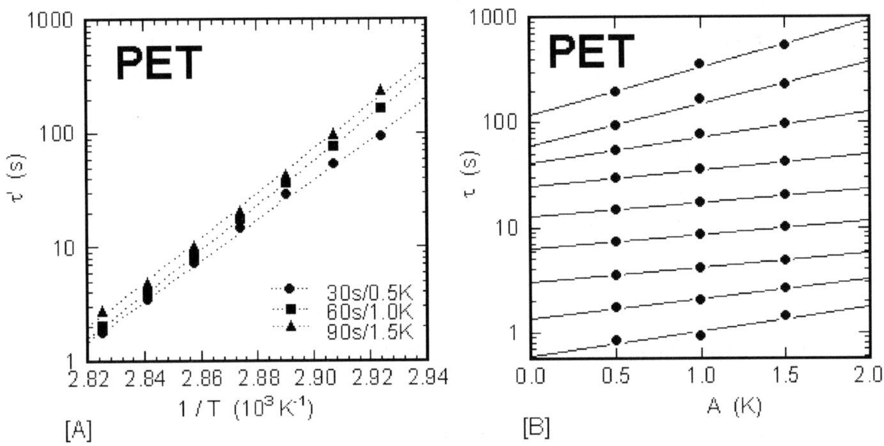

Figure 4.59. Final data analysis and computation of values for Table 4.1. [A] Calculation of the amplitude-dependent relaxation time τ'. [B] Extrapolation of τ' to zero-modulation amplitude to eliminate the temperature dependence of τ resulting from modulation.

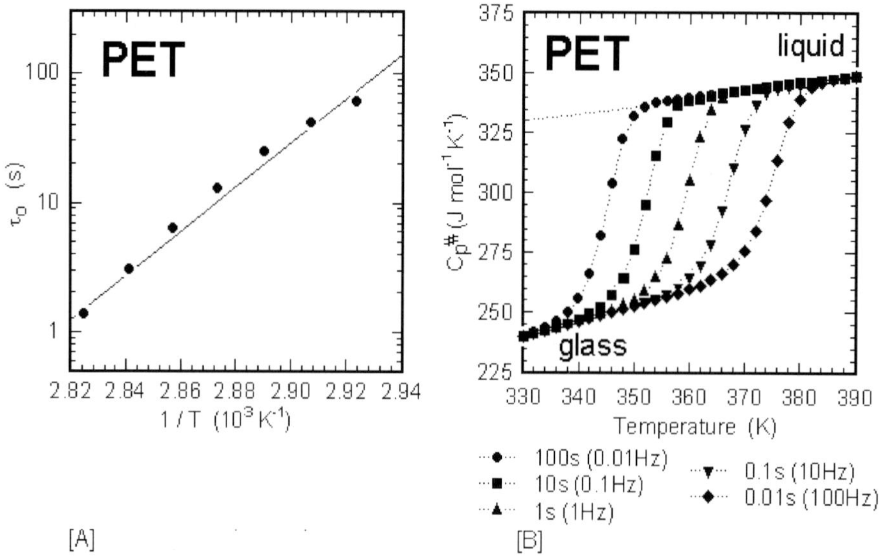

Figure 4.60. [A] Plot of the temperature dependence of τ to illustrate the computation of the Arrhenius activation energy ε_j and the pre-exponential factor B. [B] Computed heat capacity at frequencies beyond the measurement, using data from Table 4.1 for amorphous PET.

linear viscoelasticity). In MTDSC, in contrast, even changes in temperature as small as 1 K move the experiment out of the range of linear response. It is, thus, necessary to extrapolate the data to zero-modulation amplitude first, as illustrated in Figure 4.59[B]. Figure 4.60[A] depicts the extrapolation of the relaxation times at zero amplitude to give ε_j and the pre-exponential factor B.

With B and ε_j, the apparent heat capacity can be calculated for any frequency, as shown in curves [B] of Figure 4.60. Carrying out this analysis for a number of samples of different crystallinity gives the parameters B and ε_j listed in Table 4.1 [67]. Table 4.1 illustrates a large change of ε_j with pre-treatment, correlated to the common observation that crystallisation and drawing broadens the glass transition region. Another observation is that

Table 4.1. Glass transition parameters for PET

Sample (type and treatment, w_c = crystallinity)	**ε_j (kJ mol^{-1})**	**B (s)**
PET, amorphous, melt-quenched	328.19	5.59×10^{-49}
PET, 8% w_c by cold crystallisation 1 h at 370 K	350.57	2.76×10^{-52}
PET, 17% w_c by cold crystallization 1.5 h at 370 K	329.74	3.98×10^{-49}
PET, 26% w_c by cold crystallization 2 h at 370 K	173.31	2.55×10^{-25}
PET, 44% w_c by cooling from the melt, 5 K/min	152.85	2.45×10^{-22}
PET, film, biaxially drawn, 42% w_c	78.44	1.78×10^{-10}

the activation energies and the pre-exponential factors are strongly coupled. They can be written for PET as:

$$\tau = \tau_1 e^{\varepsilon_j \left[\frac{1}{RT} - \frac{1}{RT_1}\right]} \tag{23}$$

where $\tau_1 = 132.5$ s and $T_1 = 341.1$ K. The temperature parameter T_1 is only little lower than the lowest measured glass transition (346.5 K). From the simple kinetics, one expects close correspondence only in the vicinity of equilibrium. The empirical corrections developed over the years to account for the asymmetry of approach to equilibrium and for the co-operativity of the large-amplitude motion need, at present, too many parameters to be fitted quantitatively to the experimental data (TNM, VF, WLF, KWW equations [68]). Qualitative agreement has been achieved with assumed parameters [69,70].

4.3.4 *Model Calculation for the Glass Transition with an Underlying Heating Rate*

In standard MTDSC, an additional underlying heating rate complicates the analysis of Eq. (20). It takes now the form [72]:

$$\left(\frac{dN}{dt}\right) = \frac{N_0^*(1 + A_N \sin\omega t + q_N t) - N_0}{\tau_0(1 - q_\tau t - A_\tau \sin\omega t)} \tag{24}$$

with the two new parameters describing the changes due to the underlying heating rate written as: $q_N = \langle q \rangle \alpha / N_0^*$ and $q_\vartheta = \langle q \rangle \varepsilon_j / (RT_0^2)$. Although possible, the solution of Eq. (24) is rather cumbersome and numerical solutions are more convenient. Figure 4.61 shows the numerical integrations of the changes of N with time and temperature for amorphous PET for the given MTDSC parameters as expressed in Eq. (24). The curve ΔN is the change in N per second, the step of the numerical integration, and N^* is calculated from the equilibrium, given in Eq. (19) [71].

Figure 4.62 shows the first harmonic of the solution of Eq. (24), which represents the reversing heat capacity as normally computed (see Figure 4.29). The heavy line is the total C_p (curve B of Figure 4.29). Outside of the glass transition, this heat capacity is equal to the heat capacity measured with a standard DSC of the same cooling rate $\langle q \rangle$. In the glass transition region, it only approximates the apparent heat capacity of the standard DSC because of contributions of the type seen in Eq. (21). The reversing C_p decreases at a higher temperature than the total C_p because of its faster time scale of

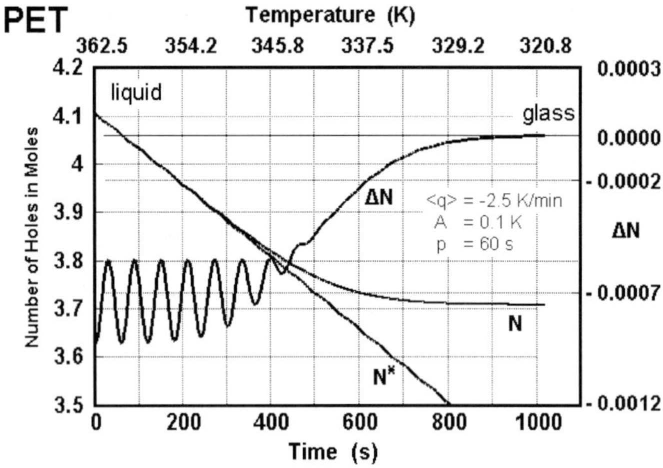

Figure 4.61. Calculated values for N^*, N and the change of ΔN (per second) using Eq. (20) with the kinetic parameters derived from Figures 4.58–4.60.

measurement. The bell-shaped curve is the difference between the total and the reversing heat capacities called the non-reversing C_p.

Figure 4.63 illustrates that the second harmonic is a minor, but not negligible, correction to the total heat capacity. Of additional interest are the remaining small ripples of the various C_p plots. The deconvolution should have removed all periodic contributions of frequency ω and also higher harmonics. Inspection of Eq. (24) shows, however, that the underlying heating rate causes a small frequency shift of the type of a Doppler effect, as found in the analysis of sound from moving sources, quantitatively assessable

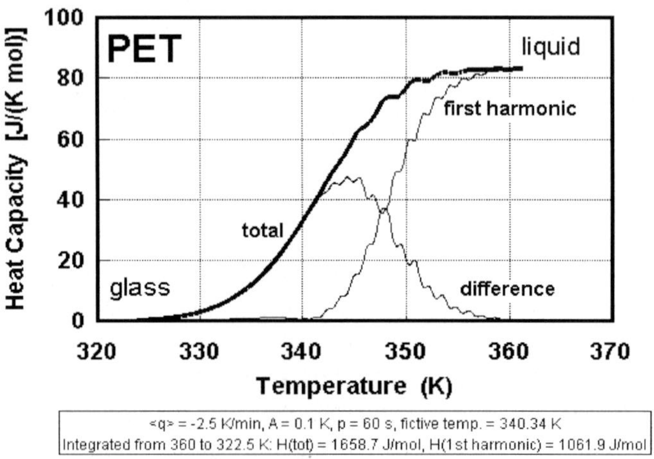

Figure 4.62. Total apparent heat capacity and the first harmonic (reversing heat capacity) of the solution of Eq. (24). The difference represents the "non-reversing contribution".

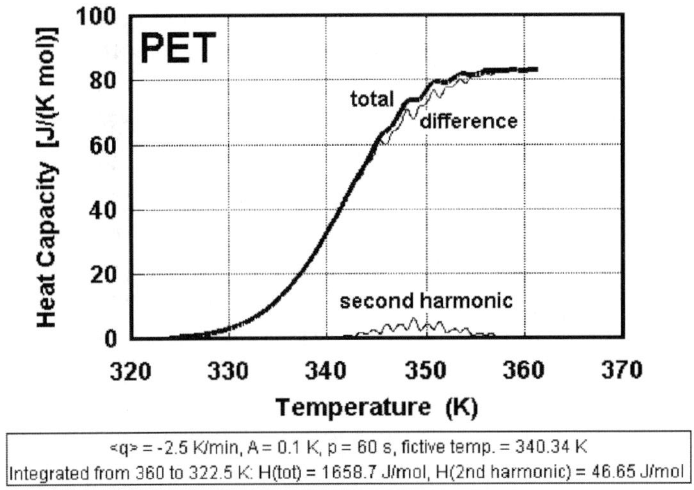

Figure 4.63. Total apparent heat capacity and the second harmonic of the solution of Eq. (24). The difference represents the error in the analysis (but note that even higher harmonics may not be negligible).

through the model calculations [25]. On heating, the oscillation frequency of the heat flow into the sample is higher than ω, because of the shorter modulation period caused by the underlying heating rate $<q>$. The reverse is true on cooling [72]. Trying to represent the heat flow with the Fourier series as shown in Eq. (14) with the slightly different frequency ω gives rise to the observed ripples. Experimental data may not show the ripple because of an additional smoothing by the commercial software, omitted in the presentation of the model calculations in Figures 4.62 and 4.63.

Figure 4.64 shows two sets of experimental data on heating and cooling, compared with the quasi-isothermal measurements derived as for

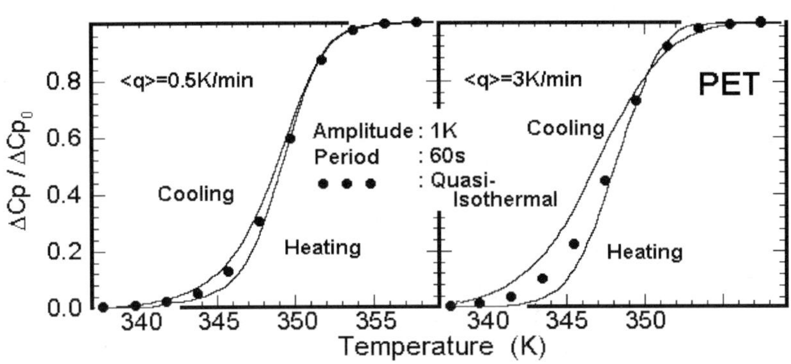

Figure 4.64. Comparison of the experimental quasi-isothermal apparent reversing heat capacity (•) and the heat capacity measured with an underlying heating rate on heating and cooling. The upper disparities of the curves can be modelled with Eq. (24), but the bottom ones cannot.

Figure 4.58. At low $<q>$, the reversing C_p approaches the quasi-isothermal data. At increasing $<q>$, the results from cooling and heating experiments separate increasingly. Close to the liquid state, model calculations, as in Figures 4.61–4.63, correspond closely to the experiment. The cross-over at larger distance from the liquid (equilibrium) state is, however, not modelled [72]. It is caused by an "autocatalytic" effect on heating and a "self-retarding" effect on cooling, as has been found also by DMA and volumetric experiments about the kinetics of the glass transition [68]. This is clear evidence of the co-operative nature of the glass transition that needs to be corrected by a more detailed kinetics expression than given in the present description.

A similar effect that illustrates the need to introduce a co-operative kinetics, which uses a relaxation time in Eq. (20) that depends not only on temperature, but also on the number of frozen high-energy conformations, is shown in Figure 4.65. The curves represent MTDSC traces on heating with a fixed underlying heating rate. The different glasses were produced by annealing at several temperatures in the transition region for various times. This process changes the number of frozen high-energy configurations, N, at the beginning of the heating experiment. One can clearly see that the better annealed samples of smaller N show a higher glass transition

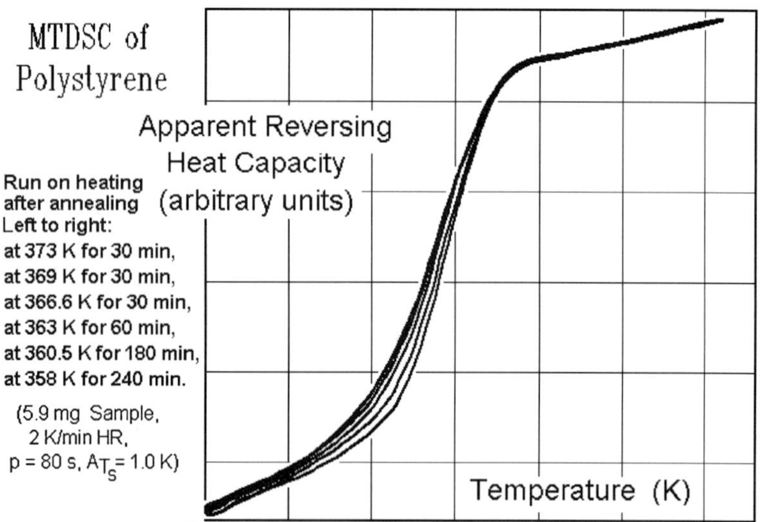

Figure 4.65. Apparent reversing heat capacities for polystyrene samples with different thermal histories. Note that to describe such reversing heat capacity, Eq. (20) would need to use a relaxation time that depends on N.

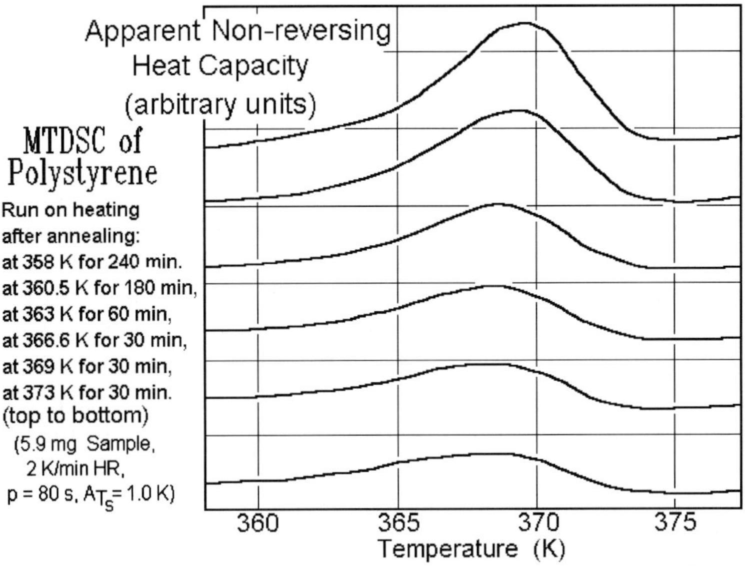

Figure 4.66. Apparent non-reversing heat capacities for polystyrene samples with different thermal histories (enthalpy relaxation). See also Figure 4.37 for the separation of the reversing and non-reversing heat flow rates for PET with enthalpy relaxation.

temperature. The differences disappear, as equilibrium in the form of the liquid state is approached, and overall, the changes are relatively small.

The apparent total heat capacity contains the enthalpy relaxation in the form of an endotherm, as shown in Figure 4.37. This well-known hysteresis effect can be separated to the degree of precision of the representation by Eq. (20) using the MTDSC software, as is shown in Figure 4.66 for the example of polystyrene [73]. The apparent, non-reversing heat capacity is the total heat capacity minus the apparent, reversing heat capacity, shown in Figure 4.65. The major contribution of the endotherms arises from non-modulated relaxation of N on heating, as long as ΔN of Figure 4.61 is much larger than the modulation-caused changes, and the glass-transition temperature is approached. With this detailed analysis, the quantitative capabilities of MTDSC of PET have been probed and details, which go beyond present day theories, have been extracted from the measurements displayed in Figures 4.37 and 4.50–4.53, which are less quantitatively analysed.

4.3.5 Kinetics of Transitions with a Latent Heat

The kinetics of processes with a latent heat, such as crystallisation and melting, can be measured either directly by isothermal or non-isothermal measurement of the latent heat or by observation of the change of the heat capacity in case liquid and solid have a substantial difference in heat capacity. The latter can be seen in Figures 4.49 and 4.51, for example. Doing the experiment quasi-isothermally over a long time interval, the kinetics can be read directly from the changing apparent heat capacity.

The measurement of latent heat is more commonly carried out by isothermal calorimetry or non-isothermally with a standard DSC on cooling. The complications on observing latent heats with MTDSC and their analysis via the first harmonic of a Fourier series as an apparent heat capacity in the case of a reversible transition are illustrated in Sections 3.4 and 4.1. For fully irreversible processes, there may be little or no advantage to use MTDSC over DSC. In fact, the total heat capacity of the MTDSC may be unduly broadened by the deconvolution procedure, so that the standard DSC or isothermal calorimetry is often preferable. The advantage of the MTDSC develops as soon as there are several simultaneous processes of different degrees of reversibility, as was seen in Section 4.2 in the discussion of the apparent heat capacity of semicrystalline polymers when approaching the melting transition (Figures 4.49–4.54).

Figure 4.67[A] shows a typical isothermal experiment carried out with a DSC. Similar experiments could be carried out with isothermal calorimeters, dilatometry and other techniques sensitive to crystallinity changes. After attainment of steady state at point 0, the experiment begins. At point 1, the first heat flow rate is observed, and when the heat flow rate reaches 0 again, the transition is complete. The shaded area is the time integral of the heat flow rate, and if there is only a negligible instrument lag, it represents the overall kinetics. In case of an excessive heat flow-rate amplitude, lag calibrations with sharply melting substances of similar thermal conductivity may have to be made (see Figure 4.22). Processes faster than about 1 min

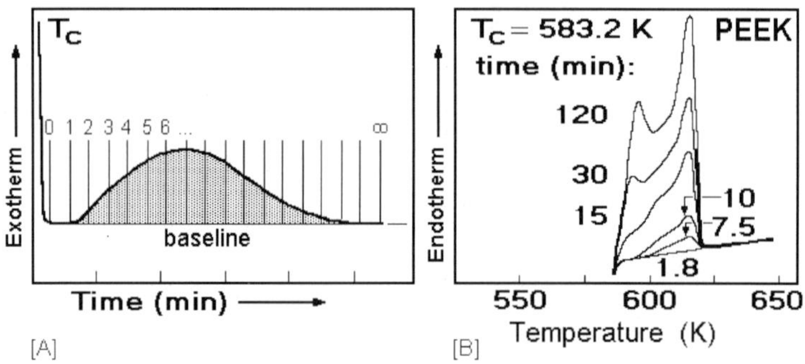

Figure 4.67. Analysis of the crystallisation kinetics. [A] Isothermal analysis with a DSC by quick cooling to the crystallisation temperature and analysis after steady state has been reached. [B] Step-wise analysis by heating after different times of crystallisation of PEEK by direct measurement from the crystallisation temperature.

are difficult to assess, since the times to reach steady state are too long and interfere with the measurement. Slower processes than perhaps 30 min are too slow for many calorimeters, particularly DSCs, to build-up a measurable heat-flow rate. Figure 4.67[B] illustrates how the isothermal method can be extended for slow processes. The curves in [B] represent heating curves with a standard DSC, starting at the crystallisation temperature, after the indicated lengths of isothermal crystallisation. The analysed polymer is poly(oxy-1,4-phenylene-oxy-1,4-phenylenecarbonyl-1,4-phenylene) (PEEK) [74]. In this analysis, it can be seen that three differently melting polymers grow in sequence. The initial crystals have an intermediate melting point and seem to improve with time to yield the main, high melting peak. Secondary crystallisation begins at a distinctly later time. The overall crystallinity has reached 48% after 120 min.

The interpretation of the calorimetric data is rather difficult because of the various effects observed by thermal analysis of polymers, detailed in Section 2.5. A review of the mechanisms of primary and secondary crystallisation, as well as annealing, has been prepared [5]. The primary crystallisation, which may already be complicated by crystal perfection after initial growth (see Figure 4.67[B]), is usually described by a crystal nucleation followed by growth. Figure 4.68 illustrates, schematically, the progress of the growth of spherulites as seen on a hot stage under a polarizing microscope. It is obvious that, in the chosen example, the number of crystal nuclei is constant ($= N_0$),

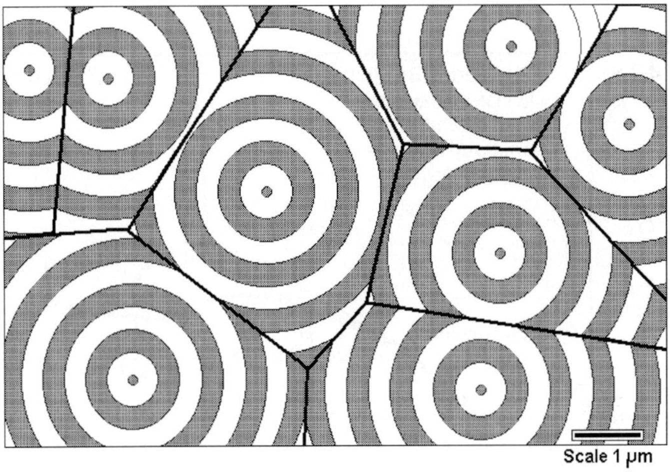

Figure 4.68. Schematic of the crystallisation of spherulites from a given number of (heterogeneous) nuclei. In case the spherulites contain (as usual) amorphous inclusions between the lamellar crystals, the heat for fusion ΔH_f^o needs to be corrected by the crystallinity reached at time infinity.

and the growth rate v is constant, as also seen in the data of Figure 4.5, and the spherulites are assumed to be space filling. All conditions are needed to produce the linear lines of impingement of neighbouring spherulites. Neither N_0 nor v is available from thermal analysis, but only if v and N_0 are known and a spherulitic morphology is proven, one can use the common Avrami equation to describe the crystallisation kinetics from the heat flow rate [5]:

$$\frac{\Delta h'_{\mathrm{f}}}{\Delta h^{0'}_{f}} = 1 - \exp\left[-\frac{4\pi N_0 (vt)^3}{3}\right] \tag{25}$$

where $\Delta h'_{\mathrm{f}}$ represents the heats of fusion per unit volume ($\Delta h_{\mathrm{f}} = \Delta h_{\mathrm{f}} \rho_{\mathrm{c}}$).

Modifications of Eq. (25) have been derived for other growth situations, but are not reliable, or not sound mathematically. Similarly, non-isothermal kinetics analyses are rather uncertain unless they are supported by structural data and isothermal thermal analysis. If the latter is available, however, non-isothermal data are not needed, except, perhaps, for quality control.

Figure 4.69 illustrates the very complicated case of crystallisation of PEcoO also known as linear-low-density polyethylene. More details about the annealing and the fraction of reversible melting and crystallizing of this polymer are discussed in Sections 4.4 and 5 [24,29]. An initial analysis by standard DSC shows a sharp, largely irreversible crystallisation peak, followed by a broad, even larger secondary crystallization, which reaches

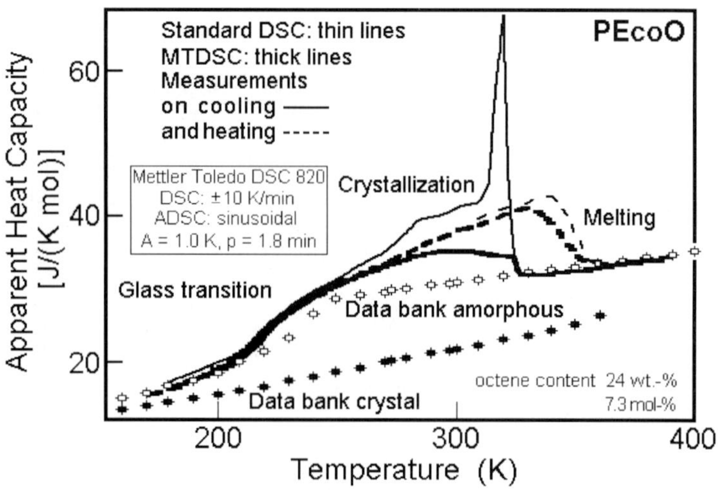

Figure 4.69. Crystallisation of PEcoO as measured by DSC and MTDSC in comparison to melting and data bank information on the two heat capacities [75].

all the way to the glass transition of polyethylene at about 237 K. The crystallisation peak is mainly the primary crystallisation, at about 250 K. However, it joins the reversing apparent heat capacity. The reversing heat capacity starts at the irreversible DSC peak as secondary crystallisation. It stays always above the heat capacity of the liquid and indicates a broad crystallisation range, i.e. at all temperatures between the initial crystallisation peak and the glass transition, crystallisation continues. A comparison with the heating curves shows that the total heat and reversing heat flows stay the same up to 310 K, and then the total heat flow is only a little larger than the reversing one. Such traces need special techniques to be separated into the six possible thermal effects identified in Section 2.5, and analysed in terms of the kinetics of primary and secondary crystallisation, a task only possible by MTDSC, as is shown in Section 4.4.

4.3.6 Mesophase Transitions

Figure 4.70 addresses the question of disordering of mesophase polymers [76]. An example of MTDSC of a liquid crystal and a condis crystal is shown. The analysis of low molar mass liquid crystals was discussed in Figures 4.42–4.44 as an example of a sharp, reversible transition and of low latent heat of transition of a sample of low thermal conductivity (Section 4.1). The liquid crystal-forming polymer has a much broader isotropisation

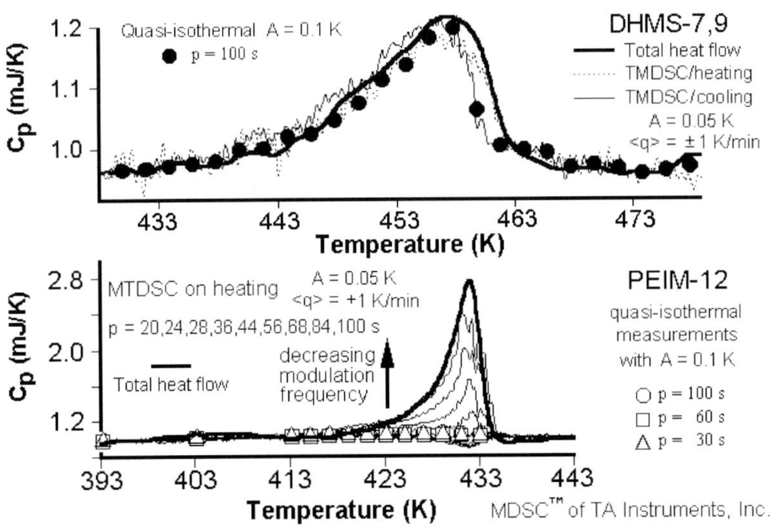

Figure 4.70. Analysis of the isotropisation/ordering transition of two mesophase polymers [54]. [A] The reversible liquid crystal DHMS-7,9 and [B] the irreversible condis crystals of PEIM-12.

temperature range, but a similarly low heat of transition and thermal conductivity. The sample is a polyether, synthesized by coupling 4,4′-dihydroxy-α-methylstilbene (DHMS) with a 1:1 molar mixture of 1,7-dibromoheptane and 1,9-dibromononane. The name of this random copolymer is abbreviated to DHMS-7,9. It forms a nematic liquid crystalline phase between 400 and 470 K. The molecular structures of the two repeating units have the following formula with $n = 7$ and 9:

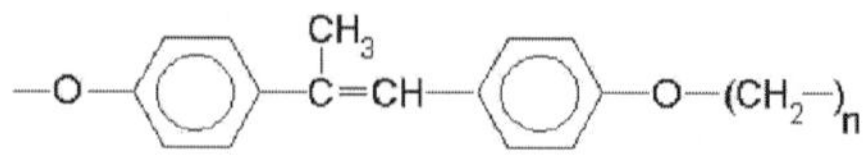

The condis crystal forming polymer is poly(4,4′-phthaloimidobenzoyl-dodecamethyleneoxycarbonyl) (PEIM-12) with the following repeating unit:

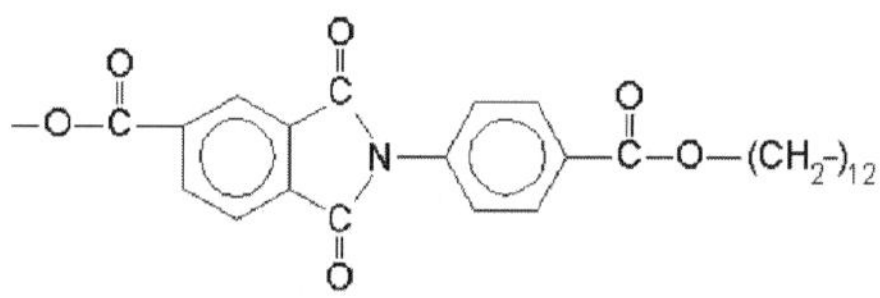

Because of the broad isotropisation transition, the DHMS-7,9 has, in contrast to the low molar mass liquid crystal, no sizeable instrument lag. All analysis methods yield the same transition peak, indicating a fully reversible transition. The condis crystal, in contrast, is fully irreversible, as can be seen by comparing the transition in the total apparent heat capacity from standard DSC, and the missing transition in the quasi-isothermal MTDSC data. On cooling, the transition shows a supercooling of 27 K, and the reversing components of the MTDSC with underlying heating rate, as well as the quasi-isothermal experiments, have practically no contribution from the latent heat.

4.3.7 Analysis by MTDSC Kinetics of Transitions with a Latent Heat

Continuing with the discussion of the isotropisation of PEIM-12 from the last paragraph, one notices that the MTDSC with an underlying heating rate approaches, at low frequency, the DSC result for the liquid crystal. Such changes may be due to difficulties of deconvolution if too few modulation cycles cover the transition (see Figures 4.30–4.35) or due to changes in the irreversible isotropisation rate with temperature. In the last case, one should be able to derive the kinetics from the frequency dependence of the MTDSC data. A simple model for the analysis of such changes of contribution of irreversible melting and crystallisation to the reversing apparent heat capacity on heating has been developed by Toda and has been applied to the irreversible melting of crystals of polyethylene and PET [56]. The development or loss of crystallinity is assumed to be directly proportional to the crystal growth or melting rate v, multiplied with the total area of the growth surface, E_{total}. This should be true at any instant, independent of the crystal morphology. The heat-flow rate is then:

$$\frac{\mathrm{d}Q_{\text{transition}}}{\mathrm{d}t} = \dot{Q}_{\text{transition}} = \Delta h_{\text{f}}' v \sum_{\text{total}} \tag{26}$$

where $\Delta h_{\text{f}}'$ is the heat of fusion per unit volume and v is the temperature-dependent linear crystal growth or melting rate (see Figure 4.45). The area of the growth faces, Σ_{total}, on the other hand, does not change significantly during the chosen small modulation amplitudes. As a result, one can write for the temperature dependence of the latent heat flow rate:

$$\frac{\partial \dot{Q}_{\text{transition}}}{\partial T} = \Delta h_{\text{f}}' \sum_{\text{total}} \frac{\mathrm{d}v(T_{\text{s}})}{\mathrm{d}T} \tag{27}$$

Combining Eqs. (26) and (27) and substituting the phase angle, H, corrected for instrument changes, as outlined below, then results in

$$\frac{\mathrm{d}\ln v(T_{\text{s}})}{\mathrm{d}T} = \frac{\partial \dot{Q}/\partial T}{\dot{Q}} = -\frac{\omega A_{C_{\text{p}}} \sin \varsigma}{\dot{Q}} \tag{28}$$

The phase angle ς is derived from the phase shifts $\varepsilon - \delta$ of the sample temperature T_{s} and the temperature difference Δ. The phase lags are measured relative to the phase of the reference. In order to find the lag due to the sample only, one must subtract $\varepsilon - \delta_0$, the interpolated value measured for the reversible baseline:

$$\varsigma \equiv (\varepsilon - \delta) - (\varepsilon - \delta_0) \tag{29}$$

During the derivation that led to Eq. (11) [31], it was also found that:

$$\cos(\varepsilon - \delta) = \frac{K A_{\Delta}}{A_{T_{\text{s}}} \omega (C_{\text{s}} - C_{\text{r}})} \tag{30}$$

These relationships allow the evaluation of Eq. (28) by finding $(\varepsilon - \delta)$ for the transition range of interest and subtraction of the appropriate reversible

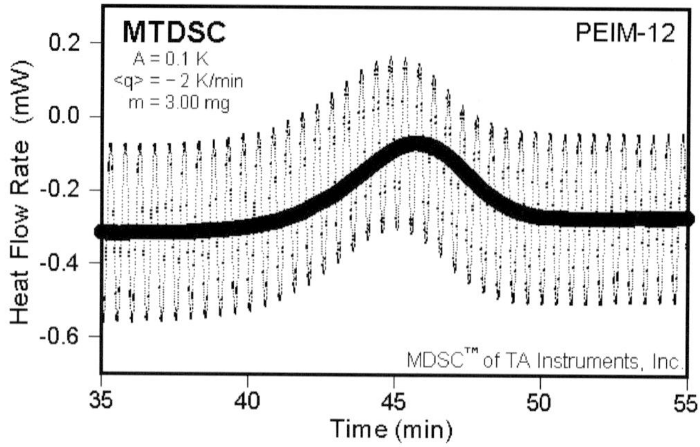

Figure 4.71. Reversing and total heat flow rates of PEIM-12 to show the ordering from the melt to the condis crystals.

baseline (by extrapolation from reversible segments of the baseline of the solid and liquid at lower and higher temperatures, respectively).

The ordering of PEIM-12 can be used as an example of the application of Eq. (28). Figure 4.71 illustrates the heat-flow rate response of the liquid PEIM-12 on cooling through the transition region to the condis phase (ordering). Figure 4.72 illustrates the rather small transition peak obtained for the three chosen frequencies after calculation of the apparent reversing

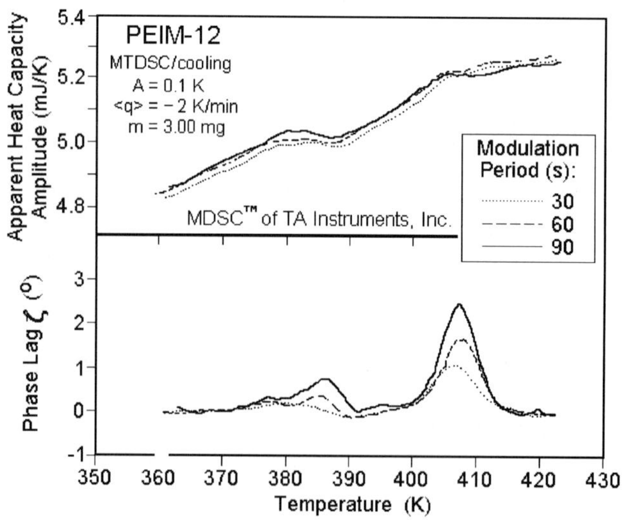

Figure 4.72. Reversing apparent heat capacity and phase lag on ordering of PEIM-12. Evaluation of the data in Figure 4.71 with Eqs. (11) and (30).

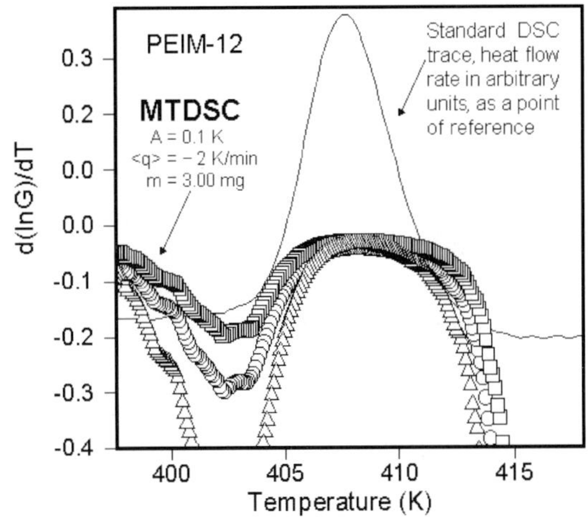

Figure 4.73. Change of the logarithm of the linear growth rate of the condis crystals of PEIM-12 with temperature. Evaluation of the data in Figure 4.71 with Eq. (28). The thin line represents a superimposed ordering peak by standard DSC, for reference.

heat capacity using Eq. (11). Finally, Figure 4.73 displays the results for the condis-crystal growth rate. Only in the centre of the transition peak, from 407 to 412 K, the phase lags are sufficiently large to yield a credible result of identical, slowly decreasing values for all three frequencies. More details can be found in the papers on crystallisation of polymers [55,56], but the interpretations must always take into account the possible crystal improvements during analysis.

4.3.8 Summary

In Section 4.3, it is shown with Figure 4.55 that the heat of fusion and its calibration to 100% crystallinity can be best accomplished by standard DSC, but the baseline is best checked or established by MTDSC. A well-established baseline of heat flow rate of the liquid is sufficient if the temperature dependence of the heat capacity is known (see Figures 4.23, 4.25 and 4.57). A detailed, simple description of the kinetics of the glass transition of semicrystalline samples is illustrated in the example of PET (Figures 4.58–4.60). Both frequency of measurement and the existing crystallinity affect the appearance of the glass transition as can be seen from the data in Table 4.1.

It is shown that the first harmonic is only an approximate measure of the reversing apparent heat capacity. There are higher harmonics and constant contributions that are not properly assessed. Model calculations by numerical integration can lead to a full evaluation, so that the reversing heat flow rate in the glass transition region can be used to extract the basic kinetic parameters (when free of instrument lags), which in turn can be used to describe properly the sample response, as shown in Figures 4.61–4.63.

Finally, the present limits are traced to an incomplete understanding of the co-operative nature of the glass transition. This result was already seen in the analysis of the enthalpy relaxation by standard DSC [77] (see also Figures 4.64–4.66).

The measurement of the kinetics of transitions with a latent heat, such as crystallisation and ordering or melting and isotropisation, is briefly reviewed for the case of standard DSC (Figures 4.67 and 4.68). The major problem of separation of the six different caloric effects that can contribute in the temperature region between the glass transition and melting point, introduced in Section 2.5, is illustrated with linear low-density polyethylene (Figure 4.69) and appears to varying degrees in the melting ranges of all polymers analysed. The secondary crystallisation, various annealing effects and contributions of reversible transitions must, thus, be removed first as discussed in the earlier-mentioned sections.

Mesophase transitions may be good examples of transitions that have only minor lag problems for their analysis with MTDSC. Examples are given in Figures 4.70–4.73, including the quantitative assessment of the kinetics using the method of Toda.

4.4 DETERMINATION OF ANNEALING AND REORGANISATION

The annealing and reorganization, as shown in Figures 4.7 and 4.10, was derived from classical calorimetry and structure analysis [3,6]. It has seen progress by the introduction of MTDSC [1]. Even qualitative analysis can identify overlapping exotherms and endotherms, but the quantitative analysis is still lacking for most of the polymers.

Especially, puzzling proved a reversing contribution to the apparent heat capacity that remains even after prolonged quasi-isothermal analysis for almost all polymers that have been analysed to date [78]. After removing all instrument effects, there remains an apparent, reversible heat capacity that is larger than the vibrational and large-amplitude contributions to the reversible heat capacity [effects (1) and (2) of Section 2.5] and also cannot be connected with the irreversible, primary crystallisation [effect (6) of Section 2.5, see Figure 4.45]. In a unique mixture, secondary crystallisation and crystal perfection [which lead to "annealing peaks" at higher than the crystallisation or annealing temperature, effects (4) and (5) of Section 2.5] are interwoven with some truly reversible melting. Local equilibria are the reason for this reversible melting, which is similar in its appearance to the melting of low molar mass paraffins [effect (3) of Section 2.5, see Figure 4.45]. In this section, an attempt is made to separate and discuss the effects (5) and (4) from the effect (3), which will be discussed in Section 4.5.

4.4.1 Annealing of PET

Poly(ethylene terephthalate) was the first polymer analysed with quasi-isothermal MTDSC, as displayed in Figure 4.50. Even this qualitative figure shows that the higher melting range of the quasi-isothermal analysis must have its origin in the annealing which occurred in the protracted analysis time. This time dependence was further analysed by extending the time of quasi-isothermal analysis from the common 20 min to 6 h, as illustrated in Figure 4.74.

The apparent reversing heat capacity shows two distinct kinetic processes of almost equal magnitude, but with time-scales differing by a factor of about 10. The data fit a double-logarithmic function and can be extrapolated to infinite time. Figure 4.75 reveals that this extrapolation reduces the apparent heat capacity to the reversible heat capacity of a semicrystalline polymer of about 15% crystallinity, ▲, but not the 35% crystallinity that was still present after the experiment. This difference in heat capacity must be a reversible latent heat.

A more detailed analysis needs first to clarify that the changes in the sample were not enhanced or even caused by the modulation itself. For this purpose, PET was analysed with three different calorimeters using various methods of modulation control and control, and experiments were performed with different lengths of modulation-free annealing. Figure 4.76

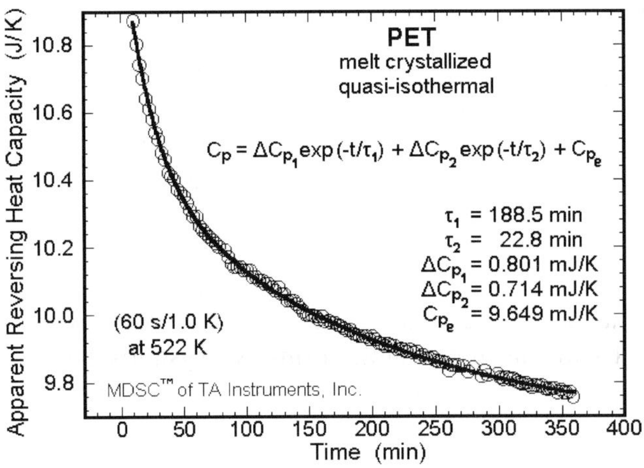

Figure 4.74. Extended quasi-isothermal MTDSC of PET in the melting range. Sample as in Figure 4.50 [79].

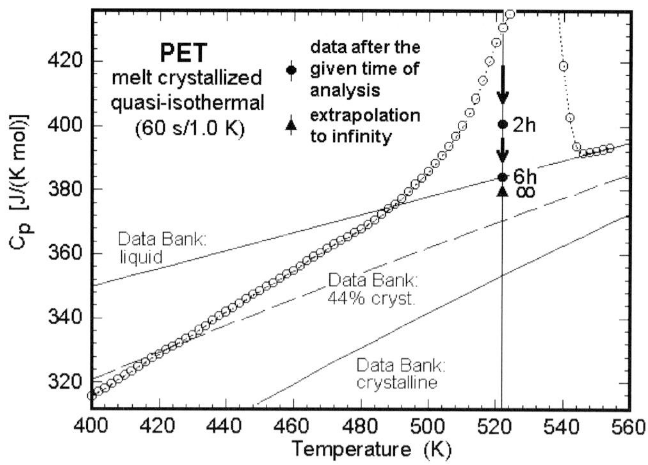

Figure 4.75. Quasi-isothermal MTDSC of PET as in Figure 4.50 with one set of long-term modulation experiments at 522 K.

shows the data from a sinusoidally modulated power-compensated DSC [80]. No differences could be observed over the whole modulation time, proving that the annealing was not influenced significantly by the modulation itself. The change in the nature of the polymer after several quasi-isothermal runs was analysed by quenching after the run and analysis by standard DSC, as illustrated in Figure 4.77. As expected, the standard DSC traces reveal typical annealing peaks about 10–15 K above the analysis temperatures. Such annealing of semicrystalline polymers, in general, is well known from standard DSC [6], but the MTDSC permits a more quantitative analysis.

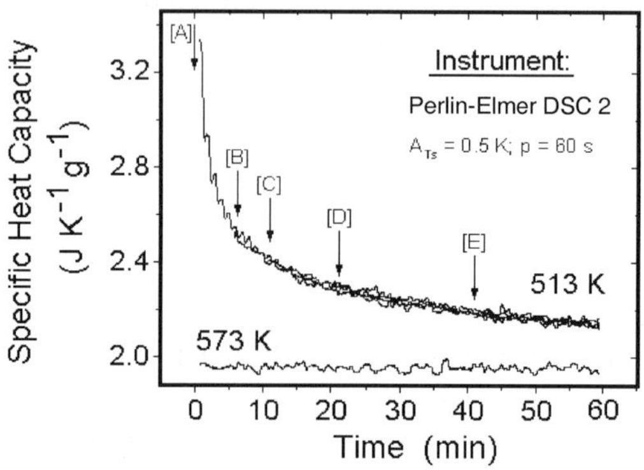

Figure 4.76. Extended quasi-isothermal MTDSC of PET in the melting range, similar to Figure 4.74, but with a different calorimeter. At 573 K, the sample is melted, and at 513 K, it is in the melting range. For the five experiments at 513 K, the modulations were started at times [A]–[E].

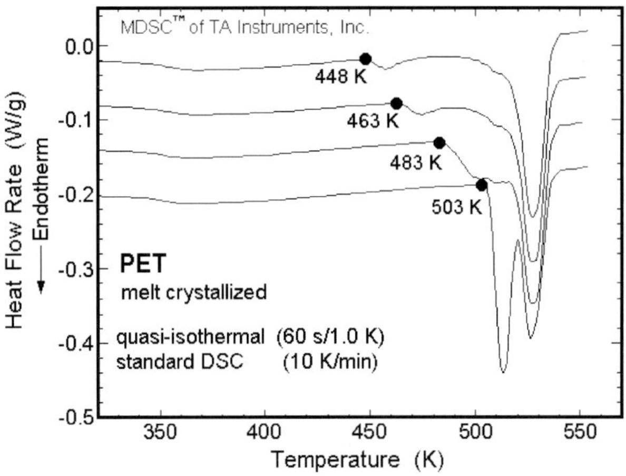

Figure 4.77. Standard DSC of PET after 20 min quasi-isothermal runs at the indicated temperatures. Before the DSC experiment, the samples were quenched to 250 K. Runs without quenching before analysis were almost identical to the traces shown here [79].

Inspection of the Lissajous figures of the analyses show a quick achievement of steady-state ellipses, so that the step-wise annealing does not seem to involve melting and recrystallisation, as is seen on single-step annealing [81], but a continuous, irreversible crystal perfection. The decreasing signal shows, furthermore, that there are some of the contributing latent heat sites lost as the experiment proceeds. This could be caused by either occasionally more perfect crystallisation, which would remove the site from melting in the next cycle, or it would lead occasionally to complete melting of a molecule, which then would need to undergo new molecular nucleation and require a supercooling beyond the modulation range. For PET, the precision of the modulation was insufficient to distinguish between excess exotherms or endotherms as a function of time. Note that both of these deviations give rise to a positive excess reversing heat capacity in plots such as Figures 4.74 and 4.76. Only a direct analysis in the time domain can resolve this question.

4.4.2 Annealing of PTT

Poly(trimethylene terephthalate) has similarly been analysed for long-term annealing. The quasi-isothermal MTDSC of the original sample is seen in Figure 4.56. Figure 4.78 shows the reduction of the apparent reversing heat-capacity amplitude with time and the extrapolated reversible apparent heat capacity. Again, a sizeable latent heat is contributed to the reversible heat capacity, and the annealing effect stretches over several hours. Figure 4.79 illustrates the heat flow rate in the time domain. After the initial

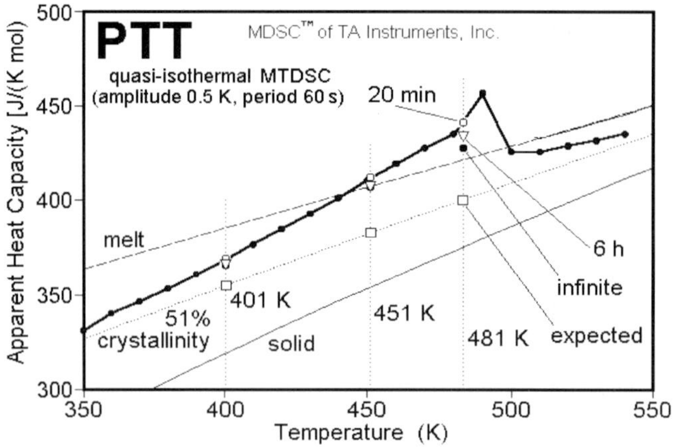

Figure 4.78. Quasi-isothermal MTDSC of PTT as in Figure 4.56 with three sets of long-term modulation experiments [51].

approach to steady state of the calorimeter, the modulation curves and the Lissajous figures are again close to symmetric. The ultimate approach to a constant ellipse seems to be more gradual on the endothermic side. If this bears out after more precise experiments, the occasional complete melting of polymer molecules would be the predominant reason for the slow kinetics of annealing.

Similarly, the fast process seems to be coupled to melting. This again indicates that the major annealing is an irreversible process that shows up

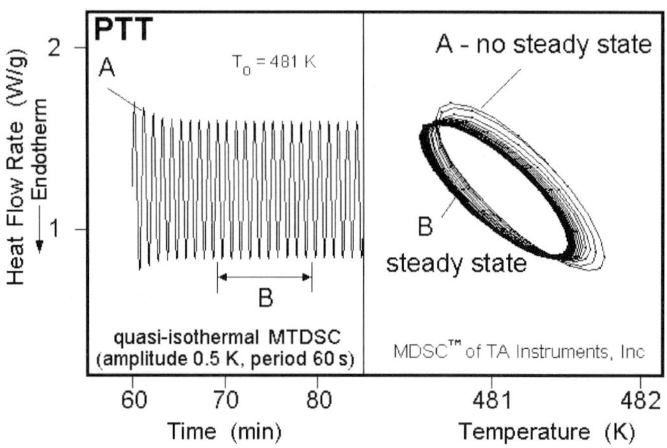

Figure 4.79. Reversing heat flow rate and Lissajous figure of the quasi-isothermal analysis at 481 K of Figure 4.78.

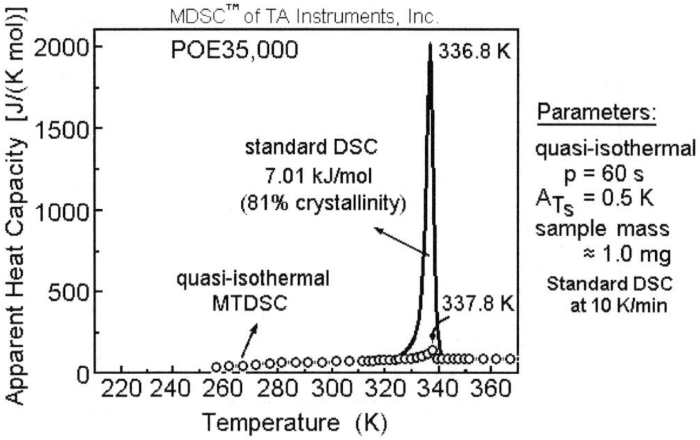

Figure 4.80. Apparent heat capacity measured by DSC and MTDSC for PEO35000 crystallised by cooling from the melt at 10 K min^{-1}. The solid curve was obtained by DSC at 10 K min^{-1}, and the open circles represent the apparent reversing heat capacity, obtained by a series of quasi-isothermal experiments [52].

mainly in the total heat flow rate, so that both total and reversing heat-flow rates should be studied simultaneously.

4.4.3 Annealing and Reversible Recrystallisation in Low Molar Mass PEO

Poly(ethylene oxide) of high molar mass behaves similarly to the PET and PTT and other polymers analysed, although special effects are seen for many analysed polymers [78,82]. Figure 4.80 represents an example of PEO of a molar mass of 35,000 Da. As before, at low temperature, standard DSC and quasi-isothermal MTDSC give the same result. Most of the melting is irreversible and shows only in the total apparent heat capacity. A small amount, however, is reversing. The irreversible melting occurs at a temperature expected for 4 folds per molecule [52].

The behaviour of the PEO with lower molar mass is different. The sample with molar mass 5000 Da, PEO5000, shows very little reversing melting. Figure 4.81 represents a magnification of quasi-isothermal MTDSC in the melting region of POE5000 Da that is compared to the standard DSC in Figure 4.47. The reversible melting peak at 333.7 K is very small, but the heat capacity still has a considerable reversible latent heat contribution over a wider temperature range. On subsequent cooling, there is a supercooling of about 20 K, as expected from the crystal growth-rate data in Figure 4.5. The subsequent reheating repeats the first heating. The melting temperature agrees mainly with that of the extended chain crystals [57,58].

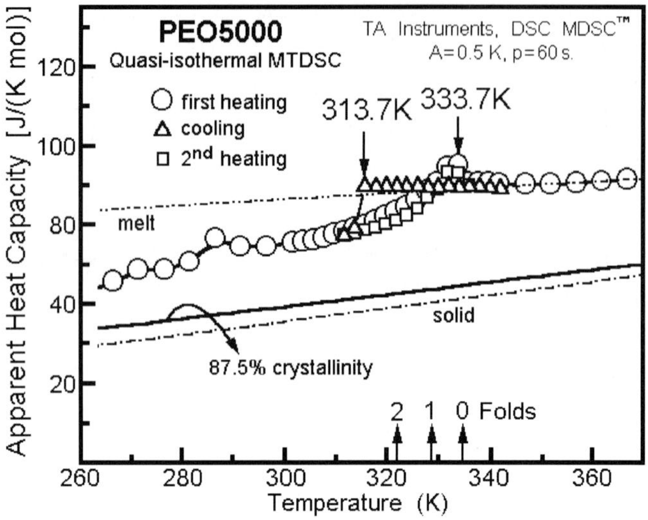

Figure 4.81. Enlarged part of the apparent reversing heat capacity of PEO5000 in the melting range. For the full trace of the initial sample, see Figure 4.47. This was followed by a quasi-isothermal set of runs at decreasing temperatures and completed by a third set of runs at increasing temperatures as for the first.

The oligomer PEO1500 of about 1500 molar mass, when quenched, shows a small reversing contribution as depicted in Figure 4.82, but it occurs at the low temperature side of the standard DSC curve, in contrast to the POE35000 in Figure 4.80. Molecules of such small molar mass are only about 10 nm long. A detailed analysis of the melting range under quasi-isothermal conditions is given in Figure 4.83. A rather large reversing

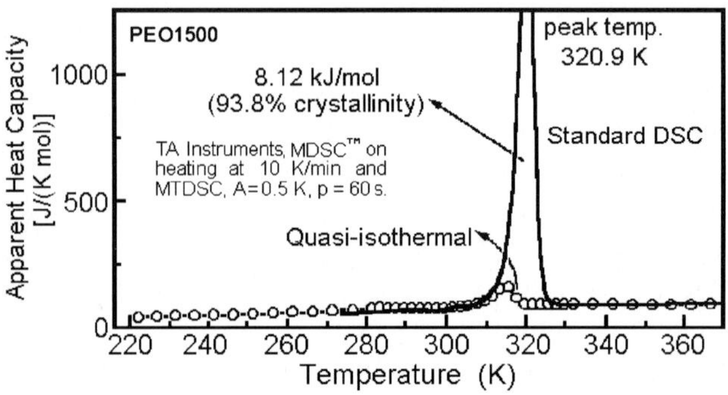

Figure 4.82. Apparent heat capacity measured by DSC and MTDSC for PEO1500 crystallised after quick cooling to 300 K. The solid curve was obtained by DSC at 10 K min^{-1}, and the open circles represent the apparent reversing heat capacity obtained by quasi-isothermal experiments [52].

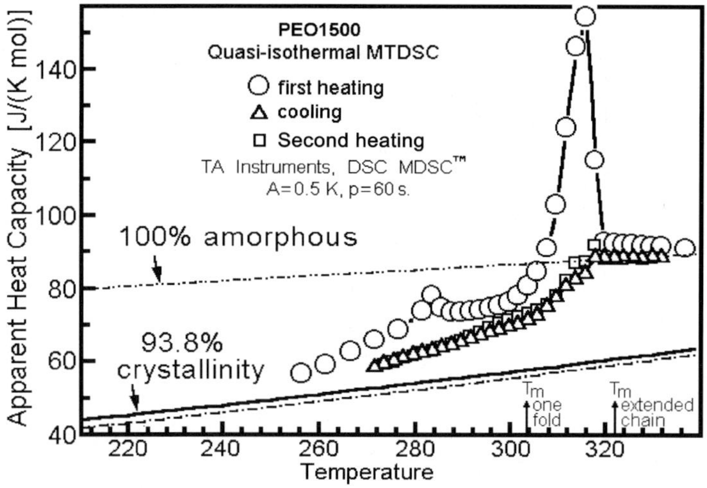

Figure 4.83. Enlarged part of the apparent reversing heat capacity of PEO1500 in the melting range. For the full initial trace, see Figure 4.82. This was followed by a quasi-isothermal set of runs at decreasing temperatures and completed by a third set of runs at increasing temperatures as for the first.

melting peak is produced by quenching, while on slow cooling, as carried out on the quasi-isothermal cooling experiment, the melting peak is even smaller than for PEO5000. In addition, the supercooling has decreased to about 4 K. As before, the apparent heat capacity reaches the expected value for 93.8% crystallinity only at much lower temperature. To clarify the nature of the reversing heat flow rate in the melting region of the quenched sample, a series of quasi-isothermal experiments was made throughout the melting range and is summarised in Figure 4.84. Each series of runs was made on a sample newly quenched from the melt to T_0, the base temperature of the quasi-isothermal experiments. In addition, parallel standard DSC traces were done on a similarly treated sample at the beginning of each series of experiments and at the end of all experiments at 315.6 K, close to the peak in Figure 4.83. The interpretation of the data is as follows. At 305.6 K, close to the melting temperature of crystals of once-folded chains, the crystallisation was practically complete and the apparent heat capacity is only a little below that of the liquid, represented by the dashed line. At 319.6 K, no crystallisation occurred, and modulation up to ±3 K also does not produce any crystallisation. The equilibrium melting temperature of POM1500 is 322.3 K, and Figure 4.83 indicated first crystallisation at about 316 K. At 310.6 K, the initial crystallinity is somewhat less, but with increasing modulation amplitude, a larger latent heat contribution can be seen. This effect increases for the experiments at 313.6 and 315.6 K. In the last case,

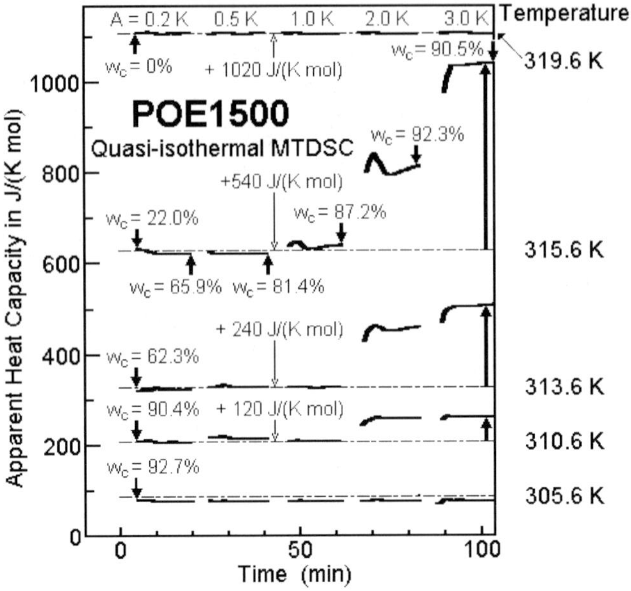

Figure 4.84. Apparent heat capacity measured by quasi-isothermal MTDSC for PEO1500 with successively higher modulation amplitudes, crystallised after quick cooling to the indicated first analysis temperatures. The indicated crystallinities were obtained by parallel experiments with standard DSC on identically treated samples measured at 10 K min^{-1}. Successive data are displaced upwards by the listed amounts [52].

the crystallinity had developed during the quenching to 22% only, but increased by the end of the first 20 min run to 65.9% (fully irreversibly since no reversing heat capacity is seen). In fact, there is a small decrease in heat capacity, which is an indication of an increase in crystallinity. This trend continues in the second run, but changes drastically with the subsequent runs, where now a sizeable increase in the apparent reversing heat capacity occurs with little further increase in the crystallinity. The Lissajous curves for the data at 315.6 K are drawn in Figure 4.85 and show the reversing part of the crystallisation and melting. The curves with an amplitude of 2 K are shown in Figure 4.86 as a function of time. The modulated heat flow rates are not sinusoidal, and by subtracting the also-drawn curve for 319.6 K after adjustment to the proper phase as a reference free of transitions, one can compare the lightly shaded melting areas to the heavily shaded crystallisation areas. The two transitions are separated by a region of approach to steady state, in contrast to the data gained on indium (see Figure 4.40). The total transition involves a crystallinity of about 10%. Following the endotherms and exotherms in Figure 4.85, one can see that the crystals that exist for a shorter time before reaching the melting point are poorer

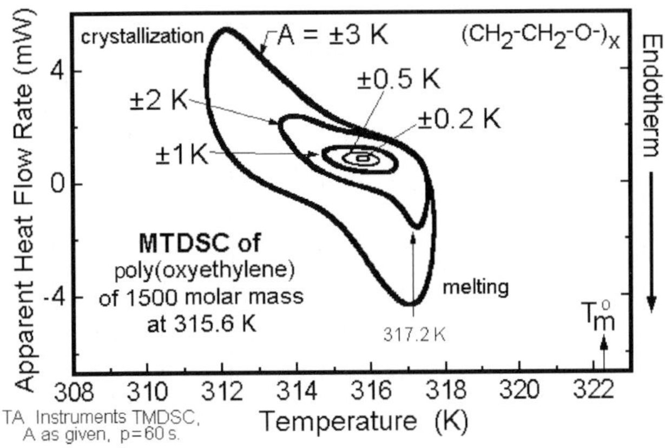

Figure 4.85. Lissajous figures of the runs at 315.6 K with different modulation-temperature amplitudes, as seen in Figure 4.84, showing melting and crystallisation at the larger amplitudes.

and melt at lower temperature. To explain the melting and crystallisation, one can assume that the poorer crystals can, for the case of the low molar mass oligomers, bridge the molecular nucleation gap in Figure 4.5, as indicated by the temperature scale. It is thus possible by combining DSC and MTDSC to study the annealing and separate it from the reversible melting. It is expected that much more details can be gathered in the future by using

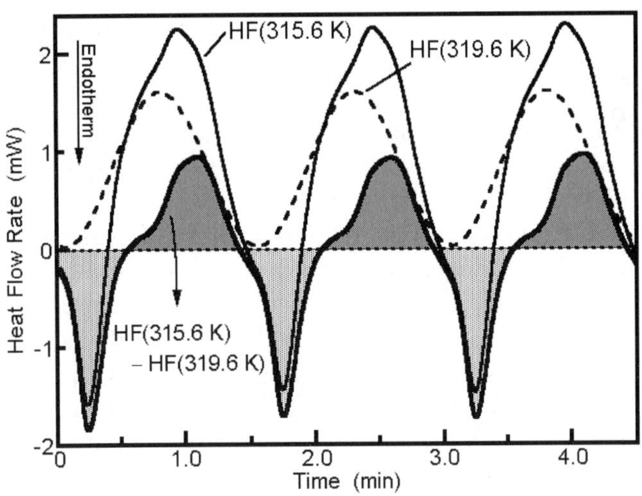

Figure 4.86. Heat flow rate in the time domain for PEO1500 at 315.4 K with an amplitude of 2.0 K as shown in Figures 4.84 and 4.85. The heavy line results on subtracting the lag-corrected liquid heat flow rate curve (at 319.6 K). The lightly shaded areas indicate the heat of fusion, and the densely shaded areas indicate the heat of crystallisation.

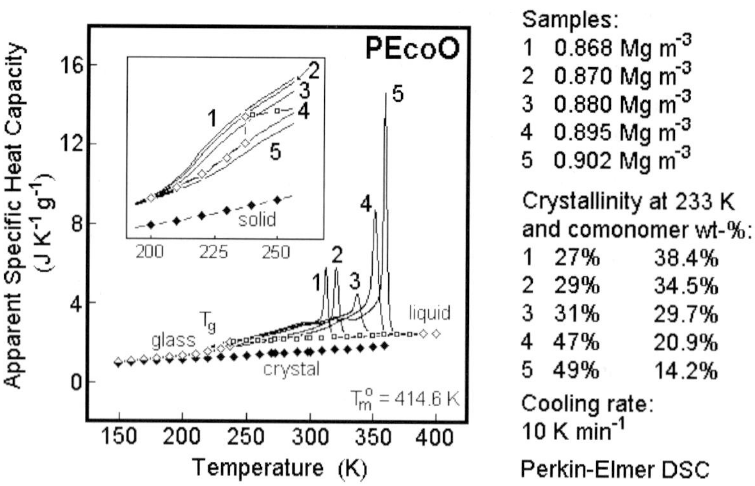

Figure 4.87. Crystallisation of PEcoO as measured by standard DSC in comparison to the data bank information on the heat capacities [24]. See also the more extensive data on sample no. 2 of this series in Figure 4.69.

different modulation profiles with enclosed isotherms and linear ramps of appropriate lengths (see Figures 4.26 and 4.27).

4.4.4 Annealing in PEcoO

Poly(ethylene-*co*-octene-1) is shown in Figure 4.69 of Section 4.3 as an example for the complicated kinetics of crystallisation of many polymers. Figure 4.87 illustrates how the broad second step of crystallisation that follows the initial, irreversible crystallisation peak increases in relative magnitude when the concentration of non-crystallisable comonomer increases. The comparison of DSC and simple MTDSC traces in Figure 4.69 allowed the qualitative identification of the irreversible crystallisation peak, common in polymers, as effect (6) of Section 2.5. The broad exotherm on cooling, which becomes increasingly more reversing as temperature decreases, is more difficult to identify. It is a superposition of secondary crystallisation, annealing and reversing crystallisation and melting [effects (5), (4) and (3) of Section 2.3], which naturally must be separated from the reversible heat capacity of the baseline [effects (2) and (1)]. The latter two effects are indicated by the glass, crystal and melt heat capacities marked in Figures 4.69 and 4.87. To analyse, a number of thermal analysis techniques must be applied to separate the effects. The PEcoO is a good example for such analysis because it has similar fractions of all effects.

The experiments to find the annealing effect are displayed in Figure 4.88. They involved stopping the standard DSC cooling trace at a given

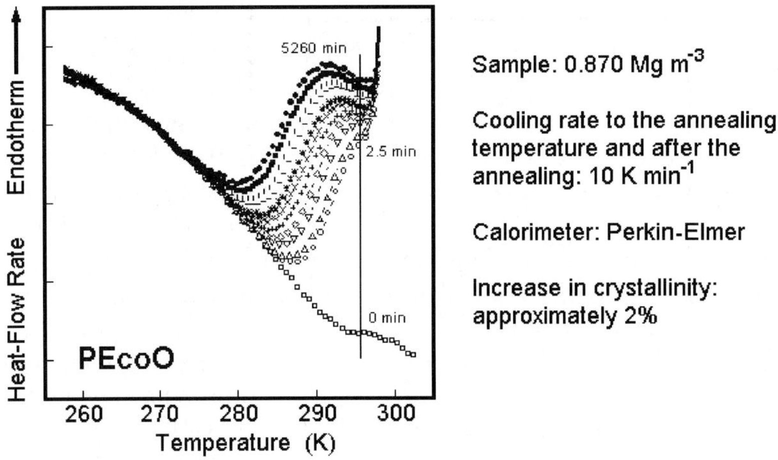

Figure 4.88. Standard DSC cooling scans of sample no. 2 of PEcoO of Figure 4.87 after annealing at 299 K for times between 2.5 and 5260 min. The trace at 0 min leads to an identical apparent heat capacity as displayed in the cooling trace of Figure 4.69. The vertical line indicates the point of the analysis of the annealed samples when steady state is reached by the calorimeter for the subsequent cooling.

temperature (299 K) for different lengths of time to affect annealing, and then continue the cooling trace. Note that the increasing heat flow rate on cooling corresponds to the decreasing apparent heat capacity in Figure 4.69, i.e. after annealing, the apparent heat capacity has decreased considerably, but regains its high value by cooling to about 280 K. Figure 4.89 shows

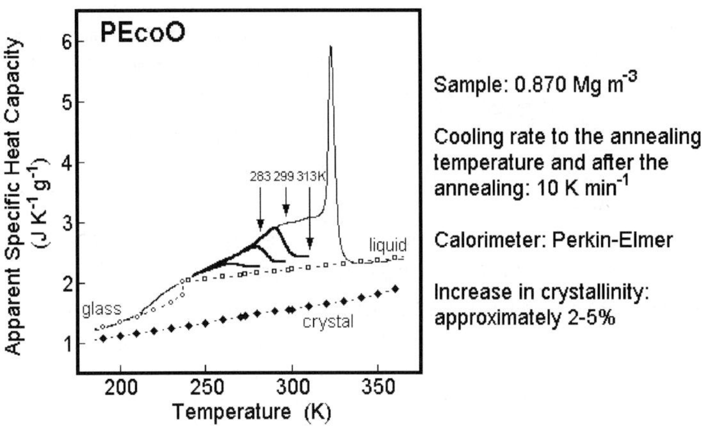

Figure 4.89. Apparent heat capacity by standard DSC on cooling of four PEcoO samples of density 0.870 Mg m^{-3}. The thin line represents the cooling trace of sample no. 2 in Figure 4.87. The other three are taken after annealing for more than 5000 min at the indicated temperatures.

the apparent heat capacities after annealing for over 5000 min at three different, marked temperatures. After this length-of-time, no further changes are expected. Much, but not all, of the latent heat has been removed by the annealing, but only for a short temperature range. Annealing at one temperature affects only the crystals growing over a narrow temperature range, so that the material crystallizing at lower temperature without annealing is unchanged.

These observations are supported by the picture developed by Flory [83] of initial copolymer crystallisation, which suggests that the uninterrupted sequences of crystallizable ethylene units of varying length crystallise in sequence of their decreasing equilibrium melting temperatures. The kinetic path leads initially to larger crystals, limited in size by the occurrence of copolymer units and the limited time available to fractionate the long sequences. The system becomes, thus, metastable before equilibrium is reached, and the subsequent secondary crystallisation is hindered by the network of the larger, primary crystals [84–86]. Both the primary and secondary crystals are, thus, not in equilibrium and able to anneal.

At the temperatures chosen for annealing in Figure 4.89, the primary crystals remain unchanged, as is shown in Figure 4.90 which is a superposition of melting traces for the samples that were made by annealing, as illustrated in Figure 4.88. Clearly, the annealing has transported the crystals

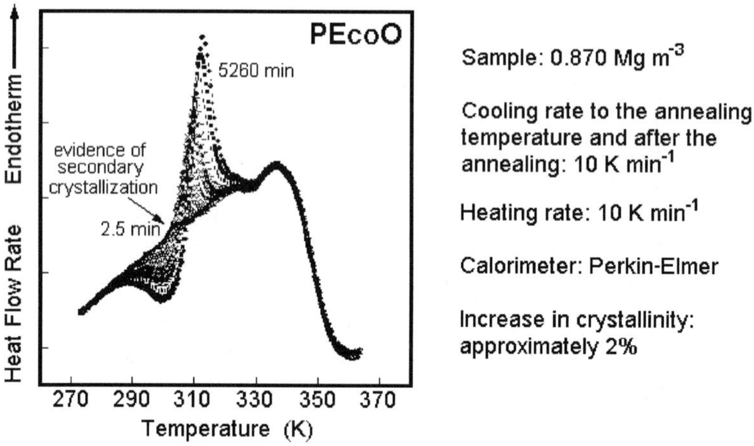

Figure 4.90. Standard DSC heating scans of sample no. 2 of PEcoO of Figure 4.87 after annealing at 299 K for 2.5–5260 min followed by further cooling to low temperature. The trace at 0 min (without minimum and maximum) leads to an identical apparent heat capacity as displayed in the heating trace of Figure 4.69. The arrow indicates a small endotherm, not compensated by a minimum at 299 K.

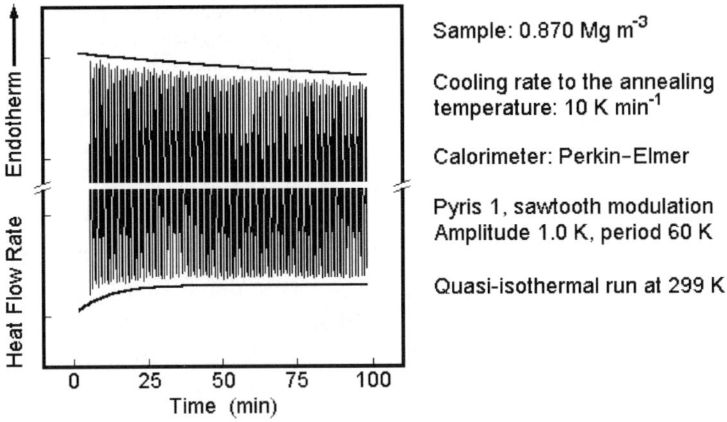

Figure 4.91. Heat flow rate and contour lines (slightly displaced for clarity) of sample no. 2 of PEcoO of Figure 4.87 on quasi-isothermal MTDSC as a function of time at 299 K.

that should have melted at about 299–312 K in an annealing peak. This annealing peak overcompensates the decrease in melting about the annealing peak. During the annealing for 500–1000 min, the crystallinity of the samples of Figure 4.89 increases by 1–5%. In addition, the arrow in Figure 4.90 points at a small melting endotherm not compensated by a minimum at the annealing temperature that points at continued secondary crystallisation at the beginning of the annealing.

Further analysis is possible by applying MTDSC to the analysis of the annealing kinetics. Figure 4.91 is a record of the heat flow rate in the time domain for the first 100 minutes of annealing. The envelopes, slightly displaced for clarity, show that the exotherm is initially larger than the corresponding endotherm, i.e. the initial process is not fully reversing, but involves also continuing secondary crystallisation. As time continues, the modulation becomes almost symmetrical, but with decreasing amplitude, describing the annealing of the sample, as seen by the moving of low sample melting in the vicinity of the annealing temperature into the annealing peak, with a slight increase of crystallinity. Figure 4.92 illustrates the uncorrected reversing specific heat capacity calculated from the MTDSC data. Clearly, two processes are separated by the double-exponential fit. As before for PET in Figure 4.74, a fully reversible contribution can be extrapolated, to be discussed in Section 4.5. The main processes, however, remain the secondary crystallisation and the annealing, documented by the standard DSC trace. The changes of the apparent total heat capacity with time both by standard DSC of Figure 4.89 and by MTDSC of Figure 4.92 yield relaxation times of about 5 and 100 min for the two processes at 299 K.

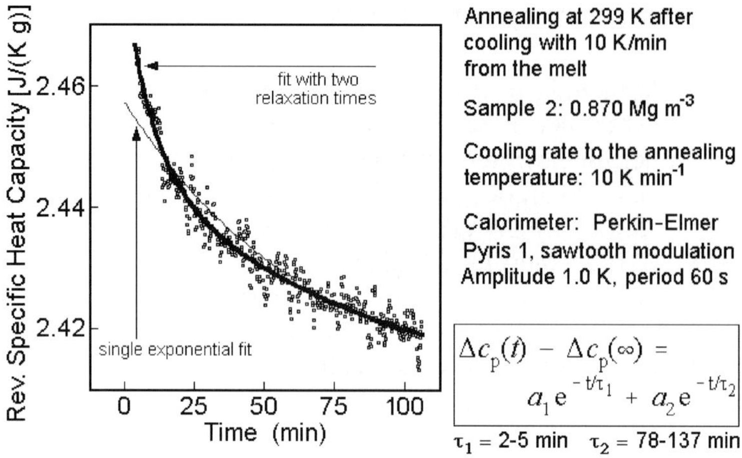

Figure 4.92. Fit of the apparent, reversing specific heat capacity generated from Figure 4.91 with a single- and a double-exponential function (thin and thick curves, respectively). The range of relaxation times refers to the different annealing temperatures shown in Figure 4.89 and refers to data gained from standard DSC and MTDSC.

4.4.5 Summary

The annealing effect predominates the thermal analysis of semicrystalline polymers between the glass and melting transitions. The thermodynamic driving force for the processes is given schematically in Figures 4.7 and 4.10, and the listing of the effects to be observed with their degree of reversibility is given in Section 2.5. In Section 4.4, the separation of annealing from secondary crystallisation and truly reversible melting is discussed. The examples of PET and PTT illustrate mainly the irreversible and partially reversing contributions to the annealing, and their separation from instrument lags and the small reversible effect to be discussed in Section 4.5 (Figures 4.74–4.79). Poly(oxyethylene) is treated as an example for the changes that occur when the chain length of the molecules gets reduced to the oligomer length (see Figure 4.5). Apparent heat capacities when compared with the true heat capacities (Figures 4.47 and 4.80–4.84) and the shape of Lissajous figures and time-domain heat-flow rates (Figures 4.85 and 4.86) gained by quasi-isothermal MTDSC allow a detailed description of the different thermodynamic perfection. Finally, PEcoO copolymers give an example of ultimate analysis of annealing and prove the special value of parallel standard DSC and MTDSC (Figures 4.87–4.92). Of even greater value for full characterisation is the simultaneous analysis of the structure of the sample by X-ray diffraction, electron, light and atomic force microscopies and solid-state NMR, as it is possible today for a full thermal analysis of materials [1].

4.5 REVERSIBLE MELTING

The reversible melting of a small portion of polymer crystals is one of the most important new observations of thermal analysis by TMDSC. For the main part of the polymers, crystallisation and melting of a sufficiently high molar mass is thermodynamically irreversible due to three reasons. A. The chain-folded crystal morphology causes non-equilibrium crystals (see Figure 4.7). B. As for most liquids, there is a need for nucleation of a crystal before it can grow (see Figure 4.68). C. Each macromolecule additionally needs to undergo molecular nucleation before it can add to a crystal (see Figure 4.6). As a result, at a given temperature, crystal and melt do not coexist in dynamic equilibrium [5].

Because of this thermodynamic irreversibility, it was expected that polymer crystallisation and melting show only a response in the non-reversing part of the modulated heat flow rate as is approximated by PEO in Figure 4.80. Exceptions were found when the amplitude of the temperature modulation is sufficiently large to bridge the region of metastability of the polymer melt due to nucleation. As indicated in Figures 4.5, 4.85 and 4.86, this became possible for oligomers with a smaller temperature range of metastability in Figure 4.5 and was helped by the initial growth of defect crystals. Furthermore, MTDSC may show erroneous reversing contributions due to distortions of the modulation caused by annealing, irreversible melting processes and accidental components of frequency ω in abrupt changes in the total heat flow rate, as indicated by Figures 4.30 and 4.31. These errors can be avoided by using quasi-isothermal analyses and waiting sufficiently to complete irreversible processes, as illustrated in Figure 4.91. The remaining reversible apparent heat capacity can then be compared with extrapolations from outside the transition region.

Truly reversible processes have been quantified for melt-crystallised PET and PTT in their melting regions, where the crystallinity changes by about 0.05% per Kelvin (see Figures 4.74–4.76 and 4.78, respectively). Further, polymers analysed in the laboratory of Schick are poly-ε-caprolactone, crystallised isothermally at 328 K with a reversible crystallinity change of about 0.1% per Kelvin [87], and poly(ether ether ketone) at 600 K, with a reversible crystallinity change of $<$0.25% per Kelvin [88]. Finally, the data for PEcoO-1 show reversible contributions in the entire temperature range between the onset of crystallisation during cooling and the glass transition. Figure 4.93 shows data which have been corrected for the frequency dependence as exemplified for the liquid at 374 K in Figures 4.18–4.20. The reversible crystallinity change per Kelvin is about 0.1% at 299 K (see Figure 4.92) and stretches over a temperature range of about 100 K at varying, but similar levels (see Figure 4.69).

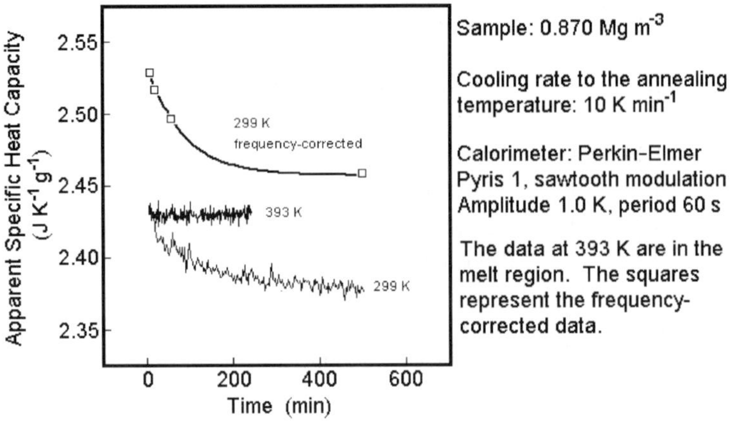

Figure 4.93. Apparent reversing specific heat capacity of PEcoO as a function of time at 299 and 393 K. The sample is in the melting region at 299 K and liquid at 393 K. The data (□) are corrected for the frequency of analysis as shown in Figures 4.18–4.20.

The temperature range of crystallisation and melting of the here-analysed PEcoO lies between the glass transition and the initial crystallisation or final melting temperatures. Such a broad range of transition for polyethylene was first documented almost 40 years ago [89]. Reversible changes in crystallinity were first suggested by structure analysis. It may be possible that the earlier-identified surface melting of lamellar crystals contributes to this phenomenon, identified by small-angle X-ray scattering experiments [90–93]. The degree of reversibility, measured in this case on linear polyethylene, is also in the order of 0.1% per Kelvin (between 323 and 373 K) and is similar to the reversibility measured on different polymer samples by MTDSC. However, in these studies, the reversible event was caused by relatively perfect, lamellar crystals of highly crystalline samples (about 80%). In case of the here-discussed PEcoO, the crystallinity is much lower, the temperature range of reversible melting is much larger and lamellar crystals are not the dominant morphology.

Figure 4.94 shows the reversing, apparent specific heat capacity as a function of time for three different crystallisation conditions. As in Figures 4.92 and 4.93, extending the annealing time, the apparent specific heat capacity becomes truly reversible, but the difference between the samples is maintained. The same copolymer reaches, thus, higher levels of reversible heat capacity for poorer crystal morphology [29]. Similarly, higher levels of copolymerisation cause poorer crystals and increase the reversible apparent heat capacity as suggested in Figure 4.87 [24]. In all cases, the final state displays perfectly reversible apparent heat capacity, as is shown in Figure 4.95 by the Lissajous figures with widely varying modulation amplitudes. There

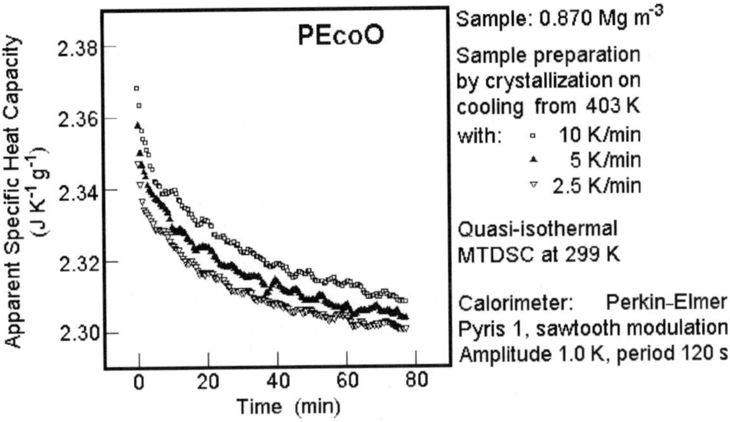

Figure 4.94. Change of the apparent reversing heat capacity of PEcoO with time for samples that were crystallised with different cooling rates at 299 K.

is no indication of a melting peak or region in the upper level of steady state, and no crystallisation shows at the lower level of steady state of the response of the sawtooth-modulated temperature.

The changes in apparent heat capacity with crystal perfection could be caused by either or both of the following contributions: (a) a reduced heat capacity and (b) a reduced latent heat contribution from a reversible crystallisation and melting. Of the three reversible contributions discussed in Section 2.5, the vibrational contribution (1) does not change significantly when changing the structure or morphology of a sample and can be omitted from the discussion (see Figure 4.1). Contribution (2), due to large-amplitude

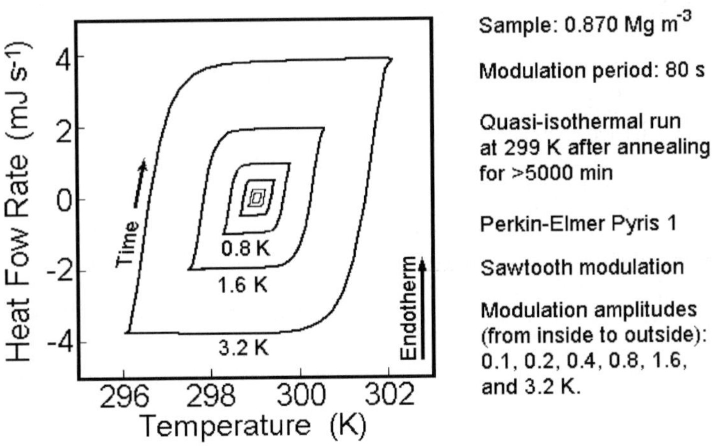

Figure 4.95. Lissajous figures of PEcoO after annealing for more than 5000 min at 299 K using the indicated amplitudes, proving full reversibility. Sample as in Figures 4.88–4.93.

motion, would go in the right direction at the temperature of interest, 299 K. The poorer crystallised sample shows an increase in the pseudo-hexagonal phase [94], which has a higher heat capacity due to a larger *gauche* concentration. One would, however, not expect a specific heat capacity higher than that of the liquid polyethylene of 2.20 J K^{-1} g^{-1}. The apparent specific heat capacities of quenched and samples cooled at 10 K min^{-1} are, in contrast, 2.50 and 2.44 J K^{-1} g^{-1}, but even the quenched samples are far from amorphous. Contributions (4)–(6) of Section 2.5 can also be excluded from further consideration because of their supercooling which exceeds the typical temperature-modulation amplitudes as can be seen from Figures 4.69 and 4.90. Based on these considerations, we identify the latent heat contribution (3) as the major cause of the increased reversible heat capacity.

Over a temperature range of 20 K, the quenched and 10 K min^{-1} cooled samples with heat capacities of 2.50 and 2.44 J K^{-1} mol^{-1} would need to change in reversible crystallinity by 2.0% and 1.6%, respectively, to account for their high heat capacity. One should note, in addition, that when calculating crystallinity, one usually uses the heat of fusion of orthorhombic crystals. However, it is likely that the pseudo-hexagonal heat of fusion may only have half the orthorhombic heat of fusion, doubling the crystallinity change to 4.0% and 3.2%.

A remaining point of discussion is to propose the scenario for the growth of a crystal morphology, which permits initially reversing and then reversible melting. Naturally, this involves a reasonable amount of speculation and points toward further experimentation needed to prove the details and give more quantitative information. Atomic force microscopy and microcalorimetry may be new tools that can pinpoint some of the morphological and structural features [95,96]. We assume that on cooling, the PEcoO sets up a network of mainly orthorhombic crystal lamellae, using the randomly occurring long sequences of ethylene in the copolymer. This early stage of crystallisation may be described as mentioned in Section 4.5 [84–86]. These lamellae are linked rather quickly by amorphous defects [4] and set up a metastable, global network of crystals. From Figure 4.87, one can guess that this type of gelation is reached at 5% of crystallisation.

After the crystal network is set up, the amorphous defects can continue locally to crystallise with very little or no long-distance diffusion. This secondary crystallisation shows the relaxation time of about 5 min in Figure 4.92 and is related to cold crystallisation, a crystallisation that is well known for crystallisation of glassy polymers close to the glass transition temperature (see Section 4.1). In the PEcoO example, this secondary crystallisation makes up more than 3/4 of the total crystallisation for the sample featured in Figure 4.69. It can involve adding chain segments to existing crystal or mesophase surfaces, or, at sufficiently low temperature, it may also involve separate crystallisation of CH_2- sequences that are sufficiently long to have reached their equilibrium melting temperature.

It is well known that, for polymers with side chains of sufficient length, the side chain may be decoupled from the polymer backbone with four to six flexible segments (CH_2- or O– groups, for example). The decoupled side chains can then crystallise at the equilibrium temperatures of the corresponding small molecules [6]. One can assume that backbone chain segments can similarly be decoupled between crystals. For a crystallisation between 310 and 250 K, this would correspond to crystallisation of sequences of 20 to 10 CH_2- groups. The fraction of crystallisable units of 10 or more methylene units in the analysed copolymer can be estimated from the molar branch concentration as being 0.47 [$= (1.000 - 0.073)^{10}$], which is the proper order of magnitude for the observed maximum mesophase crystallinity.

A similar estimate can be made for the crystallisation involving the growth faces of the already existing crystals. In this case, we assume that the attachment of the chains to the crystal precludes the need for molecular nucleation (see Figure 4.6). Taking the data of Figure 4.93, the latent heat contribution to the true heat capacity of the semicrystalline sample is about 0.28 J K^{-1} g^{-1} (the C_p of the melt is decreased by 0.06 J K^{-1} g^{-1} for the existing 10% crystallinity). This latent heat involves a crystallinity change of $0.28\ \mathrm{J\,K^{-1}\,g^{-1}} \times 1.0\ \mathrm{K} \times 14.03\ \mathrm{g\,mol^{-1}} \times 100/4110\ \mathrm{J\,mol^{-1}} = 0.10\%$. Assuming further that all of the about 10% existing crystals of the analysed samples show such reversible crystallisation and melting on their surfaces and that these crystals are isometric with a dimension of 5.0 nm, only ca. 1.5% of a monomolecularly occupied surface layer of 0.5 nm thickness need be involved in the reversible melting and crystallisation to account for the higher apparent heat capacity. If this were one single molecular segment on each of the four growth faces of the assumed crystal, it would have a molar mass of the proper magnitude to melt and crystallise at 299 K, namely 282 Da ($\approx C_{20}H_{40}$). Both estimates show that in such a poorly crystallised sample, it is possible to have local equilibria that are not restricted by molecular nucleation.

4.5.1 Summary

In the framework of the thermal effects enumerated in Section 2.5, the increased reversible heat flow rate is due to the latent heat effect (3), which may be caused by either isolated crystals, which crystallise and melt in a local equilibrium set-up within the network of primary and secondary crystals (5,6) after their rearrangement has ceased (4), or by reversible crystallisation and melting on the lateral surface areas of the crystals (5,6).

5 Recommendations

This chapter is concluded with a set of recommendations about the use of MTDSC. This new technique is a still-growing extension of the well-established standard DSC. The many examples in this chapter have shown that in almost every instance, both standard DSC and MTDSC in their various modes of operation are necessary to analyse fully the material problem on hand.

The analysis with any unknown sample should start with a trial run using standard DSC, going to the maximum temperature of interest to check if the sample pan stays closed and retains the sample. At high temperatures, many polymers become sufficiently fluid to creep out of the sealed pan or decompose and burst the pan. In this case, a lengthy and difficult cleaning of the DSC head may become necessary, which often reduces the precision of future runs and always requires a full new calibration. It is best to have an old DSC handy for the "dirty" run. This stability test can also be done in a standard oven filled with a nitrogen atmosphere.

The stability test is followed with a higher precision standard DSC runs of the delivered sample on heating, followed by measurement on cooling at a convenient rate to set a constant thermal history for comparison to other samples and completed with a second run on heating. If the first and second heatings are largely different, it may be useful to perform a third and fourth heating after identical cooling, to check on the repeatability of the thermal history. Typical heating and cooling rates can be 10–20 K min^{-1} and sample masses about 3–15 mg.

Next follows an attempt at a preliminary analysis starting with the identification of thermal events from high to low temperature. Decomposition always terminates a heating run. It is often exothermic for oxidation reaction or endothermic when accompanied by a mass loss, such as in depolymerisation with evaporation of the monomer. Since decomposition and often also loss of water from moist samples opens or bursts a sealed pan, catastrophically sharp peaks or a series of small sharp instabilities indicate such events thermally. A check of the change in weight of the pan and visual inspection of the sample after the run, which should always be routine, will confirm any sample losses, melting and decomposition. At lower temperature, one should find the melting endotherm (as well as crystallisation exotherm on cooling). Even lower is the change in baseline indicative of the glass transition: on heating, there may be possibly a small enthalpy relaxation. After inspection of these preliminary results, a decision should be made how much quantitative analysis by standard DSC is necessary, and in which temperature regions is an MTDSC analysis of value.

In order of complexity, the quantitative analysis by standard DSC can involve the following.

1) The determination of the onset, peak and end temperatures of the endotherms and exotherms. (Needs only temperature calibration, best to ±0.1 K, if fully pre-calibrated with about five strategically spaced standards, and checked weekly with the melting of indium: the preliminary run can be used for this determination.)
2) Measurement of heats of transition. (This can also be determined on the preliminary run if the proper sample mass was chosen, a good baseline was established and the area was calibrated with a standard melting substance in the vicinity of the transition temperature of the unknown sample. The calibration may change with temperature. Remember, also, when comparing heats of transition for crystallisation and melting, that heats of fusion change with temperature. If the mass in the preliminary run was not ideal to give a proper area for precision measurement, the mass and heating rate should be adjusted for highest precision in a second analysis.)
3) The measurement of the change of the heat capacity at the glass transition temperature. (This requires a mass of 10–30 mg, more than is used for the analysis of melting and crystallisation.) The increase in heat capacity after elimination of the enthalpy relaxation, i.e. between the solid and liquid baselines, can be calibrated with a standard glass transition, such as found in polystyrene.
4) For a complete analysis, a measurement of the heat capacity is necessary. [Three consecutive runs must be made at maximum precision (10–30 mg sample, 10–20 K min^{-1} heating or cooling rate).] The first is a run with two identical empty pans to establish the heat-capacity baseline and correct for asymmetry. The second run is with the sample and the same empty pan as before (with the sample pan also being matched to the one in the baseline run). The third, a calibration run with sapphire (about 30–60 mg). Outside the transition regions, this gives heat capacities that can be analysed with the ATHAS system for deviation from the vibrational heat capacity for the large amplitude motion which is coupled to the plasticity of the material. (From outside of the transition extrapolations into the transition region can be made to allow a quantitative interpretation of the transition.)

Anywhere in this course of analysis with standard DSC, one will discover transitions that should be studied further, or suspect multiple, overlapping transitions of different nature, which could be separated and studied with respect to their time dependence and reversibility with MTDSC, as described in Section 4. For an initial separation of the different transitions or transition effects, only the specific temperature ranges of interest need to be analysed.

An initial test can be made, perhaps, with an underlying heating rate of ≤ 0.5 K, a sample mass of about 0.5–5 mg (the lesser value for higher heats of transition) and a modulation amplitude of 0.5–1.5 K, coupled with a modulation period of 60 s.

As soon as the need arises to study the time dependence in the glass transition, it may be necessary to work on the quasi-isothermal analysis in its temperature range in steps of, perhaps, 1 K (larger steps below and above the actual transition). These measurements have to be made over as large a frequency range as possible and then be analysed as shown in Section 4.2. Similarly, heat capacities can be separated from non-reversing effects by quasi-isothermal MTDSC or with an underlying heating rate as shown on Section 4.2. Quasi-isothermal analyses are always necessary when there is a suspicion of calorimeter lag which falsifies the data. Even, heat capacities can be determined with much higher precision when analysed with quasi-isothermal analysis with different frequencies, so that all losses not caused by the modulation can be separated, and the contact resistances and the thermal conductivity effects can be calibrated separately for every run.

To summarise, the initial work should be done by standard DSC. As soon as this is completed, however, quality thermal analysis requires MTDSC in its many applications as documented in this book. Any up-to-date thermal analysis laboratory must, by now, be able to provide MTDSC measurements. It can be performed by the commercially available software, but it is also relatively easy to generate specific programs better suited for the problems at hand.

Acknowledgements

This work was supported by the Division of Materials Research, National Science Foundation, Polymers Program, Grant # DMR-9703692 and the Division of Materials Sciences, Office of Basic Energy Sciences, U.S. Department of Energy at Oak Ridge National Laboratory, managed and operated by UT-Battelle, LLC, for the U.S. Department of Energy, under contract number DOE-AC05-96OR22725.

References

[1] B. Wunderlich, Thermal Analysis of Materials (a computer-assisted lecture course of 36 lectures and 2879 screens, published on the Internet, downloadable without charge, including presentation software: http://web.utk.edu/~athas).

[2] E. Turi (Ed.), Thermal Characterisation of Polymeric Materials, 2nd Ed., Academic Press, New York (1997).

[3] B. Wunderlich, Thermal Analysis of Polymeric Materials, Springer, Berkin (2005).
[4] B. Wunderlich, Macromolecular Physics, Vol. I, Crystal Structure, Morphology, Defects, Academic Press, New York (1973).
[5] B. Wunderlich, Macromolecular Physics, Vol. II, Crystal Nucleation, Growth, Annealing, Academic Press, New York (1976).
[6] B. Wunderlich, Macromolecular Physics, Vol. III, Crystal Melting, Academic Press, New York (1980).
[7] W. Hemminger and G. Höhne, Calorimetry, Verlag Chemie, Weinheim, Germany (1984).
[8] H.F. Mark, N.G. Gaylord and N.M. Bikales (Eds.), Encyclopedia of Polymer Science and Engineering, 2nd Ed., J. Wiley, New York (1985–1989) (a many-volume treatise of which there is a much abbreviated version: J.I. Kroschwitz, Concise Encyclopedia of Polymer Science and Engineering, J. Wiley, New York (1990)).
[9] J. Brandrup, E.H. Immergut and E.A. Grulke (Eds.), Polymer Handbook, 4th Ed., Wiley, New York (1999).
[10] B. Wunderlich, The Athas data base on heat capacities of polymers, Pure Appl. Chem., 67 (1995) 1919 (for data see the internet: http://web.utk.edu/~athas).
[11] B. Wunderlich, Pure Appl. Chem., 67 (1995) 1919.
[12] B. Wunderlich and H. Baur, Heat capacities of linear high polymers, Fortschr. Hochpolymeren Forsch. (Adv. Polym. Sci.), 7 (1970) 151.
[13] D.W. van Krevelen, Properties of Polymers, Correlation with Chemical Structure, 3rd Ed., Elsevier, Amsterdam (1990).
[14] J. Pak and B. Wunderlich, Macromolecules, 34 (2001) 4492.
[15] G. Tammann, The States of Aggregation, Van Nostrand, New York (1925).
[16] C. Doelter, Zeitschr. Elektrochem., 12 (1906) 413.
[17] H. Baur, Thermophysics of Polymers—I, Theory, Springer, Berlin (1999).
[18] H. Baur and B. Wunderlich, J. Thermal Anal. Cal., 54 (1998) 437.
[19] B. Wunderlich, Polymer, 5 (1964) 125, 611.
[20] E. Hellmuth and B. Wunderlich, J. Appl. Phys., 36 (1965) 3039.
[21] A. Miyagi and B. Wunderlich, J. Polym. Sci. Polym. Phys. Ed., 10 (1972) 1401.
[22] B. Wunderlich, M. Möller, J. Grebowicz and H. Baur, Conformational Motion and Disorder in Low and High Molecular Mass Crystals, Springer Verlag, Berlin (1988) (Adv. Polym. Sci., 87).
[23] M.G. Broadhurst, J. Res. Nat. Bur. Std., 70A (1963) 481.
[24] R. Androsch and B. Wunderlich, Macromolecules, 32 (1999) 7238.
[25] B. Wunderlich, A. Boller, I. Okazaki and S. Kreitmeier, J. Thermal Anal., 47 (1996) 1013.
[26] M. Pyda and B. Wunderlich, Macromolecules, 32 (1999) 2044.
[27] A. Mehta and B. Wunderlich, Makromol. Chem., 175 (1974) 977.
[28] J. Pak, A. Boller, I. Moon, M. Pyda and B. Wunderlich, Thermochim. Acta, 357/358 (2000) 259.
[29] R. Androsch and B. Wunderlich, Macromolecules, 33 (2000) 9076.
[30] B. Wunderlich, Differential Thermal Analysis. In: A. Weissberger and B.W. Rossiter, Eds. Physical Methods of Chemistry, Vol. 1, Part V, Chapter 8, J. Wiley & Sons, New York (1971).
[31] B. Wunderlich, Y. Jin and A. Boller, Thermochim. Acta, 238 (1994) 277.
[32] W. Hu and B. Wunderlich, J. Thermal Anal. Cal., 66 (2001) 677.
[33] B. Wunderlich, A. Boller, I. Okazaki, K. Ishikiriyama, W. Chen, M. Pyda, J. Pak, I. Moon and R. Androsch, Thermochim. Acta, 330 (1999) 21.
[34] R. Androsch, I. Moon, S. Kreitmeier and B. Wunderlich, Thermochim. Acta, 357/358 (2000) 267.

[35] I. Hatta and N. Katayama, J. Thermal Anal. Cal., 54 (1998) 557.
[36] G.W.H. Höhne, Thermochim. Acta, 330 (1999) 45.
[37] R. Androsch, J. Thermal Anal. Cal., 61 (2000) 75.
[38] Y.K. Kwon, R. Androsch, M. Pyda and B. Wunderlich, Thermochim. Acta, 367/368 (2001) 203.
[39] M. Pyda, Y.K. Kwon and B. Wunderlich, Thermochim. Acta, 367/368 (2001) 217.
[40] J. Pak and B. Wunderlich, Thermochim. Acta, 367/368 (2001) 229.
[41] A. Boller, M. Ribeiro and B. Wunderlich, J. Thermal Anal. Cal., 54 (1998) 545.
[42] U. Gaur, H.-C. Shu, A. Mehta, S.-F. Lau, B. Wunderlich, M. Varma-Nair and B. Wunderlich, J. Phys. Chem. Ref. Data, 10 (1981) 89, 119, 1001, 1051; 11 (1982) 313, 1065; 12 (1983) 29, 65, 91; 20 (1991) 349.
[43] V.B.F. Mathot, Calorimetry and Thermal Analysis of Polymers, Hanser Publishers, New York (1994).
[44] B. Wunderlich, B.M. Bodily and M.H. Kaplan, J. Appl. Phys., 35 (1964) 95.
[45] B. Wunderlich, R. Androsch, M. Pyda and Y.K. Kwon, Thermochim. Acta, 348 (2000) 181.
[46] M. Merzlyakow and C. Schick, Thermochim. Acta, 330 (1999) 55.
[47] B. Wunderlich, J. Thermal Anal., 48 (1997) 207.
[48] M.L. Di Lorenzo and B. Wunderlich, J. Thermal Anal. Cal., 57 (1999) 459.
[49] M. Reading, B.K. Hahn and B.S. Crowe, U.S. Patent 5,224,775 (1993).
[50] B. Wunderlich, J. Chem. Phys., 29 (1958) 1395.
[51] M. Pyda and B. Wunderlich, J. Polym. Sci., Part B: Polym. Phys., 38 (2000) 622.
[52] K. Ishikiriyama and B. Wunderlich, Macromolecules, 30 (1997) 4126; J. Polym. Sci., Part B: Polym. Phys., 35 (1997) 1877.
[53] R. Androsch and B. Wunderlich, Thermochim. Acta, 369 (2001) 67.
[54] W. Chen, M. Dadmun, G. Zhang, A. Boller and B. Wunderlich, Thermochim. Acta, 324 (1998) 87.
[55] A. Toda, C. Tomita, M. Hikosaka and Y. Saruyama, Thermochim. Acta, 324 (1998) 95.
[56] A. Toda, T. Oda, M. Hikosaka and Y. Saruyama, Polymer, 38 (1997) 231, 1439, 2849.
[57] A.J. Kovacs, A. Gonthier and C. Straupe, J. Polym. Sci., Polym. Symp., 50 (1977) 283.
[58] A.J. Kovacs, C.J. Straupe and A. Gonthier, J. Polym. Sci., Polym. Symp., 59 (1980) 31.
[59] K. Ishikiriyama, A. Boller and B. Wunderlich, J. Thermal Anal., 50 (1997) 547.
[60] M. Varma-Nair and B. Wunderlich, J. Thermal Anal., 46 (1996) 879.
[61] A. Boller, Y. Jin and B. Wunderlich, J. Thermal Anal., 42 (1994) 307.
[62] K. Ishikiriyama, M. Pyda, G. Zhang, T. Forschner, J. Grebowicz and B. Wunderlich, J. Macromol. Sci. Phys., B37 (1998) 27.
[63] I. Okazaki and B. Wunderlich, Macromolecules, 30 (1997) 1758.
[64] H. Suzuki, J. Grebowicz and B. Wunderlich, Br. Polym. J., 17 (1985) 1.
[65] M. Pyda, A. Boller, J. Grebowicz, H. Chuah, B.V. Lebedev and B. Wunderlich, J. Polym. Sci., Part B: Polym. Phys., 36 (1998) 2499.
[66] B. Wunderlich and I. Okazaki, Temperature-Modulated Calorimetry of the Frequency Dependence of the Glass Transition of Poly(Ethylene Terephthalate) and Polystyrene. In: M.R. Tant and A.J. Hill, Eds. Structure and Properties of Glassy Polymers, ACS Symposium Series 710, Am. Chem. Soc., Washington, DC (1998) 103–116.
[67] I. Okazaki and B. Wunderlich, J. Polym. Sci., Part B: Polym. Phys., 34 (1996) 2941.
[68] S. Matsuoka, Relaxation Phenomena in Polymers, Hanser, Munich (1994).
[69] J.M. Hutchinson and S. Montserrat, Thermochim. Acta, 286 (1997) 263.
[70] J.M. Hutchinson and S. Montserrat, J. Thermal Anal., 47 (1996) 103.
[71] B. Wunderlich and I. Okazaki, J. Thermal Anal., 49 (1997) 57.
[72] L.C. Thomas, A. Boller, I. Okazaki and B. Wunderlich, Thermochim. Acta, 291 (1997) 85.

[73] A. Boller, C. Schick and B. Wunderlich, Thermochim. Acta, 266 (1995) 97.
[74] S.Z.D. Cheng, M.Y. Cao and B. Wunderlich, Macromolecules, 19 (1986) 1868.
[75] R. Androsch and B. Wunderlich, Proc. NATAS, 26 (1998) 469.
[76] W. Chen, A. Toda, I.-K. Moon and B. Wunderlich, J. Polym. Sci., Part B: Polym. Phys., 37 (1999) 1539.
[77] B. Wunderlich and D.M. Bodily, J. Polym. Sci., Part C, 6 (1964) 137.
[78] B. Wunderlich, Prog. Polym. Sci., 28 (2003) 383.
[79] I. Okazaki and B. Wunderlich, Macromol. Rapid Commun., 18 (1997) 313.
[80] C. Schick, M. Merzlyakov and B. Wunderlich, Polym. Bull., 40 (1998) 297.
[81] H.G. Zachmann and H.A. Stuart, Makromol. Chem., 52 (1960) 23.
[82] M. Merzlyakov, A. Wurm, M. Zorzut and C. Schick, J. Macromol. Sci. Phys., 38 (1999) 1045.
[83] P. Flory, Trans. Farad. Soc., 51 (1955) 848.
[84] H. Baur, Kolloid Z. Z. Polymere, 224 (1968) 36.
[85] H. Baur, Ber. Bunsenges., 71 (1967) 703.
[86] H. Baur, Kolloid Z. Z. Polymere, 203 (1965) 97.
[87] A. Wurm, M. Merzlyakow and C. Schick, J. Thermal Anal. Cal., 56 (1999) 1155.
[88] A. Wurm, M. Merzlyakow and C. Schick, Colloid Polym. Sci., 276 (1998) 289.
[89] B. Wunderlich, J. Polym. Sci., Part C, 1 (1963) 41.
[90] T. Albrecht and G. Strobl, Macromolecules, 28 (1995) 5827.
[91] Y. Tanabe, G.R. Strobl and E.W. Fischer, Polymer, 27 (1986) 1147.
[92] G.R. Strobl, M.J. Schneider and G. Voigt-Martin, J. Polym. Sci., Part B: Polym. Phys., 18 (1980) 1361.
[93] J.M. Schultz, E.W. Fischer, O. Schaumburg and H.A. Zachmann, J. Polym. Sci., Part B: Polym. Phys., 18 (1980) 239.
[94] R. Androsch, J. Blackwell, S.N. Chvalun and B. Wunderlich, Macromolecules, 44 (1999) 3739.
[95] I. Moon, R. Androsch, W. Chen and B. Wunderlich, J. Thermal Anal. Cal., 59 (2000) 187.
[96] B. Wunderlich, Thermochim. Acta, 355 (2000) 43.

Index

abilities of characterization methods 162, 164
adiabatic calorimetry 270, 271, 274
ageing 38, 65, 87, 151, 195 *see also* annealing
amino resins 86
amorphous 4, 21, 34, 43, 44, 47–49, 68, 86, 103, 112, 127, 155, 173, 218, 219, 222, 228, 231–235, 258,271, 272, 275, 279–281, 287, 312
amorphous phase, definition 34, 218
amplitude xi, 3–6, 12, 14, 15, 24, 26, 45, 47, 51, 56, 58, 68, 70, 75, 83, 84, 94, 101, 103, 220, 227, 232, 239, 240, 242–244, 250–254, 256–258, 261, 277–281, 286, 291, 294, 297, 301–303, 307, 309–312, 315, 316
annealing 23–26, 28–36, 38, 54, 168, 169, 172, 173, 196–198, 200, 202, 228, 230, 232, 236, 237, 259, 260, 268, 271, 272, 284, 287, 288, 294–299, 303–311
annealing kinetics 307
apparent heat capacity 50, 51, 170, 171, 232, 233, 235, 247, 248, 257, 258, 265–268, 270–274, 276–278, 280–283, 286, 289–295, 297, 299–302, 305, 306, 308–311, 313
Arrhenius 13, 16, 19, 20, 26, 29, 30, 42, 59, 64, 67, 130, 132, 278, 280
asymmetry of the calorimeter 244, 269
asymmetric interdiffusion 185, 186, 190, 192
ATHAS data bank 219, 248, 276, 315
autocatalytic effect 284
autoacceleration 107
average density 189, 190
Avrami equation 288
azoxyanisole 262, 264, 265, 268

branching 98
breadth of the glass transition region 178

cage effect 109
calibration 1, 7–9, 11, 14, 15, 50, 51, 69–71, 73, 76–79, 102, 103, 105, 171, 172, 225, 240, 244, 246, 269, 286, 293, 314, 315
calibration constant 51, 102, 171, 172
chain
  chain mobility 139
  chain motion 86
  chain segments 132, 136, 139, 140, 154, 233, 312, 313
chemical control 131, *also* kinetic control
chemically controlled 97, 116, 118, 130, 131, 134–136, 138, 139
chemical structure 141, 153, 154
chemorheology 86, 87
cold crystallization 21, 39, 45, 54, 259, 260, 268, 271, 272, 274, 276, 280, 312
complex sawtooth 244, 251, 270

M. Reading and D. Hourston (eds). Theory and Practice of Modulated Temperature Differential Scanning Calorimetry, 321–328.

computation of crystallinity 247
condis crystal 232, 237, 289, 290, 292, 293
conformational motion 220, 232, 235, 277
conversion 16, 86, 87, 93, 94, 96–101, 106–109, 112, 114, 116, 118, 124, 125, 127, 129–133, 139, 140–147, 149, 151, 155, 206, 279
conversion rate 93, 96, 100, 129
cooling rate 24–27, 29, 30, 32, 38, 55, 224, 247, 281, 311, 314, 315
co–operative kinetics 235, 284
copolymer analysis 243
copolymerisation 92, 106, 124, 140, 310
core-shell latex 196–198, 200, 201
crosslink 94, 133, 154, 204, 205, 207
  crosslink density 87, 115, 133, 143, 147, 151, 153, 154, 204
crystalline xii, 43, 44, 46–49, 55, 66, 68, 156, 173, 221, 222, 233, 235, 238, 248, 270, 275, 290, 310
crystallisation 8, 18, 19, 21, 22, 25, 27, 39, 42, 44, 45, 47, 54, 55, 66, 68, 156, 218, 219, 222, 224–228, 233–237, 245–249, 251, 252, 255–261, 263, 264, 267, 268, 271–274, 276, 280, 285–290, 293, 294, 297, 299, 301–304, 306–315
crystallisation kinetics 258, 272, 286, 288
crystal perfection 41, 232, 236, 252, 260, 261, 268, 287, 294, 297, 311
cure, experimental procedures 94
cure, experimental techniques 102
cure diagrams
  time-temperature-transformation (TTT) diagram 87, 145–148
  continuous heating-transformation (CHT) diagram 87, 145–147, 149,
cure paths 87, 93, 116, 147
  isothermal 110, 147
  non-isothermal 147
  quasi-isothermal 127, 129, 134, 135, 137, 138, 142, 147, 150
  combined cure paths 116
cure rate law 91, 94,129, 133, 139, 147
curing xii, 20, 83, 85, 86–89, 93, 97, 98, 99, 102, 103, 110, 115, 121, 132, 136, 139, 140, 145, 151, 153, 155
crystallinity 43, 46–49, 55, 68, 218, 222, 247–249, 262, 267, 272–277, 280, 286, 287, 291, 293, 295, 301, 302, 307, 309, 310, 312, 313
crystallization by standard DSC 245
crystallization kinetics 258, 272, 286, 288
crystallization of indium 224, 245, 264

Debye-Bueche neutron scattering 183
deconvolution xi, xii, 2, 4, 6, 9, 11, 12, 22, 39, 54, 55, 237, 239, 248, 252, 253, 258–260, 266, 282, 286, 290
  comments 10
  complete 9–11, 105, 259
  simple 6, 7, 9, 10, 18, 20, 23, 44, 105
deconvolution problem 237, 258
degradation 115, 143, 151
degree of segregation 162
density 61, 66, 87, 115, 116, 133, 143, 147, 151, 153, 154, 163, 187, 189, 190, 204, 228, 235, 288, 294, 305
determination of crystallinity 275
determination of heat of fusion 246–248, 275
determination of kinetics 275
devitrification 23, 112, 114–116, 121, 134, 149, 151, 154, 155, 278
dielectric thermal analysis (DETA) 94, 95, 102
differential of heat capacity 164, 175
differential scanning calorimetry (DSC) 1, 2, 6, 13, 15, 20–23, 37, 41, 43, 49, 50, 54, 55, 63, 73, 79, 83–85, 95–100, 103, 104, 110, 141, 143, 154, 162–164, 166, 169, 170, 172, 173, 177, 179, 204, 211, 217–219, 224, 227, 230, 233, 237, 238, 240–242, 245–247, 249–252, 256, 257, 259, 260, 263, 265–267, 270–272, 274–277, 281, 286–290, 293, 294, 296, 297, 299–308, 314–316
diffuse interface 163, 184–187, 189, 211
diffusion
  segmental diffusion 98, 154
  translational diffusion 93, 154
  diffusion coefficient 93, 98, 132, 140, 184–186, 189, 190, 192, 193, 196, 201
  diffusion effects 87, 97, 133
  diffusion factor 129, 130, 132, 134–137, 139, 155
  diffusion limitations 129, 140

diffusion control 93, 95, 97, 129, 131, 136, 138, 139, 141, 147, 155
diffusion-controlled 19, 87, 93, 98, 109, 114, 116, 118, 129, 131, 136, 140, 141, 143, 151, 153, 155
overall diffusion control 93, 94, 98, 131
specific diffusion control 93, 109
diffusion time 185–194
diglycidyl ether of bisphenol A (DGEBA) 102, 121, 125–127
disordering transition 232, 234, 236
dynamic mechanical analysis (DMA) 95, 112, 279
dynamic rheometry (viscometry) 84, 85, 94, 107, 139, 143, 147, 148, 150

Ehrenfest equation 178
elastic modulus 86, 93
elastomer 85
enthalpy 7, 20, 25, 27–31, 33–39, 4, 44, 46, 47, 49, 54, 61, 63–66, 87, 97, 98, 100, 101,112, 116, 118, 141, 142, 151, 183, 220–224, 228–232, 257–260, 268, 270, 274, 285, 294, 314
entropy 87, 94, 283, 203, 218, 221–223, 229–232, 236, 248, 249, 270
epoxy resins 21, 86, 88, 93, 130, 139, 155
equilibrium melting 43, 46, 48, 49, 217, 219, 222, 225, 229, 234, 235, 301, 306, 313
extended-chain 226, 227, 262, 267, 268, 299

fibre 91, 156, 218, 226, 231, 232
fictive temperature 34, 35, 37
film 22, 142, 165, 183, 189, 195–198, 200, 218, 273, 280
first harmonic 242, 250, 257, 260, 267, 278, 281, 282, 286, 293
flexibility of a polymer 223
flexible macromolecules, definition 218, 226
fold 152, 227, 299, 301, 309
four-component blend 169, 208
Fourier equation of heat flow 238, 240
Fourier transform 5, 6, 55, 56, 64, 84, 253, 257
Fourier representation 242
free enthalpy 221–224, 228–232, 270
free-radical polymerisation 93, 106, 108, 109, 140, 141, 145
inhibition 92, 109
initiation 92
propagation 90–92, 109, 141
termination 89–92, 99, 109, 124, 139, 141
transfer 105, 109
frequency *see* period

gauche conformation 235
Gaussian function 165, 166, 208
glass transition xi, xii, 1, 7, 13, 15, 16, 20–35, 37, 38, 442, 43, 47, 49, 51, 53–55, 61, 63–66, 77, 86, 94, 95, 97, 106, 114, 116, 121, 123, 133, 136, 140–147, 153–156, 162, 164–166, 169–171, 173–175, 179, 182, 183, 186, 190, 204, 207–209, 218, 220, 228, 232, 235, 248, 258–260, 268, 269, 271–276, 278–281, 284, 289, 293, 294, 309, 310, 312, 314–316
effect of annealing/ageing 34, 294, 297, 304, 308
effect of cooling rate 27
effect of frequency/period 25, 101, 114
glass transition, broadening 164, 265, 268
gel 86, 93–95, 98, 143, 144,146
gel effect 98, 106, 109, 140
gel fraction 96, 94
gel point 93, 94
gelation 86, 87, 94, 98, 106–108, 139, 143, 145–149, 204, 312, *see also* gel/sol
Gibbs-Thomson equation 229
glass
glass effect 109
glassy state 87, 93, 95, 106, 116, 121, 124, 144, 146, 149, 151, 153–155, 278
glass transition 1, 7, 13, 15, 16, 20–35, 37, 38, 42, 43, 47, 49, 51, 53–55,61, 63–66, 77, 86, 94, 95, 97, 106, 114, 116, 121, 123, 133, 136, 140, 141, 143–147, 153156, 162, 164–166, 169–171, 173–175, 178, 179, 182, 183, 186, 190, 204, 207–209, 218, 220, 232, 335, 248, 258–260, 268, 269, 271–276, 278–281, 284, 289, 293, 294, 309, 310, 312, 314–316
glass transition-conversion relationship 94, 141, 147
group vibration 220, 235, 270

harmonics 13, 55, 56, 60, 242–244, 250, 251, 261, 282, 283, 293
heat capacity 2–4, 6–12, 14, 17, 18, 20–23, 26–32, 34, 35, 37–40, 42–57, 60–64, 67–74, 76, 77, 79, 83–85, 95, 100–102, 104–108, 110–115, 117–120, 122–129, 134, 139–141, 144,145, 154–156, 164–167, 169–172, 174, 175, 177, 178, 180, 186, 187, 204, 205, 218–221, 232, 233, 235–239, 241–249, 252, 253, 257–260, 263, 265–278, 280–286, 289–295, 297, 299–302, 304–313, 315
  average 60, 68, 76
  calibration 50, 51, 102, 103, 105, 172, 243, 307
  complex xii, 6, 11, 12, 123, 62, 64, 76, 83–85
  frequency dependence 55, 61, 63, 102, 140, 141
  effect of cure 140
  kinetic 9, 11, 12, 21, 54, 60, 67
  non-reversing 7, 9–11, 14, 60, 285
  reversing 6–11, 14, 21, 23, 32, 34, 37, 40, 49, 50, 53, 57, 60, 67–69, 170, 239, 252, 253, 271, 274, 278, 281–285, 289, 295, 297, 299–302, 311
  vibrational 2, 22, 23, 30, 39, 43–48, 220, 221, 235, 271, 278, 315
heat capacity of melt 271
heat capacity of poly-p-dioxanone 270
heat capacity of sapphire 50, 269
heat capacity of solids 269, 274
heat flow xi, 1–6, 8–13, 17–23, 32, 39, 41, 50, 51, 56–62, 69, 77–79, 83–85, 95–97, 99,100, 103–107, 110–123, 126–130, 134, 141, 154–156, 164, 167, 170, 174, 237–242, 245–248, 253–255, 257–265, 267, 268, 276, 283, 285–289, 291, 292, 297–299, 301–303, 305, 307–309, 313
heat flow phase 83, 84, 104–107, 111–113, 118, 119, 122, 123, 127–129, 155, 156
heat flow rate 237–240, 242, 245–249, 253–255, 257–265, 268, 276, 285–288, 291–293, 298, 299, 301–303, 305, 307–309, 313
heat of fusion 43, 218, 219, 221, 229, 246–249, 257, 262, 263, 266–268, 271, 275, 276, 291, 293, 303, 312
higher harmonics 56, 60, 242–244, 250, 251, 261, 282, 283, 293
hysteresis 30, 33, 285

increment of heat capacity at the glass transition 164, 166, 186, 187
indium 4, 21, 22, 34–36, 40, 41
inorganic polymer glass (IPG) 103, 112, 113, 119, 121
instrument lag 217, 224, 245, 247, 256, 258, 260, 262, 268, 272, 286, 290, 294, 308
instrument problem 237
integral analysis of the enthalpy 257
interdiffusion 165, 183–187, 189, 190, 192–194, 196, 200
interdiffusion coefficient 184, 185, 189, 190
interface 161–165, 179, 182–190, 192–195, 197–200, 209–211, 275
interface overlap 179
interfacial thickness 162–164, 183, 185, 189, 190, 193, 197, 198, 200
intermediate phase, definition 218
inter-particle 196
interpenetrating polymer networks 156, 161, 167, 195, 203, 204, 211
interphase 156, 161–163, 201, 203, 204, 211
intramolecular cyclization 154
irreversible melting 227, 267, 271, 291, 299, 309
isotropisation transition 232, 262, 264, 265, 268, 289, 290, 294

kinetics
  reaction kinetics 87, 105, 116, 118, 121, 133, 140, 141, 154, 156
  kinetic analysis 98–100
  kinetic processes 55, 57, 60, 77, 84, 295
kinetic modeling
  empirical kinetic models 87, 139
  mechanistic kinetic models 87, 91, 109, 125, 133, 155
  optimisation 133, 134, 139, 270

simulation 143
software 283, 285
kinetics of the glass transition 33, 235, 274, 276, 284, 293

large-amplitude motion 220, 278, 281, 312, 315
latent heat 38, 43, 44, 46, 68, 221, 236, 237, 245, 248, 253, 256, 258, 269–271, 274, 276, 285, 286, 289–291, 294, 295, 297, 299, 301, 306, 311–313
latex particles 195–198, 201
light-heating modulated-temperature dsc 141, 142, *also* LMDSC
linearity 13, 15, 39, 73, 175
linear response 13, 15, 242, 280
liquid
liquid state 116, 243, 278, 284, 285
liquid crystal 156, 219, 262, 264, 268, 289, 290
Lissajous figure 39, 265–267, 269, 297, 298, 302, 303, 308, 310, 311
local equilibrium 235, 313

meander modulation 251
measurement of heat capacity by MTDSC 221, 269
mechanistic information 124
melamine-formaldehyde resins 88, 103, 109, 111, 124, 125
melting
reversible 8, 16–19, 30, 34, 38, 42, 227, 235, 236, 239, 252, 255, 256, 258, 261–264, 267, 268, 271, 276, 286, 288–292, 294, 295, 297, 299, 303, 304, 307–313
melting by standard DSC 245, 266, 271
melting kinetics 227
melting of indium 245, 256, 262, 264, 315
mesophase melting 219, 231, 232, 237, 289, 294, 312, 313
mesophase transitions 289, 294
metastable crystal 228, 231
methylenedianiline (MDA) 102, 121, 125, 127
methyl ethyl ketone peroxide 103
Mettler-Toledo DSC 245, 256
miscible blend 165, 174–176
mobility
chain segment mobility 86, 139, 140
co-operative mobility 140, 141
molecular mobility 108
mobility factor 102, 121, 123–125, 129, 136, 139, 147, 151, 155
modulated temperature differential scanning calorimetry (MTDSC) 2, 4, 5, 11–13, 15–18, 20–24, 26, 32, 34, 37–39, 41, 45, 47, 49–52, 55, 61, 63, 67, 80, 83–85, 100–104, 106, 108, 109, 112, 118, 127, 129, 134, 135, 137, 139, 140, 144, 145, 147–150, 154–156, 164, 165, 167, 169, 170, 172, 177, 179, 183, 185, 196, 197, 204, 209, 211, 217, 219, 224–227, 229, 233, 235, 237, 239, 241–243, 245, 248, 250–259, 261–267, 269–277, 279–281, 284–286, 288, 290, 293–300, 302–304, 307–310, 314–316
modulation xi–xiii, 2–6, 12–19, 23, 2–25, 29, 32, 34, 38, 39, 42, 46, 49–52, 55–61, 63, 66–71, 74–78, 83, 84, 101–103, 106, 140–142, 224, 233, 238–243, 248, 250–253, 255–264, 266, 267, 269, 272, 278,279, 280, 283, 290, 291, 295–298, 301–304, 307, 309, 310, 316, *see also* amplitude or period
modulation amplitude 15, 83, 101, 103, 239, 261, 279, 280, 291, 301, 302, 310, 312, 316
modulation period 51, 70, 84, 101, 102, 140, 238, 240, 242, 243, 252, 256, 266, 283, 316
modulus (elastic modulus) 6, 23, 83–85, 95, 112, 113
molecular weight 86, 93, 98, 108, 127, 140, 145, 184, 185, 194
moving interface 184
mtdsc parameters
modulation amplitude 15, 83, 101, 103, 239, 261, 279, 280, 291, 301, 302, 310, 312, 316
modulation period 25, 51, 67, 70, 84, 101, 102, 140, 142, 238, 240, 242, 243, 252, 256, 266, 283, 316, *also* modulation frequency

temperature-dependent heat capacity calibration 51, 102, 172
heating rate 3–6, 10, 14, 15, 17, 30, 39–42, 45–47, 49–52, 83–85, 100, 101, 104, 105, 112, 114, 116, 118, 121, 134, 143, 146, 147, 149, 151, 153, 164, 170, 225, 231, 232, 237–240, 245–247, 252, 256, 261–263, 265, 266, 272, 281–284, 290, 315, 316
multi–component polymer blends 161–164, 173
multiple frequencies 141, 244
multiple melting and crystallization of indium 256

nanophase, definition 218, 233, 234
network formation 85, 127, 129, 204, 206
Newton's law constant 237, 241
nomenclature 12, 49,237
nuclear magnetic resonance (NMR) 98, 103, 163, 164, 204, 232, 233, 308
nucleation of crystals 67, 224
nucleation of molecules 224

oligomers 86, 88, 94, 225, 226, 234, 300, 303, 308, 309
one–point calibration constant 171

particle size 113, 119–121, 196
peak resolution 190, 191, 207–210
pentacontane ($C_{50}H_{20}$) 45, 46
period xii, 4, 5, 13–15, 17, 24–27, 29, 30, 32, 40, 45, 47, 50, 51, 63, 84, 99, 101–103, 140, 237, 238, 240, 266, 283, 316
choice 15, 31
effect of 101
Perkin–Elmer DSC 242, 243
phase angle 9, 53, 54, 83, 84, 105, 106, 111, 121, 155, 291
phase lag 4–6, 9–12, 21, 44, 46, 51–54, 58, 70, 76, 240, 253, 292
correction 12, 18, 52, 54
phase separation 91, 127, 129, 155, 156, 180, 183, 184, 196, 199, 200, 202–205
phenolic resins 86
phenyl glycidyl ether (PGE) 102, 125–127
physical mixture 174–176
poly(4,4′-phthaloimidobenzoyl-dodecamethyleneoxycarbonyl) (PEIM-12) 70–73
poly(butyl methacrylate) 196
poly(butyl methacrylate - butyl acrylate), 196
polycarbonate 21, 47
polyethylene 1, 2, 8, 9
poly(epichlorohydrin) 179
polyester-styrene resins (unsaturated polyester resins)118, 124
polyethersulfone (PES) 127–129, 156
poly(ethylene-co-octene)(PecoO) 18,19, 20, 69, 87–95
poly(ethylene terephthalate) (PET) 11, 37, 39, 50–53, 58–64, 74–77
poly(ethyl methacrylate) 177
polymer blends 2, 34, 161–164, 166, 169, 178, 179, 184, 203, 211
polymer diffusion 196
polymer melting by MTDSC 258
polymer miscibility 162, 165, 173, 174, 178, 183
polymer networks 85, 115, 121, 125, 159, 161, 167, 203–205, 211
polymerisation
condensation polymerisation 88, 91, 109
step-growth polymerisation 86, 88, 90, 91, 98, 108, 109, 125, 129, 154
chain-growth polymerisation 89–91, 108, 109
free-radical polymerisation 90, 93, 106, 109, 140, 141, 155
heterogeneous polymerisation 95
poly(methyl acrylate) 176, 190, 208
poly(methyl methacrylate) 165
poly(oligoamide-12-altoligo-oxytetramethylene) 248, 249
poly(oxy-1,4-phenylene-oxy-1,4-phenylenecarbonyl-1,4-phenylene) (PEEK) 67
poly(oxyethylene), POE 47, 80–86
poly-p-dioxanone, PPDX 49
polystyrene 23–27, 32, 36, 165, 167–169, 172, 204, 269, 270, 284, 285
poly(styrene-co-acrylonitrile) 165, 173
poly(vinyl acetate) 176, 186, 190, 208
poly(vinyl chloride) 176, 177
post-cure 109, 116–118, 156
primary amine reaction 127
pseudo-isothermal 239

quasi-isothermal 14, 49, 83, 99, 103, 107, 111, 118, 119, 121–124, 127–129, 134, 135, 137, 138, 142, 147, 150, 151, 154, 239, 241, 242, 245, 250, 252–255, 258, 262–274, 276–279, 283, 284, 286, 290, 294–302, 307–309, 316
quasi-isothermal analysis of the glass transition 279
quasi-isothermal melting modeling of MTDSC 262, 264
quasi-isothermal melting of indium 262, 264

Rabinowitch 131
rate constant
  Arrhenius law 13, 16, 19, 20, 26, 29, 30, 42, 59, 64, 67, 130, 132, 278, 280
  activation energy 19, 20, 23, 29, 65, 100, 105, 130, 132, 278, 280
  pre-exponential factor 30, 100, 130, 132, 280, 281
reaction
  chemical 12, 16, 17, 19–23, 25, 33, 34, 49, 55, 56, 66, 87, 91, 94, 96, 97, 100, 106, 124, 131, 139, 155
  heterogeneous reaction 112, 119, 121
  order 57
  rate of reaction 105, 112, 114, 118, 137
  reaction enthalpy 97, 100, 101, 112, 118, 141, 142, 151
  reaction exotherm 103, 105, 106, 112, 118, 153
  reaction, heat capacity change 124–126
  reaction kinetics 87, 105, 116, 119, 121, 133, 140, 141, 154, 156
  reaction mechanism 87, 91, 92, 125, 127, 140, 141, 155
  residual reaction 97, 100, 118, 142
reaction-induced phase separation 127, 156
reagents
  accelerator 91, 102, 103
  catalyst 88
  initiator 89–92, 103
  inhibitor 91, 92, 103
  monomer 86, 90, 92, 93, 108, 109, 140, 153, 203, 225, 304, 314
recommendations about the use of MTDSC 314
relaxation
  relaxation phenomena 106, 121, 155
  structural relaxation 87, 116, 162
relaxation time 29, 61, 63, 66, 98, 132, 278–280, 284, 307, 308, 312
reorganisation 45, 230, 231, 252, 260, 261, 268, 276, 294
reptation 185, 189, 194, 195, 199
reversible melting 227, 252, 258, 261, 264, 267, 268, 276, 288, 294, 299, 303, 308, 310, 312, 313
reversing heat capacity 6, 7, 9–11, 14, 21, 23, 32, 34, 37, 40, 49, 50, 53, 57, 60, 67–69, 170, 239, 252, 253, 271, 274, 278, 281–285, 289, 295, 297, 299–302, 311
reversing signal 7, 8, 10, 12, 18, 20, 22–26, 29, 33–39, 42, 44–47, 49–52, 54, 55, 80, 118, 170, 239, 250, 255, 256
rigid amorphous phase (RAF), definition 218
rubber
  rubbery state 87, 95, 106, 123, 144, 146

sapphire 7, 14, 50, 51, 74, 244, 269, 315
sawtooth modulation 240–243, 251, 264
secondary amine reaction 88, 127
secondary crystallisation 236, 260, 268, 276, 287–289, 294, 304, 306–308, 312, 313
segment mobilities 139, 140, 184
selenium 3
self-retarding effect 284
sinusoidal modulation 13, 56, 74, 238–240, 242, 250, 251, 256, 260, 262, 264
skeletal vibration 235
sol 86, 146
specific interactions 175, 177, 190
spherulite 287, 288
standard DSC 63, 238, 240, 241, 245, 249–251, 257, 259, 260, 263, 265–267, 270–272, 274, 275, 277, 281, 286–288, 290, 293, 294, 296, 297, 299–302, 304–308, 314–316
stationary condition 256
step-wise modulation 250, 258, 273, 297
structural relaxation 87, 116, 162
structured latex films 165, 195

styrene 23–27, 32, 36, 90, 92, 103, 106, 109, 118, 121, 124, 140, 165, 167–169, 172, 173, 204, 269, 270, 284, 315
supercooling 224–228, 233, 234, 246, 252, 256, 263, 273, 290, 297, 299, 301, 312
superheating 218, 219, 226–228, 252
surface free energy 196, 228
symmetrical interdiffusion 185, 192

TA Instruments DSC 85, 224, 261
temperature gradient 4, 43, 73, 237, 238, 240, 241
thermal
  conductivity 52, 61, 62, 72, 79, 80, 83, 97, 106, 118, 134, 139, 151, 240–242, 246, 262, 264, 266, 268, 270, 278, 286, 289, 290, 300, 316
  diffusion 116, 142, 186, 190, 240
  resistance 1, 15, 53, 78, 80, 240, 263, 265, 270
thermal history 164, 166, 167, 170, 173, 227, 228, 275, 284, 285, 314
thermosets, see thermosetting systems
thermosetting systems 87, 88, 96, 98, 108, 109, 111, 124, 129, 133, 141, 143, 146, 148, 149, 152, 153, 156
  amino resins 86
  epoxy systems 102, 106, 121, 129, 133, 136, 140, 149–151, 153
epoxy-anhydride 89, 91, 104–106, 112, 114, 116, 117, 125, 134–137, 139, 142–145, 147–154
  epoxy-amine 88, 89, 108, 114–116, 118, 125, 126, 137, 139, 142, 147, 149–156
    melamine-formaldehyde resins 88, 103, 109, 111, 124, 125
    inorganic polymer glass 91, 94, 103, 112, 113, 119, 120, 121, 141, *also* IPG
  phenolic resins 86
  polyester-styrene resins (unsaturated polyester resins) 106, 118, 121, 124
third law of thermodynamics 221
time-scale 272, 295
topological constraints 101, 139
torsional braid analysis 95
trans-1,4-polybutadiene 12
trans-conformation 235
trial run 314
Trommsdorff 140
types of modulation 250

vibrational energy 220
viscosity 94, 107, 108, 147
vitrification
  partial vitrification 116, 121, 155
  degree of vitrification 121, 123, 155

water 88, 89, 91, 99, 109, 110, 195, 196, 217, 314
weight fraction 103, 161–164, 186–190, 192, 194, 197–199, 202, 209, 211, 275
Williams-Landel-Ferry (WLF)132, 133, 281,

zero-entropy-production melting 229–231, 236, 248

HOT TOPICS IN THERMAL ANALYSIS AND CALORIMETRY

1. M. E. Brown: *Introduction to Thermal Analysis.* Techniques and Applications. 2nd rev. ed. 2001 ISBN 1-4020-0211-4; Pb 1-4020-0472-9
2. W. Zielenkiewicz and E. Margas: *Theory of Calorimetry.* 2002 ISBN 1-4020-0797-3
3. O. Toft Sørensen and J. Rouquerol (eds.): *Sample Controlled Thermal Analysis.* Origin, Goals, Multiple Forms, Applications and Future. 2003 ISBN 1-4020-1563-1
4. T. Hatakayama and H. Hatakayama: *Thermal Properties of Green Polymers and Biocomposites.* 2004 ISBN 1-4020-1907-6
5. D. Lörinczy (ed.): *The Nature of Biological Systems as Revealed by Thermal Methods.* 2004 ISBN 1-4020-2218-2
6. M. Reading and D. J. Hourston (eds.): *Modulated-Temperature Differential Scanning Calorimetry.* Theoretical and Practical Applications in Polymer Characterisation. 2006 ISBN 1-4020-3749-X